BAN CHIANG, NORTHEAST THAILAND, VOLUME 2A: BACKGROUND TO THE STUDY OF THE METAL REMAINS

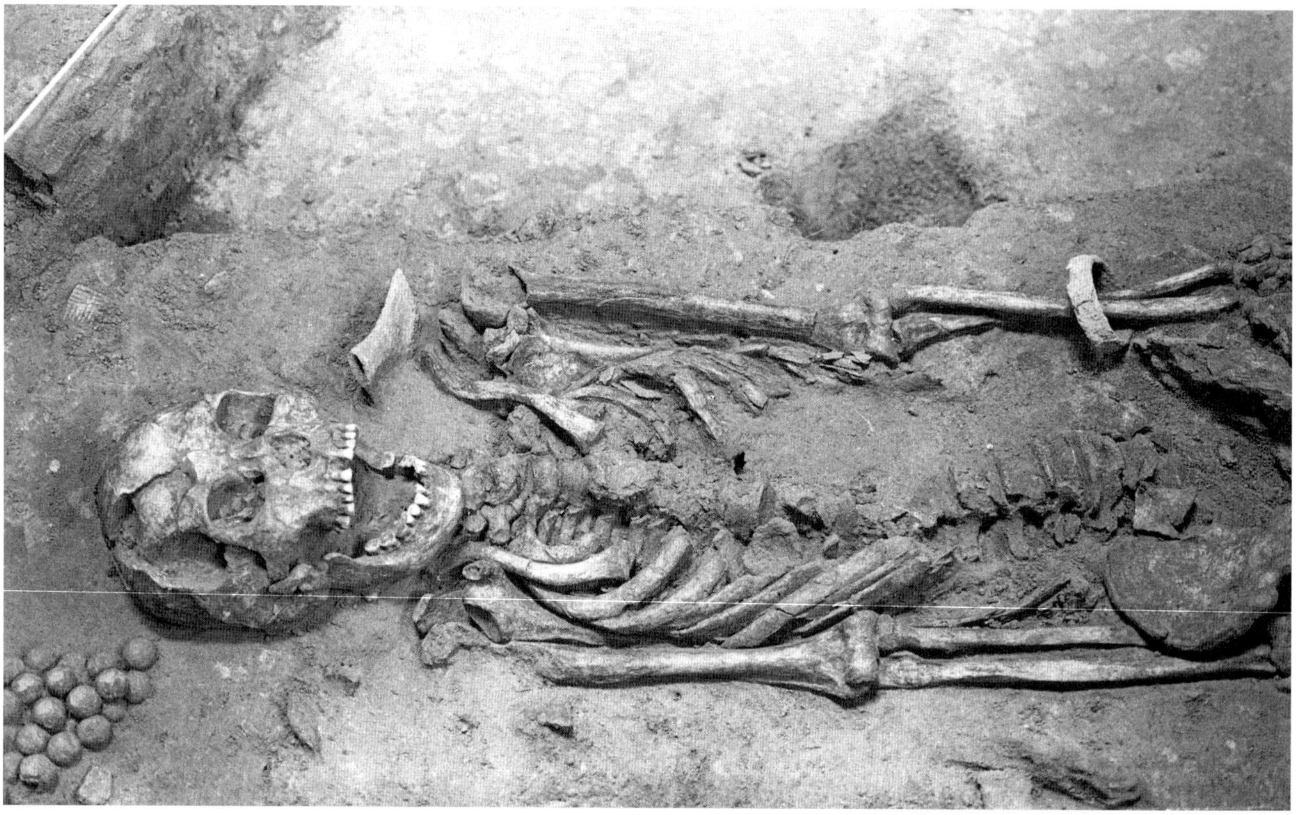

Ban Chiang BCES Burial 23, nicknamed "Vulcan," is an Early Period burial of a robust man with a bronze adze at his left shoulder and four bronze bangles on his left lower arm.

THAI ARCHAEOLOGY MONOGRAPH SERIES

Series editor, Joyce C. White

Pietrusewsky, Michael, and Michele Toomay Douglas
Ban Chiang, A Prehistoric Village Site in Northeast Thailand I: The Human Skeletal Remains, 2002.

White, Joyce C., and Elizabeth G. Hamilton, editors
Ban Chiang, Northeast Thailand, 2A: Background to the Study of the Metal Remains, 2018.

White, Joyce C., and Elizabeth G. Hamilton, editors
Ban Chiang, Northeast Thailand, 2B: Metals and Related Evidence from Ban Chiang, Ban Tong, Ban Phak Top, and Don Klang, forthcoming.

White, Joyce C., and Elizabeth G. Hamilton, editors
Ban Chiang, Northeast Thailand, 2C: The Metal Remains in Regional Context, forthcoming.

White, Joyce C., and Elizabeth G. Hamilton, editors
Ban Chiang, Northeast Thailand, 2D: Catalogs for Metals and Related Remains from Ban Chiang, Ban Tong, Ban Phak Top, and Don Klang, forthcoming.

University Museum Monograph 149

BAN CHIANG, NORTHEAST THAILAND, VOLUME 2A: BACKGROUND TO THE STUDY OF THE METAL REMAINS

Joyce C. White and Elizabeth G. Hamilton, editors

UNIVERSITY OF PENNSYLVANIA MUSEUM OF ARCHAEOLOGY AND ANTHROPOLOGY
PHILADELPHIA

CATALOGING-IN-PUBLICATION DATA IS ON FILE WITH THE LIBRARY OF CONGRESS

ISBN 13: 978-1-931707-21-3
ISBN 10: 1-931707-21-9

Distributed for the University of Pennsylvania Museum of Archaeology and Anthropology
by the University of Pennsylvania Press.

Printed in the United States of America on acid-free paper.

DEDICATION

This volume is dedicated to the memory of John V. Hastings, whose support for the Ban Chiang Project over 40 years included everything from coding amorphous metal fragments, to migrating the original database from a mainframe to a desktop computer, to funds and fundraising, to chairing the board of the Institute for Southeast Asian Archaeology. His steadfast efforts and help over the decades enabled the project to continue after the death of the late Chester Gorman in 1981. His was a life that truly made a difference. Thank you, John.

Contents

Figures

Tables

Contributors

Joyce C. White received her Ph.D. in Anthropology in 1986 from the University of Pennsylvania, with a dissertation on the chronology of Ban Chiang. She has been Director of the Penn Museum's Ban Chiang Project since 1982 as well as a Senior Research Scientist and Associate Curator for Asia at the Museum, before founding and becoming the Executive Director of the Institute for Southeast Asian Archaeology in 2013. She is also currently an Adjunct Professor in Penn's Department of Anthropology and a Consulting Scholar at the Penn Museum, where she continues her lifelong passion to foster scholarship of the prehistory of Southeast Asia with special emphasis on Thailand and now Laos.

Joyce C. White
Executive Director
Institute for Southeast Asian Archaeology
3260 South St.
Philadelphia, PA 19104-6324
joycewhite@iseaarchaeology.org

Elizabeth G. Hamilton received her Ph.D. in Anthropology from the University of Pennsylvania in 1995, with a dissertation on the development of the copper-working industry in late Iron Age and Roman Period Gaul. She has been working on the analysis of Southeast Asian metals since 1999. She is currently a Research Associate at the Institute for Southeast Asian Archaeology and a Consulting Scholar at the Penn Museum.

Elizabeth G. Hamilton
ISEAA/Ban Chiang
hamilt@upenn.edu

James D. Muhly studied Classics and Ancient History, receiving his B.A. from the University of Minnesota in 1958. He was awarded a Ph.D. from Yale University in Assyriology and Near Eastern Archaeology in 1969. He taught at the University of Minnesota from 1964 to 1967, moving to the University of Pennsylvania in 1967. He retired as Professor Emeritus from the University of Pennsylvania in 1997, serving as Director of the American School of Classical Studies at Athens from 1997 to 2002. James has taught and published widely on various aspects of the history and archaeology of the Aegean and Near East, with special emphasis on metallurgy. He is currently studying the metal objects of the Petras cemetery. His Foreword to the Ban Chiang metals study is found in TAM 2A.

James D. Muhly
Proteos 36
Palio Faliron, Athens 175 61,
Greece

Samuel K. Nash received his Sc.D. in Metallurgy from the Massachusetts Institute of Technology in 1951. After 26 years as a physical metallurgist attached to the US Army Frankford Arsenal, he retired as the Chief of the Metals Engineering Branch. From 1953–1995, he served as Adjunct Professor and then Lecturer in Materials Science at Drexel University. After his retirement, he volunteered as a metallurgist at MASCA, the Penn Museum's former Museum Applied Science Center for Archaeology, and in 2009 became a Consulting Scholar at the Museum. He retired from the Museum in 2016. Samuel Nash co-authored with Elizabeth Hamilton the metallographic and elemental analyses presented in TAM 2B, chapter 4.

Samuel K. Nash
1420 Locust St.--9A
Philadelphia, PA 19102

William W. Vernon received his Ph.D. in Geology from Lehigh University in 1964 and an M.S. in Anthropology from the University of Pennsylvania in 1984. He specialized in the analysis of minerals, rocks, and ceramics using microscopic methods. He was co-director of Archaeological Excavations in NY State from 1972–1980 and Project Geologist for the study of an ancient copper mine at Phu Lon in northeast Thailand in 1985 and 1988. He was also a member of MASCA and provided geological analysis for the studies of crucibles at Ban Chiang and Phu Lon. He retired as Emeritus Professor of Geology and Anthropology from Dickinson College in 1996. William Vernon conducted the study of the Ban Chiang crucibles presented in TAM 2B, chapter 5.

William Vernon
wwvernon@comcast.net

Vincent C. Pigott holds a Ph.D. in Anthropology (1981) from the University of Pennsylvania, and since 1984 has been the co-director of the Penn Museum's Thailand Archaeometallurgy Project. He is currently a Consulting Scholar in the Museum's Asian Section. Previously he served as the Museum's Associate Director and as a Senior Research Scientist in MASCA. After MASCA, he spent a decade as a Visiting Professor at the Institute of Archaeology, University College London. He is focused on the prehistory and archaeometallurgy of mainland Southeast Asia and maintains a strong interest in the origins, transmission, and societal impact of metallurgy across Eurasia. Vincent Pigott's summary of prehistoric copper production evidence in Thailand and Laos is found in TAM 2C, chapter 2.

Vincent C. Pigott
vcpigott@aol.com

T. O. Pryce trained in archaeology and archaeological sciences at University College London and Sheffield University, completing his Ph.D. in Southeast Asian archaeometallurgy in 2009. He completed a Leverhulme Trust Early Career Fellowship at the University of Oxford and a Senior Post-Doctoral Fellowship with the Institut de Recherche pour le Développement before being recruited by the Centre National de la Recherche Scientifique (C.N.R.S.) in 2013. Pryce is currently a Researcher at the C.N.R.S, UMR 7528 Prétech and UMR 3685 NIMBE. Oliver Pryce's lead isotope analyses of the Ban Chiang copper-base assemblage are presented in TAM 2C, chapter 3.

T. O. Pryce
opryce@gmail.com

Foreword

The story has it that a Harvard student, Steve Young, living in a village in northeast Thailand, tripped over the root of a kapok tree while out walking in the summer of 1966 and fell face first on some buried pots. These pots immediately became famous, prized by collectors and museums the world over. This represents the discovery of the site of Ban Chiang, and the pottery, which Joyce White calls the "idiosyncratic flamboyant Ban Chiang Late Period red-on-buff style pottery," found its way to all the major museums of the world. Four museums in California, as well as museums in New York City, Boston, Washington, Cleveland, and Minneapolis, built up major collections. Ban Chiang became one of the most famous archaeological sites in Thailand. But it was obvious that the looting had to be controlled and proper excavations carried out at the site.

In 1974 and 1975, the University of Pennsylvania Museum of Archaeology and Anthropology (formerly shortened to the University Museum but now called the Penn Museum) joined Thailand's Fine Arts Department to conduct excavations at Ban Chiang, led by Chester Gorman and Pisit Charoenwongsa. Froelich Rainey, then the director of the Penn Museum, had visited the site in 1973 and had confirmed the presence of significant numbers of metal objects in copper and iron and was interested in the possibility of a very early development of bronze metallurgy in northeastern Thailand. This possibility had been raised by earlier excavations at the nearby site of Non Nok Tha, discovered by Chester Gorman. It was excavated in 1966 by his Ph.D. adviser, Wilhelm Solheim of the University of Hawaii and the "Father" of modern Southeast Asian archaeology, and again by his graduate student Donn Bayard in 1968. On the basis of a series of thermoluminescence (TL) dates, Solheim concluded that bronze metallurgy had already been developed in northeastern Thailand in the 4th millennium B.C. This conclusion was elaborated in a famous article by Solheim (1968) on "Early Bronze in Northeastern Thailand." It was this article that brought the Penn Museum to Ban Chiang. Gorman, by 1973, was an assistant professor of Anthropology at the University of Pennsylvania. In an unprecedented move, the Thai Fine Arts Department permitted all the cultural materials excavated from this site, including from 142 burials, to be sent to the Penn Museum for study and analysis.

As the editor of *Expedition* (the magazine published by the Penn Museum) at the time, I arranged for the publication of a special issue devoted to the Northeast Thailand Archaeological Project (NETAP), published as volume 18(4) for 1976. This issue contained an article by Gorman and Charoenwongsa on the excavation itself entitled "Ban Chiang: A Mosaic of Impressions from the First Two Years." More important for the issues being discussed here is an article by Tamara Stech Wheeler and Robert Maddin on "The Techniques of the Early Thai Metalsmith." At the time this issue was published, the early dating for bronze metallurgy in Thailand at ca. 3000 B.C. was still accepted.

Gorman died in 1981 at the very young age of 43, and Joyce White, a graduate student in Anthropology at the University of Pennsylvania who had worked with Gorman on the analysis of Ban Chiang since the arrival of materials at Penn in 1976, took on the task of final publication. As regards the metal materials, this she has now accomplished, with the help of other scholars, notably Elizabeth Hamilton and Vincent Pigott.

What has been produced is the most comprehensive monograph ever published in the field of archaeometallurgy. It is not only a complete analysis of all the metal finds from Ban Chiang and from three other contemporary sites, Ban Tong, Ban Phak Top, and Don Klang, but it is also an introduction to the discipline of archaeometallurgy. While discussing the different methods of metal analysis now in use, the text gives a clear, detailed description of the basic technology involved in each technique. In dealing with iron, the text provides an excellent explanation of all the technical terms involved, including carburization, quenching, and tempering. The chapter on geology provides the best summary of the geology of Thailand ever written. The chapter on chronology offers the best-available account of all the problems and complications connected with radiocarbon dating, including a discussion of the controversies connected with the type of material being dated. The chapter on Lead Isotope Analysis by Thomas Oliver Pryce represents the first serious use of this new technique in Southeast Asian archaeology (see also Pryce et al. 2014). Anyone who reads this publication is going to have an excellent understanding of everything involved with the discipline of archaeometallurgy.

For White, "This volume's objective to present the early metallurgical evidence from Ban Chiang and related sites in light of a contemporary framework aims to contribute to advancing the current global discussion of the development of early metallurgy." Archaeometallurgy is, by its very nature, an interdisciplinary project. As White puts it, "the study of ancient metals is an exemplar of collaborative scholarship that integrates the hard and social sciences, and frequently the humanities as well, in order to address in an effective manner the meaning and role of metals in past societies." For the project leader this is an awesome task, for it involves coordinating and pulling together the work of many different scholars, from a variety of academic disciplines, in order to produce a coherent final publication. This White has accomplished in an exemplary manner.

The present study is published as four parts of the Penn Museum's Thai Archaeology Monograph Series (TAM 2A–2D). The first book, TAM 2A, has seven chapters dealing with background theory, geology, and metallurgy. This is followed by TAM 2B, which presents the metal-related evidence from the four sites involved in the study. Data and issues of the larger regional context are presented in TAM 2C. The remarkable thing is that, over the past 100 years since the first discussion of metal finds from Thailand, the general concept of a "Southeast Asian Metallurgical Province" has remained remarkably stable. Finally, TAM 2D comprises catalogs of metal and related remains.

The most important feature of the present study is the large number of metal objects involved, comprising a total of 639 well-provenienced prehistoric and protohistoric artifacts excavated in 1974–1975, with the entire collection shipped to Philadelphia in 1976. Of these, 176 artifacts were subjected to metallographic analysis, and over 50 to elemental analysis. This analytical work represents a core value of the present publication. As Hamilton puts it in her chapter 2 in TAM 2B:

> Every effort was made to select samples of sound metal from a full range of artifacts, strata, and sites, which provides confidence that as much as possible of the copper-base, and to a lesser extent, iron technology practiced at Ban Chiang, Ban Phak Top, Ban Tong, and Don Klang, has been sampled.

To my knowledge, such a comprehensive and well-organized research project is without precedent in the history of archaeometallurgy.

For anyone without a good background in the intricacies of metal analysis, TAM 2B, chapter 2 by Hamilton, on the methods of metal analysis, will probably be the most rewarding. It provides the information that every reader needs in order to evaluate the analytical results being presented. I know of no better discussion on the relative merits of analysis by proton-induced X-ray emission spectroscopy (PIXE) as compared with X-ray fluorescence (XRF). It even provides Table 2.2 comparing results of the two methods on nine artifacts. This chapter also contains an excellent discussion of analysis by Scanning Electron Microscopy (SEM) and Optical Emission Spectroscopy (OES). Finally, it includes an explanation of conducting hardness testing by using Vickers microhardness testing. This represents one of the many reasons why I consider this publication to be our best introduction to the study of archaeometallurgy.

White considers the impact of metal technology in Southeast Asia to have been very different from that documented in the ancient Near East or the eastern Mediterranean world. Metal age Thailand never developed an "imperial" metallurgy, with the production of metal artifacts at the service of a powerful national empire. The metallurgy that developed seems to have been one dominated by the needs and wishes of small village consumers. Objects were made for daily use, not for imperial display. This influenced the very nature of metal production. Copper smelting seems to have been carried out in crucibles, producing small amounts of molten metal to be used to cast small ornamental and utilitarian copper-base artifacts. Thailand never produced anything resembling the copper production of Late Bronze Age Cyprus, a technology involved in producing thousands of copper oxhide ingots, each weighing up to 30 kgs.

This, in turn, influenced the very nature of the metal objects being produced. The best way to appreciate the difference is to compare Thai metal artifacts with the use of metals, precious and base, documented in the book by Marian H. Feldman (2006), *Diplomacy by Design. Luxury Arts and an "International Style" in the Ancient Near East, 1400–1200 BCE*. Here we have a magnificent array of art objects created, it would seem, to be used as diplomatic gifts to be exchanged among the imperial powers of the day. Nothing like this exists in prehistoric metal age Southeast Asia. There, major imperial powers never developed during the bronze or iron periods. There is no metal age evidence for the existence of imperial conquerors or empire builders. We have no Sargon of Akkad or Naram Sin, no Tuthmosis III or Ramses II. It was a different world, one dominated by village organization, with metal artifacts made at the local level and designed for local consumption.

There is one other very important difference. White, in her understanding of the development of metallurgy in prehistoric Thailand, according to the anthropology of technology, places great emphasis upon being able to know who made the metal artifacts in question and where they were made. Nothing could be further from our understanding of the luxury metal artifacts studied by Feldman, who admits that (2006:12):

> Attempts to identify where these objects were produced remain unsuccessful. From stylistic analysis and distribution patterns of excavated examples, we can make a few observations concerning their production, but we cannot say anything precise regarding locations of manufacture for individual pieces.

The interesting thing is that this lack of knowledge does not concern scholars working in the Bronze Age eastern Mediterranean. They are interested in the objects themselves, as examples of the international style of the age. It is the artistic style of the object that is important. For White, working from an entirely different cultural and intellectual background, it is the anthropological, the social background of the Thai metal objects that is important.

White is skeptical of absolute dates and absolute chronology. There are too many chronological problems. Better to emphasize relative chronology, what she describes as "multi-interregional relationships." When she took over the Ban Chiang project in 1982, she had already realized that a date of ca. 3000 B.C. for bronze metallurgy at Ban Chiang was simply not possible. She felt the 18th century B.C. was more plausible, and this is what she argued in her doctoral dissertation, completed in 1986. Today she would accept a beginning date of ca. 2100 B.C. for Ban Chiang burials and metallurgy appearing by the early 2nd millennium B.C. It has to be remembered that the 1960s and 1970s were a period when there was much discussion of multiple independent inventions, with basic discoveries, especially the smelting of copper ore, being made independently in many different parts of the world. This general atmosphere certainly influenced the ready acceptance of an early chronology for bronze metallurgy at Ban Chiang. But such developments certainly did not take place in Thailand. The earliest evidence for the use of metal at Ban Chiang indicates that the technology arrived fully formed. There is no evidence for the existence of preliminary stages or experimental practices.

It is, however, still not possible to determine the origin of this metallurgy. White believes that, to the three basic questions of where, when, and how, it is best, at present, to set aside the first question and to concentrate upon the when and the how. What she hopes to be able to determine in this Ban Chiang study is: "By what social processes was metal technology transmitted." And the emphasis upon "social processes" is one of the defining characteristics of the entire volume. This is central to her application of the new archaeometallurgy paradigm employing the "anthropology of technology," an approach to understanding human technological development with a "focus less on the grand-narrative and more on detailed processes occurring at the societal level."

It is, of course, the "grand-narrative" that has dominated scholarship in prehistoric archaeology since the days of V. Gordon Childe. For White, "To truly understand the spread of technologies, we must look not at the spread of finished products but at the spread of *chaînes opératoires*." But all arguments regarding the spread of metal technology have, in fact, up until recently been based upon the distribution of finished products. It was such evidence that Colin Renfrew used in his famous 1969 article, showing that metal finds from the Balkans were much older than anything known from the Aegean. It is interesting that this argument has now been given an unexpected twist. Scholars now argue that the evidence for early metallurgy in the Balkans is so impressive that the smelting and casting of copper metal must have developed first in the Balkans, in the 5th millennium B.C., and spread from there to the rest of the Old World.

This question of multiple inventions of metal technology is clearly one of great interest to White. It appears again and again throughout the chapters of this volume. The modern discussion goes back to the work of Childe who was convinced that things developed first in the ancient Near East and spread from there throughout the Old World. These ideas were given a more sophisticated, theoretical background in a famous 1964 article by Theodore Wertime. My early work in archaeometallurgy was done together with Wertime, and together we edited a book on *The Coming of the Age of Iron* (1980). Wertime was not your usual academic scholar. Most of his professional career was spent in the diplomatic service, going back to being a member of the team that negotiated the cease-fire between North and South Korea. Wertime organized many important field expeditions and international conferences. He wrote two books on the history of steel and many articles on the cultural background of early metallurgy. But White is one of the few scholars I have worked with who has a serious understanding of the great importance of Wertime's work in archaeometallurgy. This is just another reason why I consider this Ban Chiang publication to be such an important accomplishment.

The same could be said about White's appreciation of the work of Childe, another very idiosyncratic scholar. Although Childe's basic ideas are no longer considered acceptable, White understands Childe's remarkable achievement. On the basis of what would now be considered a very limited body of evidence, Childe fashioned a scholarly paradigm, what White calls the "Conventional Paradigm," which dominated the field of prehistoric archaeology for over 50 years and is still of great influence today.

I conclude this foreword with a brief discussion of what is today the most significant controversy in Southeast Asian archaeology, the debate over chronology and general archaeological interpretation between White and Charles Higham. Each scholar's interpretation is based fundamentally upon the material from different excavations: Ban Chiang for White and Ban Non Wat for Higham. Higham, from the University of Otago, is a very distinguished Southeast Asian scholar and the author of many books, including *The Bronze Age of Southeast Asia* (1996a). His excavations at Ban Non Wat have been published by the Thai Fine Arts Department in Bangkok in six massive, very impressive volumes. Of central importance here is *The Origins of the Civilization of Angkor, Vol. 5: The Excavation of Ban Non Wat, Part 3: The Bronze Age*, edited by C. F. W. Higham and A. Kijngam, and published in 2012.

The differences between the two scholars are considerable. White puts the beginning of the bronze age in northeastern Thailand at ca. 2000 B.C., Higham at ca. 1000 B.C. White dates the beginning of the use of iron at ca. 800 B.C., Higham at ca. 500 B.C. How are such differences possible, at this stage of our understanding of Southeast Asian archaeology? White discusses the problem at numerous points in the Ban Chiang volume. She has also written a review article of volumes 4, 5, and 6 of Higham's Ban Non Wat publications in *Antiquity* (2013). Higham provides a detailed discussion of his position in the section of his volume 5 devoted to "The Copper-base Industry." He has also written a special article in *Antiquity* on "Debating a Great Site: Ban Non Wat and the Wider Prehistory of Southeast Asia" (2015a).

It is obviously not possible, in this context, to discuss all the issues involved in this debate. Much of the chronological disagreement centers upon the evidence from TL dates and from the use of shell, bone collagen, and rice-tempered pottery in obtaining radiocarbon dates. White reviews non-chronological issues and summarizes her critique of Higham's approach in TAM 2C, chapter 6. The issues are not going to be resolved any time in the near future. As White puts it at the very beginning of her Chapter 1: The data and perspectives presented here are likely to be a part of scholarly conversations about early metallurgy for years to come.

James Muhly

Preface

The Thai Archaeology Monograph Series volume 2 (TAM 2A–D), which focuses on the metals and related evidence from the site of Ban Chiang, Thailand, is the product of decades of work by scores of people. Its purpose is to present and interpret only one portion of the evidence excavated in 1974 and 1975 from the archaeological site of Ban Chiang, Thailand, the evidence for prehistoric metallurgy. Ban Chiang metallurgy is widely considered, however, to be one of the seminal contributions of this important site to global archaeology.

The lens through which Ban Chiang metallurgy can be viewed in the 21st century is a different lens than was available in the mid-1970s. The long journey to publication has in the end enabled a much richer and more nuanced presentation than would have been possible if the final study of metals was published a year or two after excavation. With the passage of time, the assemblage was organized and studied more thoughtfully and thoroughly, the significance of data once thought not relevant (such as small metal fragments) clarified, sampling could be more rigorous, and technical studies more comprehensive. Interpretive frameworks have evolved tremendously since the mid-1970s and '80s, and the ones available today are vastly more suitable for interpreting the Ban Chiang metals assemblage than the ones available forty years ago. Also of benefit is that metals assemblages from several other sites in Thailand and in fact Eurasia add regional perspectives that only recently have been possible.

We hope this study provides a foundation upon which many future studies of the archaeometallurgy of Southeast Asia and many future Ph.D. dissertations can build. Numerous lines of future inquiry are suggested throughout the text. For present and future scholars of prehistoric technology, a new baseline is set in terms of both the kinds of data needed to address contemporary questions about past metallurgy and the essential techniques needed to study prehistoric metals assemblages. The questions that can be addressed with up-to-date archaeometallurgical methodologies have never been more interesting and relevant to contemporary archaeology. Proper study of metallurgical data offers some of the best means to uncover the social and economic networks of past societies and to document their shifts over time. Such fine-grained investigations of archaeological data are the best means by which archaeologists can contribute to social science perspectives on how past societies navigated the challenges they faced ranging from developing regional economies to resolving social conflicts.

It is not every archaeologist who has the opportunity to undertake scholarship involving a UNESCO World Heritage Site. It is an honor, a privilege, a challenge, and an inspiration. But a work like this cannot be accomplished by the lone wolf scholar. Teams of participants, at least some of which must engage in a sustained, years-long effort are required. Expertise from multiple types of individuals and scholars must be recruited, organized, and their work brought to bear on the endeavor—participants ranging from metallurgical scientists to archaeological illustrators to database managers. The funds to support the participants' work in a years-long or in this case decades-long project also must be raised.

The major funders of the excavations and associated field and specialist activities at Ban Chiang in 1974 and 1975 include the late Dorothy and Otto Haas, the Penn Museum, the Fine Arts Department of Thailand, and the Ford Foundation. The National Science Foundation Grant BNS-77-14425 funded the early stages of the analysis of the Ban Chiang materials in Philadelphia, including developing the original database of finds and the initial studies and conservation of Ban Chiang metal artifacts.

Several funders supported the research program after Chester Gorman passed away in 1981, including the late Betty Starr Cumin, Bruce and Peggy Mainwaring, the Penn Museum, and the University of Pennsylvania Research Foundation. In 1992, Joyce White founded the Friends of Ban Chiang to facilitate fundraising for ongoing work. John V. and Christie Hastings sponsored two "Thai Archaeology Challenge" grant programs during the late 1990s and early 2000s. In 1999, the Henry Luce Foundation provided a grant to the Penn Museum to support publishing Thai archaeology monographs. Receipt of this grant was facilitated by Leslie Laird Kruhly, then Director of Development at the Penn Museum. In addition to funding the editing of the first volume on the Ban Chiang excavations concerning the skeletal data, that Luce grant also allowed the hiring of Elizabeth Hamilton to systematize the metals data and reanalyze the metals assemblage. The beginning of drafting this volume and the posting of the first open access database on the Ban Chiang metals on the Southeast Asian Archaeology Scholarly Research website was also undertaken with that program.

A second grant from the Luce Foundation to the Penn Museum in 2008 for "Strengthening Southeast Asian Archaeology" supported aspects of the metallurgical work, in particular the migration of the data into the current version of the program's database as well as ongoing drafting of the current volume. The steadfast support of Helena Kolenda, Program Officer at Luce for Asia, for the Penn Museum's research and publication programs regarding Southeast Asian archaeology is acknowledged with profound gratitude.

After the Penn Museum ceased support of Penn's Southeast Asian archaeology research program in 2013 in order to prioritize gallery reinstallation and undergraduate education, several donors provided funding to establish the new Institute for Southeast Asian Archaeology (ISEAA), in order to continue the research and publication programs for the archaeology of Ban Chiang and Laos formerly conducted under the auspices of the Penn Museum. These donors included John V. and Christie Hastings, Criswell Gonzalez, Susan Catherwood, Bruce Mainwaring, Gretchen Hall, John Hover, and many other contributors to the ISEAA. The Hastings, Cris Gonzalez, Leslie Laird Kruhly, and Vincent Pigott became the board of the new Institute, and their belief in the program is deeply appreciated. A bequest to the Ban Chiang Project from a former volunteer, the late Ruth Brown, has been instrumental in funding the final stages of the monograph preparation, including final drafting, editing, illustration preparation, and layout. Elizabeth Van Horn provided additional funds to support editing. Finally, the Royal Thai Embassy in Washington, DC supported the upgrading of the online open access Ban Chiang metals database to provide an enhanced interface for scholars and the public interested in the metallurgical data from Ban Chiang.

The individuals whose efforts have made this study and monograph possible are also numerous. It is impossible to properly thank the scores of individuals who contributed to this study and this volume in one way or another. Many names are lost in the sands of time. In some cases, only categories of participants can be acknowledged. Nonetheless heartfelt thanks are directed to every individual who has ever assisted in the Ban Chiang excavation, post-excavation data organization and analyses, and publication preparations. The authors of this volume stand on your shoulders.

For the excavation, deep gratitude is extended to the Fine Arts Department of Thailand and to the late Chester Gorman and Pisit Charoenwongsa, co-directors of the excavations at Ban Chiang in 1974 and 1975. They and the excavation team developed a recording system that enabled a meticulous documentation of all metal finds recovered from all contexts from the site. They recruited and led a multi-national team where many of Thailand's future archaeologists got their early experiences in field archaeology.

William Schauffler, a graduate student at the University of Pennsylvania's Department of Anthropology, had several roles beyond field participant in 1974. He directed the excavations at Ban Tong, Ban Phak Top, and Don Klang, and he designed the original database in which the Ban Chiang metals and other data were recorded, again facilitating their meticulous examination, including application of current analytical and interpretive protocols.

The late Tamara Stech and Robert Maddin, pioneers in the field of archaeometallurgy, not only undertook the first metallographic and elemental analyses of metals excavated from Ban Chiang, they appreciated early on the significance of the Ban Chiang finds for global archaeometallurgy. While a professor at Penn's Laboratory for Research on the Structure of Matter, Bob Maddin also began teaching a course at Penn on archaeometallurgy, continued by Vincent Pigott after Bob moved to Cambridge. Ban Chiang, Ban Tong, Ban Phak Top, and Don Klang provided samples for several Penn students' projects, including those of young Southeast Asian archaeologists who were funded by the Ford Foundation for getting their M.S. degrees in Penn's Anthropology Department. This course facilitated the sampling and preliminary technical study of scores of metal artifacts from the four sites in our study corpus. Timbul Haryono from Indonesia, Surapol Natapintu from Thailand, and Christine Abiera conducted metallographic analysis on 26 copper-base prehistoric artifacts from the Ban Chiang study corpus. Rikke Giles analyzed six iron artifacts. Alissa Hinckley analyzed 50 copper-base artifacts from Ban Tong, Ban Phak Top and Don Klang, of which 20 were analyzed with SEM/EDS as well for her Master's study at Bradford University. This course also introduced Elizabeth Hamilton to the study of ancient metals.

Post-excavation analysis was assisted by innumerable students in the University of Pennsylvania's work-study program as well as by volunteers. These individuals labeled, coded, bagged, organized, catalogued, drew, tracked the metal and metal-related finds through lab protocols and various other tasks. One work-study student, Alissa Hinckley had a major impact on the course of study of the metals when she recognized the contribution of the fragmentary metal pieces, 80% of the collection, and reclassified and recoded them. She ultimately took 50 samples to work on her Master's Project at Bradford University, focusing on the high-tin bronze topic. Recently, editorial assistance with the monograph's references cited and proof-reading was undertaken by our work-study students Chloe Kaczvinsky, Meerabelle Jesuthasan, Sophie Spalding, and Anna Pugsley.

The contributing authors to this monograph, Elizabeth Hamilton, Vincent Pigott, William Vernon, Sam Nash, and T. Oliver Pryce, are gratefully acknowledged for their scholarship and for hanging in on the extremely long process that brought this study to completion. Elizabeth Hamilton has been the pillar upon which this volume stands. She conducted the reanalysis, wrote (and rewrote) chapters, guided the intellectual approach to the study and interpretation of the metals assemblages, organized and cleaned the data and databases, and worked closely with illustrators, authors, and editors.

Vincent Pigott contributed to this volume far beyond writing TAM 2C, chapter 2. While at MASCA (Museum Applied Science Center for Archaeology) he fostered the continuation of the analysis of the study collection after the initial studies by Tamara Stech and Bob Maddin. When he taught the archaeometallurgy course at Penn, he encouraged students to study samples from the collection, which enabled sampling, mounting, and the initial assessment of the metallography including photographic records of the microstructures. He mentored Southeast Asian students who were at Penn under Ford Foundation sponsorship who studied metallurgy. And Vincent served on the dissertation committees for both Elizabeth Hamilton and Joyce White at Penn and was the primary supervisor for T. Oliver Pryce's Ph.D. at University College London. He worked closely with both Sam Nash and William Vernon over the course of their long association with Penn Museum's archaeometallurgy research.

Vincent also read and provided detailed comments and edits to every chapter in this volume and sometimes to more than one draft. His contributions to TAM 2A, Chapter 7, and TAM 2C, chapters 4 and 5, were particularly extensive and helpful.

Ardeth Anderson has put years of work into the illustrations, graphics, maps, and other imagery in this volume. She also came to us as a work-study student from the Penn Department of Fine Arts, and spent several years doing measured artifact drawings and site plans. She has also supervised Penn work-study students and Ban Chiang Project volunteers over the years in the stippled drawings of metal artifacts.

Volunteers Dan LoMastro and Vivian Wolovitz have undertaken scanning of slides and drawing of metal-related artifacts that have contributed to this volume.

Several other scholars provided feedback on early drafts of particular chapters and sections, including Robert Giegengack on Chapter 6, Christopher Thornton on Chapter 4, and David Welch and Judy Voelker on portions of TAM 2C, chapter 4. Marie-Claude Boileau was very influential in formulating the underlying theoretical framework applying anthropology of technology and communities of practice perspectives on craft production. She provided feedback and suggestions for an early draft of Chapter 4 and portions of Chapter 2 in TAM 2A. Janice Cross and Page Selinsky, scholars in their own right, undertook the editing of the volume, and their sensitive and astute feedback, which improved the volume in numerous ways, is gratefully acknowledged. We are also extremely grateful to Jim Muhly for providing his unique historical and disciplinary perspective in the volume's Foreword.

Joyce C. White

1

Ban Chiang, Northeast Thailand, and the Archaeology of Prehistoric Metallurgy

Joyce C. White

Ban Chiang and Early Metallurgy

The discovery of prehistoric bronze artifacts at Ban Chiang, Thailand, was one of those archaeological finds that excited both scholars and laypeople. Even the *Washington Post's* political op-ed writer, Joseph Alsop, devoted a column entitled "Rewriting Human History" (1975) to musing on the implications of the finds for tearing apart the established narrative for the development of civilization. Although the initial astonishing dating of Ban Chiang bronze to the 4th millennium B.C. (Gorman and Charoenwongsa 1976) was rejected 35 years ago (White 1982, 1986), the rethinking of the role and significance of early metallurgy for the development of human society in Eurasia very much continues, and Ban Chiang and Southeast Asia are still key contributors to the conversation (e.g., Roberts 2011:143; Sherratt 2006; Thornton and Roberts 2014:4).

The Ban Chiang finds signaled the beginning of the unraveling of a 20th century archaeological paradigm that assumed technological progress was the driver of global human social change (e.g., Muhly 1976; Wertime 1973:876; see also Rainey 1992:244, 248; cf. Dunnell 1988). Although the details of the Ban Chiang finds and their significance have altered considerably since the 1970s, this book is still about paradigmatic change in how archaeologists view the human past and ancient metallurgy generally, with Ban Chiang and Southeast Asia as a key case study. The unfolding of the real significance of Ban Chiang and Southeast Asian metallurgy over the decades since the discovery has coincided with such changes in not only the study of ancient metallurgy (Thornton 2009b; Thornton and Roberts 2014), but also in the larger discipline of archaeology as a humanistic and social science (Hodder 2012a, 2012b).

Discussions of paradigms, methods, and theories are needed for the assessment of the significance and meaning of the finds within a contemporary framework. However, the primary focus of volume 2 in the series of monographs presenting the Ban Chiang excavations conducted by the University of Pennsylvania and the Fine Arts Department of Thailand (hereafter Penn/FAD) is on the archaeological and technological details of the metal finds from Ban Chiang and three contemporaneous sites in northeast Thailand (Ban Tong, Ban Phak Top, and Don Klang). It is these details that provide the evidence for reconstructing the role and meaning of metals in this particular part of the world during prehistory. The finds are also discussed in their wider regional context, especially the drainage basins of the Chao Phraya and middle Mekong Rivers. This chapter presents a broad overview of the issues.

Why Care about Early Southeast Asian Metallurgy?

The history of the discovery, the mastery, and the social ramifications of metal technologies has intrigued students of humanity's past for more than two millennia. The ca. 8th century B.C. Greek poet

Hesiod may have been among the first in western civilization to associate metals (in his case gold, silver, bronze, and iron) with "Ages"—temporal periods characterized by distinct sociopolitical, as well as spiritual and moral, qualities. In the ensuing centuries, "Ages" named after stone and metal technologies became a hallmark of European conceptualization of societal development from the earliest times, framing a western "origin narrative" (Grayson 1983). For the discipline of archaeology, this conceptualization became known in the 19th century as the "Three Age System." As the scientific discipline of archaeology gradually emerged over the 20th century, when, where, how, and why metallurgy began and then spread to other regions became central questions, with most attention focused on the Near East and Europe (e.g., Childe 1930; see reviews in Roberts et al. 2009; Wailes 1996a; and Chapter 3).

Southeast Asia was not thought to have had any noteworthy role in the investigation of early Old World metallurgy until the late 1960s and 1970s, when excavations at Non Nok Tha and Ban Chiang brought northeast Thailand into the discussions (Bayard 1971, 1972; Gorman and Charoenwongsa 1976; Solheim 1968). Since then, debates on the dating and significance of the early bronze and iron in this region have raged and shape-shifted as chronologies were revised, new data emerged, and intellectual trends changed. Volume 2 of the Thai Archaeology Monograph Series is published at a juncture of hot debate and paradigm change not only within Southeast Asian archaeology (e.g., Higham et al. 2011a; Higham et al. 2011b; Pigott and Ciarla 2007; Pryce et al. 2010; White 2017; White and Hamilton 2009), but also in the study of early metallurgy generally (Killick and Fenn 2012; Thornton 2009a, 2009b; Thornton and Roberts 2014; see also Chapters 3 and 4). The data and perspectives presented here are likely to be a part of scholarly conversations about early metallurgy for years to come.

The Unexpected Place

Excavations at Ban Chiang by the Penn Museum and the Fine Arts Department of Thailand in the mid-1970s came at a time when most archaeologists thought that metallurgy first emerged in the Near East, with tin bronze appearing in some locations after 3000 B.C. (Clark 1969:109; see Chapter 3 for full discussion). The initial interpretation of the dating for the earliest bronze artifacts excavated at Ban Chiang (Gorman and Charoenwongsa 1976) at first seemed to raise the possibility that alloying copper and tin to form bronze in Thailand may have preceded similar metallurgy not only in China but even in the Near East and Europe. It was suggested that Thailand might even have been their source for the technology (Muhly 1976:12) or for the tin used in the Near East (Selimkhanov 1984, 1996–97).

The significance of these possibilities stimulated articles in the *New York Times*, *Washington Post*, and *Time Magazine*. As Muhly noted (1976:11), "we are not just talking about where and how ancient man learned to make an alloy of copper and tin. What is involved here is the nature of one of the basic advances in man's technology *and the challenge to a long-cherished paradigm of historic development*" (emphasis added). This paradigm saw technological discoveries as primary indices of advance in human civilization, as indicators of ever increasing mastery over the physical world, as *progress*. Moreover, the origins and early development of metal technology was seen as intimately intertwined with, and an important causative factor of, other advances in civilization such as states, writing, social hierarchies, and economic systems dependent on divisions of labor among classes. The origin of metallurgy was also entwined with an "origin myth"—that western civilization began in Mesopotamia, and technological and other advances spread from there to other parts of the Old World. The development of bronze and iron metallurgy was seen as intimately interrelated with the development of organized warfare and its association with potentially oppressive and often imperialistic states (see fuller discussion in Chapter 3).

Revision by White (1982, 1986) of the original dating of the earliest bronze at Ban Chiang from 3600 B.C. (Gorman and Charoenwongsa 1976) to ca. 2000 B.C. and iron from 1600 B.C. to after 800 B.C. relegated the argument for Southeast Asian metallurgical primacy to the archaeological trash bin. Nevertheless, even with a date in the early 2nd millennium B.C., bronze product manufacturing in Southeast Asian villages placed the technology at a surprisingly early age and in an unexpected place and social context. Remote from early urban societies, with no associated evidence of warfare or class society, in an area that in recent times has been impoverished, it was difficult for scholars of human history to explain with the paradigms of the time how and why these societies had such mastery over a complicated technology such as bronze metallurgy. Of course, at the time Ban Chiang was excavated, there were mammoth gaps in the archaeological record of Asia and particularly Southeast Asia (Clark 1977). Since then, more data from Thailand and other parts of the Asian Old World have accumulated to fill in at least some of the gaps in archaeological understanding (e.g., Chernykh 1992; Frachetti 2012; Higham 2002; Linduff 2000). New dates from Ban Chiang and other sites have further complicated the picture (Higham et al. 2011a, 2015; J. White 2008). Yet scholars are still grappling with how to interpret the prehistoric metallurgy of Ban Chiang, Thailand, and Southeast Asia in general (Higham et al. 2011b, 2015; Pigott 2012a; Pigott and Ciarla 2007; Pryce 2009; White 2015a, 2015b; White and Hamilton 2009, 2014).

Southeast Asia's challenge to entrenched interpretations of the "Bronze and Iron Ages" remains (Muhly 1988:16; Pigott 1998, 2012a; White and Hamilton 2009). Prehistoric metallurgy and metal age societies continue to be vigorously investigated in Thailand, and understanding the significance of metals in prehistoric Thailand has provided one of the most important stimuli for recent expansion of archaeological research in the region (e.g., Ciarla 2007a; Natapintu 1988, 1991; Pigott and Natapintu 1988; Pigott et al. 1992; Pryce 2012; Pryce et al. 2014; Sayavongkhamdy et al. 2009). Scholarly disagreements and perspectives have evolved as new sites have been dug (Higham and Kijngam 2012a, 2012b; Higham et al. 2011b), chronologies have been revised (Rispoli et al. 2013), socioeconomic evidence has been amassed (White and Pigott 1996), and archaeology of neighboring regions such as southern People's Republic of China (hereafter PRC) has advanced (Ciarla 2007b). Fundamental new data will continue to accumulate, as large portions of core areas, such as Laos and Myanmar (Burma), as well as much of southwest and southern PRC are still virtually archaeological *terra incognita* for the relevant time periods. Debates and research on this topic persist up to the present (Higham 2002:116; Higham et al. 2011a, 2015; Pigott 2012b; Rispoli et al. 2013; White 2015a, 2015b; White and Hamilton 2009, 2014). Scholars specializing in other parts of the Old World observe the data coming out of Thailand, anticipating that it will shed light on their own regions (Mair 1998:20–21; Mei et al. 2015; Pigott 2012a; Roberts 2011; Sherratt 2006). Southeast Asia is one of the geographic areas with so far only "meagre" archaeometallurgical data that Rehren and Belford (2013:6) think offers a fascinating opportunity "to study mechanisms and effects of the creolization of technology." In short, the Ban Chiang metals remain relevant not only for regional archaeology but also for global archaeology.

Aims of this Study

This study has four main objectives: (1) to lay out detailed excavated and analytical evidence about metal technologies from four prehistoric sites in northern northeast Thailand: Ban Chiang, Ban Tong, Ban Phak Top, and Don Klang (Fig. 1.1); (2) to place those data in a larger regional context of the middle Mekong and Chao Phraya Basins, using comparative metallurgical data from other sites in northeast Thailand, central Thailand, and Laos; (3) to present current theoretical perspectives that can position the empirical data in their social, economic, and political contexts; and (4) to contribute to the ongoing debate between Joyce White and Charles Higham regarding the importance and interpretation of prehistoric metallurgy at Ban Chiang specifically and in Thailand and Southeast Asia generally. The evidence and discussions are published in four parts of the Penn Museum's Thai Archaeology Monograph (TAM) series: TAM 2A background; TAM 2B data from the four study sites; TAM 2C regional perspectives and syntheses including a summary critique of the Higham metal age model; and TAM 2D catalogs of the metals, crucibles, and prills.

The four sites discussed in detail were excavated in 1974 and 1975 by the Northeast Thailand Archaeological Project, or NETAP, a joint Penn/FAD research program. These four sites seem to be mixed mortuary/occupation sites. That is, the deposits resulted from a combination of mortuary activities that were predominantly human burials, and occupation activities resulting from village life. Although evidence for small-scale metal product manufacturing activities at these four sites is presented in TAM 2B (chapter 5), in general these sites represent primarily "end users" of metal products; their metallurgical remains shed light primarily on metal consumption.

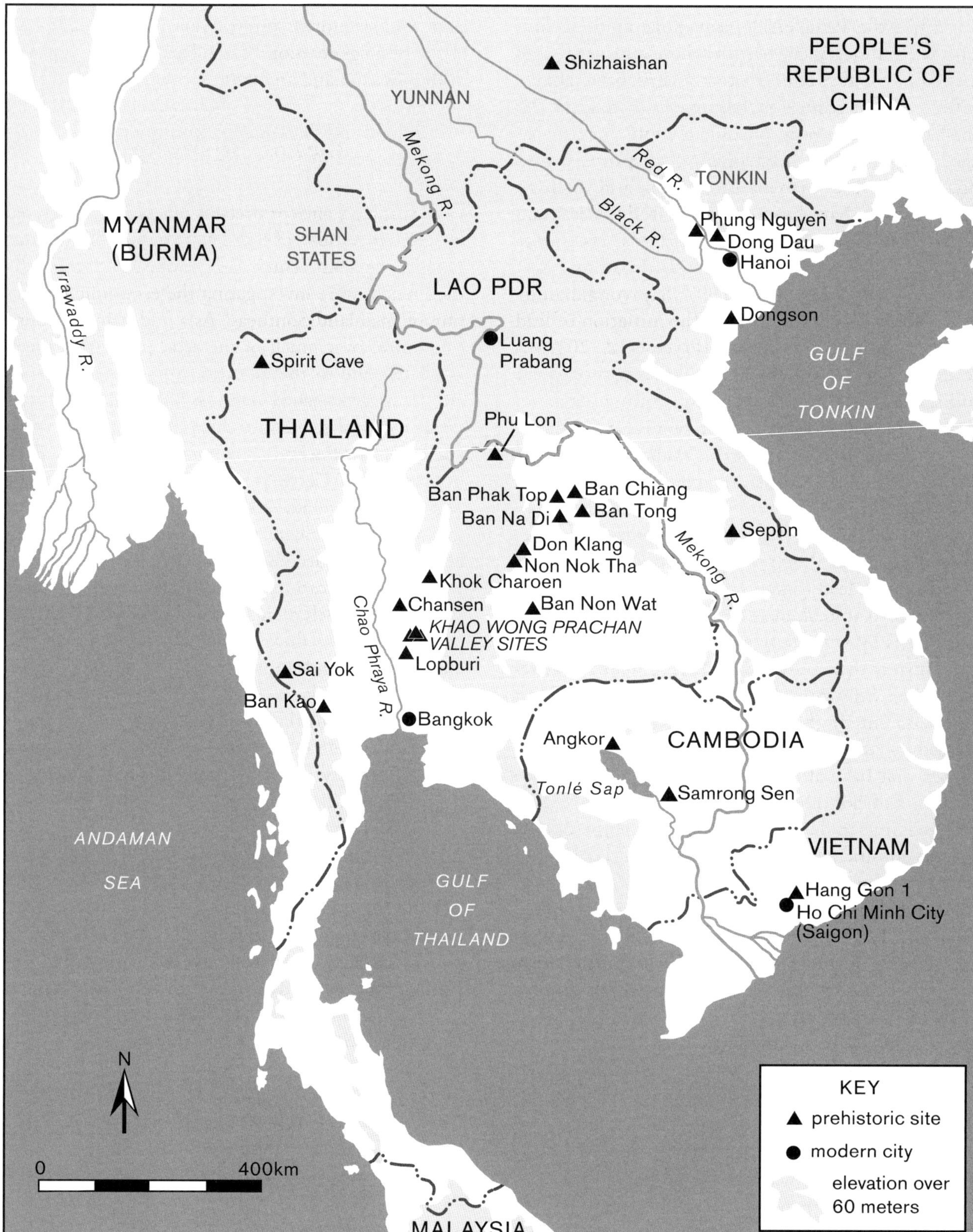

Figure 1.1 Location of Ban Chiang and other sites in Southeast Asia mentioned in Chapter 1.

Since the Penn/FAD excavations at these four sites were undertaken, work at more consumer sites as well as metal producer sites in Thailand enables a fuller regional picture to be presented in TAM 2C than would have been possible a couple of decades ago. Sites that had been involved in primary production activities, namely the mining and smelting of metal ores, have been excavated in northeast and central Thailand and Laos (Ciarla 2007a, 2008; Pigott and Natapintu 1988, 1996–97; Pigott and Weisgerber 1998; Pigott et al. 1997; Sayavongkhamdy et al. 2009; Tucci et al. 2014). The initiation of lead isotope studies in the region (Pryce et al. 2014) is revealing specific relationships between prehistoric metal producing and consuming communities and, thus, greatly enriching the regional picture.

In addition to documenting data, this study also strives to place the prehistoric metal production and consumption in Thailand into a theoretically informed contemporary interpretive framework. Data do not speak for themselves. Mere accumulation of data does not lead to understanding; rather the continual efforts to make sense of data, old and new, are the means by which understanding unfolds. Therefore, just as this study seeks to document metal-related data from four sites, this volume also strives to document the background and issues that have shaped its creation, including changes in intellectual trends that have affected how the metals from Ban Chiang and Southeast Asia have been perceived by scholars both in the past and currently. Appreciation for the empirical evidence will begin in this chapter with a review of a history of perspectives on prehistoric metals from mainland Southeast Asia prior to the Penn/FAD excavations at Ban Chiang. Subsequent parts of the chapter set the interpretive stage for understanding the Ban Chiang metals from a contemporary point of view.

Finally, an important aim of this study is to present a reinterpretation of the role and place of metallurgy during Southeast Asian prehistory in contradistinction to the prevalent interpretation advocated primarily by Charles Higham over the past several decades (e.g., Higham 1984a, 1996a; Higham et al. 2011a, 2011b). The methodologies, theoretical perspectives, global evidence for early metallurgy and empirical data from Ban Chiang and other prehistoric sites in Thailand presented in this study are synthesized in TAM 2C (chapters 4, 5, and 6). A new socioeconomic interpretation is proposed for the coming generation of scholars to debate and test with new data and analyses.

The Lost Bronze Age

How have ancient metal artifacts been perceived by past generations of Southeast Asian prehistorians? Late in the 19th century as Europeans, particularly the French, were investigating the economic geography of mainland Southeast Asia and gaining political control over much of the area, prehistoric metal artifacts came to the attention of several explorers. Bronze objects such as bangles, points, axes, and fish hooks were collected from the ground, from markets, and as gifts by the French at least as early as the 1870s in Cambodia (Corre 1879a, 1879b; Noulet 1879). Socketed bronze adzes were collected from Yunnan and the northern Shan states and displayed at the British Museum (Cartailhac 1890:646; Lefèvre-Pontalis 1904:24). Publications from the *Mission Pavie* (Pavie 1904), the French commercial and scientific expedition to what was then known as "Indo-Chine," discussed similar finds in several articles. Grouped into a section entitled "Préhistoire," these articles document several collections that included bronzes, along with greater numbers of polished stone adzes, compiled by 19th century Europeans throughout "Sud-Est de l'Asie" (Fig. 1.2). In one article, for example, Massie (1904) presented a catalogue of 138 stone and 20 bronze artifacts collected in the vicinity of Luang Prabang, Laos, between 1888 and 1892 (Fig. 1.3). The bronze artifacts comprised eleven axes (haches),[1] one needle (aiguille), two fish hooks (hameçons), four chisels (ciseaux), one lance, and one agricultural implement. By the end of the 19th century, a similar range of bronze objects, especially socketed adzes, had been recovered from the territory extending from Burma (now Myanmar) east to Tonkin, Vietnam, and from Yunnan in southwest China south to the Mekong Delta (Lefèvre-Pontalis 1904).

Initial Applications of the Three Age System

What did the scholars of the day make of these early finds, lacking as they did any direct means of

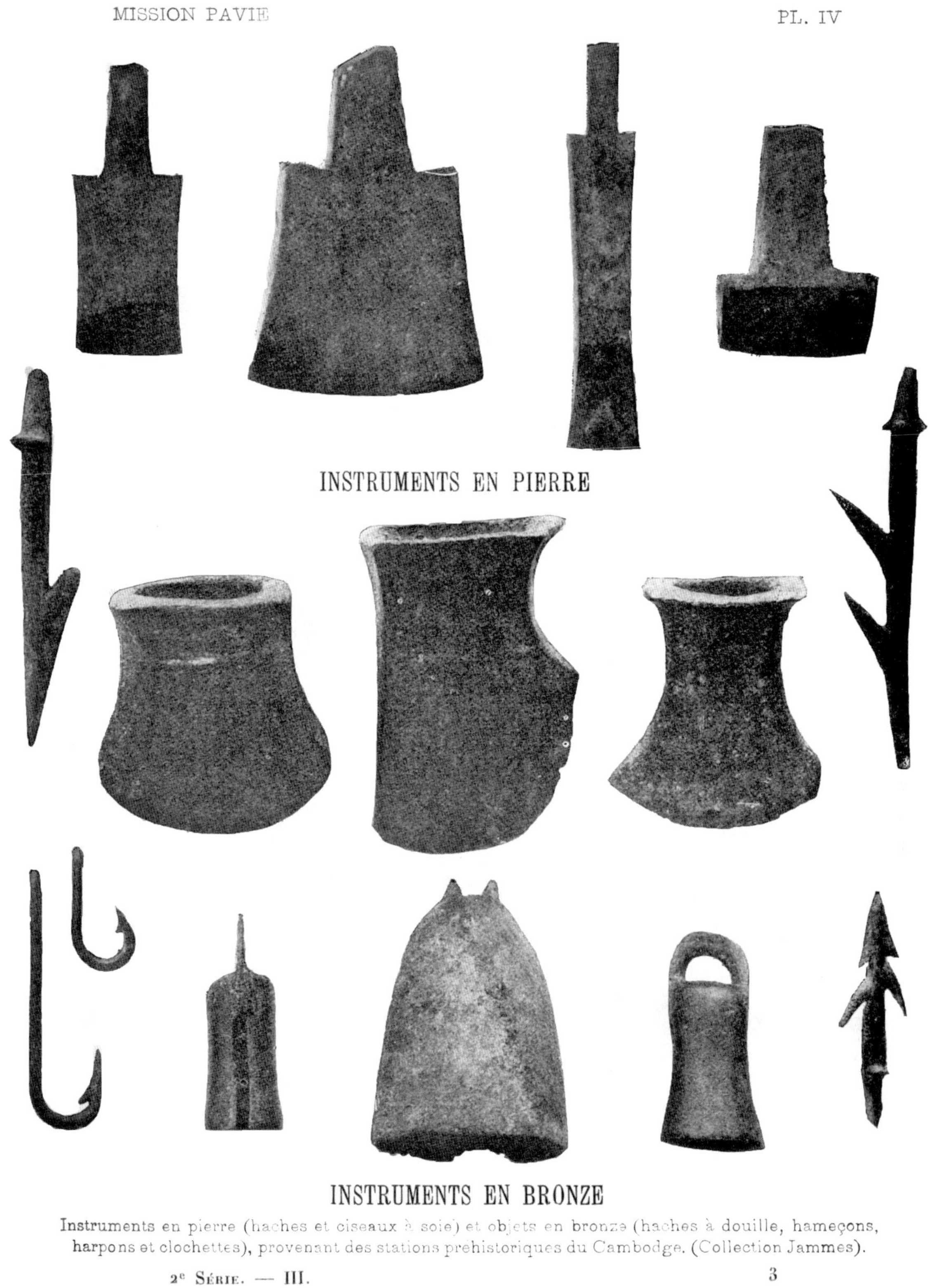

Figure 1.2 Finds of bronze and polished stone recovered from Cambodia by Jammes and reported by Massie (1904).

MISSION PAVIE PL. III

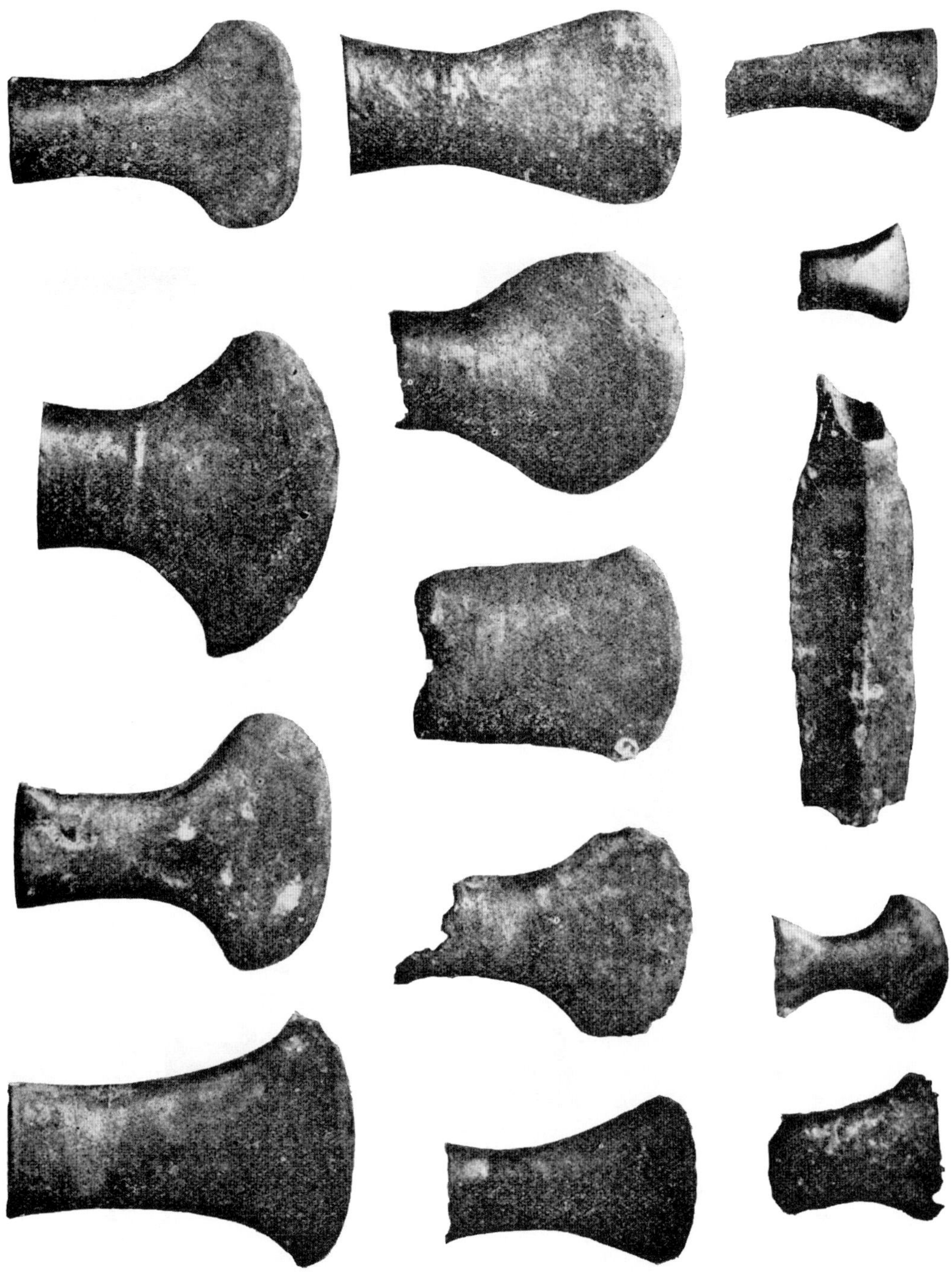

Haches et ciseaux à douille, en bronze. (Collection Massie.)

Figure 1.3 Copper-base socketed adzes and a spear point collected by Massie in Laos during the Mission Pavie (Massie 1904).

determining their absolute age, or even relative age via in situ stratigraphic contexts? Indigenous explanations of the bronzes described them as products of bolts of lightning or other supernatural conceptions, similar to older explanations of similar artifacts found on European soil (Pavie 1904:5). The Europeans, however, attributed the artifacts to societies that predated the historic Southeast Asian civilizations such as the Khmer at Angkor (e.g., Cartailhac 1890:644). The authors in the Pavie (1904) volume generally placed the bronze artifacts in the context of "l'âge du bronze" (bronze age) that emerged from "l'âge de la pierre" (stone age), or more specifically the later part of that phase termed "l'âge de la pierre polie" (age of polished stone). It is clear that they are using these "ages" in the sense of "technologically-based periodization" (Bacus 2004:258; see also Massie 1904:10; Pavie 1904:1–2). They assumed that the same sequence of technological advance and progress articulated for Europe could be applied here, i.e., the materials from which tools were made began with stone, then progressed to bronze, then to iron, and that the Three Age System applied equally well to Southeast Asia as to Europe. That most of these early collections came from surface finds, gifts, or purchases from local inhabitants helps explain why they are minimally mentioned in more recent archeologically based discussions of early metallurgy in Southeast Asia (except see Higham 1996a).

For a few decades in the late 19th and early 20th centuries, the copper-base finds noted above were discussed as typical of a Southeast Asian bronze age (Noulet 1879; Pavie 1904). Of course, the temporal depth of a purported bronze age could only be guessed, since at that time radiocarbon and other chronometric dating techniques were not yet available. Nonetheless, it is noteworthy that the basic geographic and typological parameters of what more recently has been termed the "Southeast Asian Metallurgical Province" (White 1982:48, 1988) were delineated more than 100 years ago.

Many discussions compared the Southeast Asian finds to artifacts that were considered prehistoric from Europe and elsewhere (e.g., Mansuy 1920; Noulet 1879). The ubiquitous adzes and axes intrigued the early observers. Cartailhac (1890:647), for example, concluded after discussing five bronze adzes found by Moura at Samrong Sen, "on voit que ce sont, en somme, de menus objets qu'il serait imprudent de classer soit parmi les armes, soit parmi les outils d'usage vulgaire" (we see that they are, in fact, small objects and it would be imprudent to classify them as either arms or utilitarian tools). Much attention was paid to comparing the shapes of stone and bronze adze/axes to other Eurasian examples. Interestingly, some of the early authors (e.g., Patte 1936:299) compared socketed bronze axes to similar ones found in Russia and Siberia (an observation later pursued by White and Hamilton [2009, 2014]). Although iron artifacts were not noted among the earliest Southeast Asian collections, it was assumed that iron technology eventually replaced bronze in the area at least for tools (Lefèvre-Pontalis 1904:25).

Although similar copper-base objects were found throughout the region, some variation in types and frequency of finds were noted. Bells and harpoons were only noted in Cambodia, for example. More than one collector (Verneau 1904:38; see also Lefèvre-Pontalis 1904:23) observed that "Dans le Sud de la péninsule, le bronze est plus rare que dans le Nord." (In the south of the peninsula [referencing the time when Indo-china, Thailand, and Malaya together were called the "golden peninsula," Keyes 1977], bronze is rarer than in the north.) A collector of prehistoric items in Cambodia, Jammes, perhaps disappointed in his recovery of bronzes, even went so far as to call "Laos, *le pays du bronze*" (Lefèvre-Pontalis 1904:23; emphasis in original). [Laos, the country of bronze.] Verneau's (1904) assessment of the regional distribution of bronze finds and their abundance in the northwest relative to east and south of the region concluded, "que les importateurs de l'industrie métallurgique ont pénétré par l'angle Nord-Ouest dans la péninsule indo-chinoise" (the importers of metallurgy penetrated the Indochinese peninsula at an angle from the northwest).

"L'âge du bronze" was not used initially to assert implications regarding warfare, states, or variations in wealth or social classes. Over time, however, some scholars began to examine relationships between these technologically-based periods of time and "social stages," (e.g., Worman 1949a) although neutral voices can also be found throughout the history of discussions about the societal implications of the prehistoric bronzes (e.g., Mansuy 1920). Nevertheless, the problems in applying the European Three

Age System to Southeast Asia quickly emerged once it appeared that at a site called Samrong Sen polished stones and bronzes were apparently contemporaneous.

Samrong Sen

Samrong Sen was the first prehistoric site discovered in Cambodia, and it became a destination location for colonial bureaucrats and travelers during the late 19th and early 20th centuries. Numerous collections were undertaken near this huge shell midden, which had at least six meters of deposit peppered with mortuary interments. The finds contributed to much speculation about the region's prehistoric past. Heine-Geldern (1946:164–165) gave a sense of Samrong Sen's perceived potential when he stated, "If we except the Mediterranean countries and the ancient Near East, there are few prehistoric sites in the world which could have equaled its enormous wealth in pottery and in weapons, tools and ornaments of stone, bone, shell and bronze.... Had it been systematically excavated, it could have given us a complete chronology and sequence of cultural phases for a large part of the neolithic and bronze age periods of Indo-China. Indeed, it could have played a similar role for the prehistory of Southeast Asia, on a somewhat smaller scale, a role similar to that which Troy holds in the prehistory of Anatolia and southeastern Europe." Unfortunately, this clearly important site was largely destroyed in the early decades of the 20th century as its shell was mined for making lime, although recently some remnants have been excavated by Ly (2001). Nevertheless, this site, its dispersed collections, and the early discussions about it have remained of interest to recent prehistorians of Southeast Asia (Demeter et al. 2002; Heng 2007; Ly 2001; Murowchick 1988a, 1988b).

Due to an apparent association of bronze artifacts with an overall stone age context, the site introduced some Three Age confusion into the discussions of Southeast Asian prehistory. Numerous visits, collections, and "excavations" (with the methodologies of the times) took place at the site between 1876 and 1890 (reviewed in Worman 1949a, 1949b), recovering numerous polished stone tools, many as grave goods. A smaller number of bronze rings, bells, chisels, hooks, projectile points, and socketed axes were collected from local villagers (Cartailhac 1890; Corre 1879b). One of the Europeans who dug pits at the site, Jammes, claimed that bronze and stone artifacts came from the same level (Cartailhac 1890:644). However, Mansuy also excavated at the site in about 1901 and again in 1923 but did not find bronze in situ. Although the precise provenience of the bronzes associated with Samrong Sen has always been highly uncertain (even Cartailhac [1890:644] complained of the lack of detail relating to provenience provided by Jammes), the circumstantial evidence was convincing enough for some Europeans to consider that the bronze, though rare, came from contexts where stone tools predominated (Patte 1936:298). The implication was that bronze was in use by an essentially stone age society (Patte 1936:280).

Three Age Struggles

Beginning in the late 19th century but continuing throughout the twentieth, discussions of prehistoric finds in Southeast Asia reveal an uncertainty as to how to fit the finds within a Three Age framework (e.g., Lefèvre-Pontalis 1904; Worman 1949a, 1949b). The situation was brought to the fore by Samrong Sen, where bronze tools appeared to have been rare in contexts where a lively industry of polished stone tools prevailed (see for example Patte 1936). What label should be applied to the Samrong Sen society? Should it be "Bronze Age," emphasizing the mere presence, though rare, of metal technology; or "Stone Age," acknowledging the predominant tool material (e.g., Cartailhac 1877:98); or the "Age of Polished Stone"; or alternatively "New Stone Age," Lubbock's "Neolithic" (Lefèvre-Pontalis 1904:20), emphasizing the prevalent polished stone tool technology? The most accurately descriptive "l'âge de la pierre polie et du bronze" (age of polished stone and bronze) used by Corre (1879a, 1879b) and Noulet (1879) never caught on. Other treatments wrote of a transitional phase between the stone and bronze ages (Chauvet 1928:292).

The same terminological struggles with applying the Three Age System to a Southeast Asian context recurred throughout the 20th century. Many years after the initial discoveries at Samrong Sen, Worman (1949b:327), unconvinced that any bronze was truly associated with the lithic industry and asserting there was no evidence at the site for agriculture to merit even an attribution to the Neolithic (in the early 20th century V. Gordon Childe and others had firmly linked the term "Neolithic" with the presence of agriculture), concluded that Samrong Sen was a "purely Stone Age deposit." Some publications emphasizing the prevalence of the stone technology noted that bronze appeared initially in "late Neolithic" societies, sometimes termed "Néolithique supérieur," or even "Enéolithique" (Saurin 1971:35). Struggles about Three Age terminology recur today, decades later (see Higham 1989:28, 1996a:7; Nguyen et al. 2004; White 2002, 2017). Worman (1949b:328) would not be the last to exclaim, "It [the confusion pertaining to the Southeast Asian bronze age] clearly shows the need throughout Indo-China for a careful reexamination of early archaeological work and for a well-planned program of modern excavation by professional archaeologists" (cf. Higham 2015a:1214).

The early literature also reveals disagreements as to age and duration of a Southeast Asian bronze age (e.g., Pavie 1904:26), with appropriate caveats reflecting the realization that they did not know the actual age of the artifacts and, thus, could not at that time prove their various propositions. Both a "short" chronology of a few centuries before the Christian era (Fuchs 1882:365) and a "longer" chronology for an advanced civilization of thousands of years of duration prior to the Angkorian civilization (Jammes as discussed by Lefèvre-Pontalis 1904:26) were proposed. Although the specifics on dates have changed, the age, duration, and significance of the bronze age in Southeast Asia have remained controversial up to the present among regional archaeologists, even with metal artifacts now recovered from in situ, presumably datable, contexts (e.g., Higham et al. 2011a, 2011b, 2015; Pigott 2012a; Pigott and Ciarla 2007; J. White 2008, 2015b; White and Hamilton 2009).

In the early decades of the 20th century, archaeological principles of stratigraphy, assemblage analysis, and typology came to be applied more systematically to prehistoric finds in Southeast Asia. Mansuy (1920) provided a synthesis of the industries of polished stone and bronze tools, making many observations that still hold today. He noted, for example, that bronze and stone tools appeared to be contemporaneous and presumed that bronze appeared in the region when a polished stone tool industry was at its apex. He also noted that bronze and stone weapons are rarely found, in contrast to their prominence in European prehistoric assemblages (Mansuy 1920:11). He characterized the bronze finds in Southeast Asia as having a simple range of forms and narrow inventory of objects, with predominance of a single tool type—the socketed adze in a limited range of sizes, and of fragile construction (i.e., they are small, light, and thin at least in comparison to, for example, European copper/bronze shaft hole adze/axes). Overall his comparison of European and Southeast Asian bronzes led him to conclude (Mansuy 1920:10) they were so different in so many aspects that "ces deux industries sont complètement indépendantes" (the two industries are completely independent). Mansuy (1920:12) allowed that the bronze technology in Southeast Asia could be indigenous or imported from India or China. Generally, his references to the Three Age System were sparse; he wrote mostly of the "l'industrie du bronze" seemingly to avoid "l'âge du bronze."[2]

Post-Dongson Revisionism

The perception of a bronze age in Southeast Asia shifted markedly beginning in the late 1920s with the discovery and excavation of the site of Dongson in Vietnam (Goloubew 1929). A wider range of bronze artifacts (including large and elaborately decorated drums) and greater prevalence of armaments (including swords, spearheads, daggers, and arrowheads) were recovered in situ associated with rare examples of iron. Artifacts of clear historic Chinese origin provided a means, for the first time, to "date" a prehistoric culture in Southeast Asia. Coins and armaments from the historical Chinese Han dynasty associated with Dongson contexts indicated a date of the first centuries preceding and postdating the birth of Christ (Goloubew 1929, 1937).

Subsequent discussions of Dongson culture remains, such as by Janse (1947) and Karlgren (1942), argued that Dongson could date from the 3rd or 4th century B.C. The earliest "date" proposed for the Dongson in this pre-radiocarbon period was ca. 800 B.C. by Heine-Geldern (1937). His reasoning compared Dongson motifs (e.g., double spirals) with similar motifs found in late Chou dynasty art. He asserted that these motifs originated in the Caucasus of eastern Europe and were brought to Eastern Asia by Thraco-Cimmerian tribes between about 800 and 600 B.C. (1937:191–194). His view of commonalities between Southeast Asian and European bronzes thus contrasted with Mansuy's assessment of the independence of the two traditions.

Despite the presence of iron, during the mid-20th century Dongson came to be considered by some the "typical Bronze Age" for Southeast Asia (Heine-Geldern 1935; Saurin 1971:35; Van Heekeren 1958; see discussions in Glover and Syme 1993:46 and Solheim 1982:168). At its first discovery, however, Goloubew (1932:139) did not collapse all prehistoric bronze finds in Southeast Asia into the Dongson period, stating, "Le technique du bronze était sans doute connue au Tonkin bien avant l'époque représentée par les sépultures de Đòng-sơn" (Bronze technology without doubt was known in Tonkin well before the period represented by the Dongson tombs).

However, in the mid-1930s pre-Dongson bronze-using societies in Tonkin and elsewhere in Southeast Asia ceased to be envisioned. Heine-Geldern (1935) was perhaps the first proponent of this revision, but the view is also clearly entrenched by Pearson's (1962) review and synthesis of Dongson. Thus, all prehistoric mainland and island Southeast Asian bronzes were collapsed into a Dongson culture (Heine-Geldern 1935:315) or Dongson complex (Pearson 1962), that ranged geographically from Yunnan to Vietnam to Indonesia.

This reinterpretation of early bronze in Southeast Asia reflected mid-century trends in archaeological scholarship, and it provides a good example of how a theoretical framework can influence chronological interpretation by archaeologists. At the time Dongson was discovered, Euro-American anthropological theorists had been developing a view of world prehistory that emphasized diffusion of culture traits from centers of "superior civilization" to peripheral areas. The "culture area concept" and *Kulturkreis* (culture circle) views were in part developed in reaction to 19th century cultural evolutionists. Proponents of the theory of *Kulturkreis* sought to refocus explanations of culture change away from processes of progress internal to societies (i.e., cultural evolution) and toward contacts with external societies (Rebay-Salisbury 2011). Concepts of technological progress were mixed with migration and diffusion scenarios, as well as hierarchical relations among stages of civilization. In the absence of absolute dates from coins or historic texts, these theories were used to place archaeological cultures into temporal frameworks. Combining culture circles with a sequence of "culture grades," any part of the world could be interpreted in terms of its position in a hierarchical scheme of cultural strata from primitive to civilized or superior. Each culture was viewed as a product of varying combinations of diffusion of traits and migrations of people.

The volume and splendor of the Dongson bronze finds in Tonkin matched an expectation for a "superior" civilization for Southeast Asia, in some ways comparable with Near Eastern and Egyptian Bronze Ages (ignoring the presence of iron, however). In the 1920s and early 1930s, Sir Leonard Woolly's excavations at Ur, and his uncovering of the Royal Tombs and ample finds of bronze vessels, weapons, and ornaments, helped define the Bronze Age as both a technological and sociopolitical entity. According to the *Kulturkreis* thinking of the time, the parts of Southeast Asia where small finds of bronze axes were found in essentially stone age contexts, in the absence of absolute dating, began to be perceived as derivative of the superior civilization, rather than as derivative of antecedent societies.

The discovery and excavation of Shizaishan in Yunnan during the 1950s, with its variant bronze drums, rare iron, and associated Han Chinese artifacts, further consolidated the picture of a Tonkin-Yunnan corridor defined by the Red and Black River Basins where a high "Bronze Age" existed dating to the late 1st millennium B.C. The finds of bronzes outside of the Tonkin-Yunnan corridor, such as the common socketed adze/axes collected in the Mekong drainage basin, came to be thought of as related to Dongson and of the same or younger age (Pearson 1962:34). The prehistoric bronze outside of the Tonkin-Yunnan corridor came to be seen as illustrating the spread of aspects of Dongson culture via trade (Pearson 1962:44), and only bronze objects useful in "hinterland markets," such as ornaments and tools, found their way to these areas that were seen as peripheral to "the true bronze culture" in Tonkin.

This brief foray into mid-20th century archaeological theory is more than just a review of outdated ideas. As Rebay-Salisbury (2011:56) in her discussion of hidden paradigms underlying current archaeological explanations noted, "Remnants of the *Kulturkreislehre* still influence the way in which the archaeological record and archaeological cultures are thought about and dealt with."

The Intellectual Context When Ban Chiang Bronze Was Uncovered

World War II, the French Indochina War, and the American Vietnam War curtailed the progress of archaeological research in mainland Southeast Asia and isolated the region from contact with archaeologists from developed countries and concomitantly from current method and theory. Of the geopolitically isolated countries, only northern Vietnam continued any significant amount of research into prehistory, building upon the groundwork laid during the French period (Davidson 1979).

By the mid-20th century, the possibility recognized earlier that a prehistoric Southeast Asian bronze technology could have existed that was older than Dongson and outside of the Tonkin-Yunnan corridor seems to have been lost. This mid-century viewpoint is clearly evident in the first two editions of Grahame Clark's global archaeological synthesis, *World Prehistory*, each edition having had numerous printings published between 1961 and 1977.

> Neither south-east Asia, Indonesia nor the Philippines experienced a phase of technology fully comparable with the Bronze Age in certain parts of the Old World. Yet, while stone tools continued in general use into the Christian era, a certain number of bronze artifacts, named after the rich settlement and cemetery of Dong So'n in northern Annam, found their way over these territories during the latter half of the first millennium B.C. and in the richer graves of Annam these were sometimes accompanied by objects made of iron.... Bronze working...may have been established in Annam by the sixth century B.C.... The Dongson bronzes affected only a small segment of the population, which remained as a whole in a basically Neolithic stage until the general use of iron was spread by Hindu merchants...during the early part of the Christian era. (Clark 1969:238–239)

Clark's statement, with its *Kulturekreis* overtones, makes clear that in the 1960s and early 1970s, at the time when Ban Chiang was discovered and excavated, archaeologists did not think that Southeast Asia had ever had a discrete bronze age. Clark's comment that most of the population remained at a "Neolithic stage" despite the presence of bronze technology further illustrates the scholarly tendency of the time to pigeonhole societies into stages of development, and the preference that the stages be characterized by particular levels of technology. The incongruities that resulted from applying these criteria to the Southeast Asian evidence, where societies were at a "Neolithic stage" despite having access to bronze and even iron technology, were soon to be laid bare. Even though the second (1969) edition of Clark's *World Prehistory* appeared after a handful of radiocarbon determinations from Southeast Asian contexts were first published, there is no hint in his synthesis of the revolution about to occur as chronometric dates began to be incorporated into understanding the prehistory of the region.

The Radiocarbon Revolution

It is hard for those born after 1950 who have an interest in prehistory to imagine life before radiocarbon dating. The age for anything prehistoric—lacking any known historic link even from a stray coin—was at worst pure guesswork or at best inferences based on a variety of suppositions such as those of the *Kulturekreis* school. The introduction of absolute dating into the picture was sure to upset some scholars' deeply cherished beliefs.

Radiocarbon dating, first developed in 1949, was slow to be applied to sites in Southeast Asia, with the first dates coming from Niah Cave on Borneo in the late 1950s (Bronson and White 1992:491). A handful of dates trickled out of the region in the 1960s (see references and data compiled in Bronson and White 1992), and the first radiocarbon determinations published from the prehistoric period of mainland Southeast Asia included a few from each country: Cambodia (e.g., Carbonnel and Delibrias 1968), Thailand (e.g., Gorman 1969; Sørensen 1967), peninsular Malaysia (Dunn 1966), and Vietnam (Delibrias et al. 1965:238). The new dates laid the groundwork for significant changes in prior conceptions of the ages of different phenomena, ranging from the presence of *Homo sapiens* to the appearance of horticulture (Bronson and White 1992:491).

Dates associated with bronze were also surprising. Saurin (1968:3) referred to a date of about 2000 B.C. associated with axe molds from Hang Gon 1 near Saigon (Sa-205; Delibrias et al. 1965:238). In northern Vietnam, the Phung Nguyen complex discovered in 1959 provided clear evidence for pre-Dongson bronze, according to overviews by Davidson (1975:89–90, 1979), who noted finds of copper alloy lumps associated with Phung Nguyen deposits. A date from the site of Dong Dau, just younger than the Phung Nguyen phase and undisputedly from a society that cast bronzes, was 3328±100 B.P. (Davidson 1979:117 n. 8 indicates this date was published in *Khảo Cổ Học* in 1969.) This date today would be calibrated to roughly 1600 B.C. (see also Nguyen Quang Mien 2005 for more recent Dong Dau dating). Due to the small number of these initial dates and the thinness of the accumulated archaeological record from the region, there was suitable hesitant wording in these early reports. However, after mentioning a wide range of possible interpretations of the dates associated with bronze in Vietnam, Davidson (1979:102) concluded, "the astounding implication remains that a knowledge of bronze-casting existed in Viѐt-nam in the early second or late 3rd millennium B.C." This implication was echoed from finds and dates several hundred kilometers to the west in northeast Thailand at a site called Non Nok Tha.

Non Nok Tha

While former French Indochina descended into war during the 1960s, Thailand, a country that had never been colonized, started to engage with the international archaeological community. During the 1960s, a handful of joint projects between the Fine Arts Department of Thailand and western countries undertook excavations at several sites. These projects included those conducted with the United States (University of Hawaii at Non Nok Tha and Spirit Cave, and University of Pennsylvania at Chansen), Denmark (Ban Kao and Sai Yok), and Britain (Khok Charoen and other sites in central Thailand). These projects resulted in the introduction of modern archaeology into a country on mainland Southeast Asia that had seen much less archaeological attention over the preceding decades than had the former French colonies. Thailand's openness to engaging with Euro-American archaeological scholarship and methodologies brought significant changes to the position of Southeast Asian archaeology in global scholarship. One result of this sudden concentration of multinational archaeological research was that the majority of new absolute dates from Southeast Asia came out of Thailand (Bronson and White 1992:492).

Dates pertinent to bronze came from the site of Non Nok Tha located in Khon Kaen Province in northeast Thailand. Excavated in 1966 and 1968 by Solheim and Bayard, the initial publication of ten dates, including 3rd millennium B.C. dates in association with bronze (Solheim 1968), began what became a decades-long debate. The efforts to date the site's sequence has now spanned nearly fifty years, included numerous dating methodologies, and demonstrated the history of struggles and pitfalls of the development of chronometric dating techniques and their application to real archaeological problems over the latter half of the 20th century and into the 21st century. Ultimately, the site had 70 radiocarbon dates on bone apatite (2), bone collagen (7 undertaken in the 1960s–1970s and an additional 14–16 recently; Higham et al. 2014a), charcoal (36), rice temper (11), as well as four thermoluminescence (TL) dates from sherds. Bayard's and Solheim's chronology for bronze at Non Nok Tha as beginning during the 3rd millennium B.C. was subsequently supported by reviews of the data by Smith (1979) and Spriggs (1996–97). However, Higham et al. (2014) have proposed a revised chronology for Non Nok Tha spanning the period of the 14th to the 7th century B.C. based on 14 recent "collagen" dates.

Over the years, Higham (e.g., 1996–97; see also Higham et al. 2014) and Loofs-Wissowa (e.g., 1983a), among others, have challenged the early chronology on a variety of grounds. Initially the most difficult problem was the relationship of a dated sample to its archaeological context in heavily disturbed deposits. Another difficult problem, especially in the early days, were discrepancies between radiocarbon dates and dates from the new TL dating technique. Thermoluminescence, which was developed around 1960, had a rocky history in its early applications to directly dating pottery in Southeast Asia (see review in White 1986:281–285).

Non Nok Tha provided an early example of TL dating of excavated pottery, and the results appeared to mesh reasonably well with at least some of the radiocarbon dates from the site (Bayard 1972; Spriggs 1996–97:943). However, what can only be described as chaotic results from some early TL dates, particularly those run on unprovenienced pottery from Ban Chiang, exposed the technique to profound questions about its reliability (White 1986:281–285). After criticizing various logical underpinnings as well as "public relations" aspects that occurred with the early bronze dating, Loofs-Wissowa (1983a) rejected the validity of ^{14}C dates in general as well as all TL dates from Non Nok Tha, but then advocated the reliability of several TL dates run at Australian National University on unprovenienced sherds from "black Ban Chiang ware" (Loofs-Wissowa 1983a:9–10). He interpreted these as supporting a bronze age at Non Nok Tha and Southeast Asia generally as no older than the 1st millennium B.C.

Higham's critiques have generally rejected TL dating (although recently he is receptive to TL dates from selected sites such as Khok Charoen [e.g., Higham and Thosarat 2012:90; Higham et al. 2011a:245–246]), and he initially focused on the stratigraphic and consistency problems of the Non Nok Tha charcoal dates. For many years he championed AMS (Accelerator Mass Spectrometry) dates on rice temper from burial vessels as providing the solution to the problem of dating Non Nok Tha and other sites with complex mortuary sequences (e.g., Higham 1988b:75, 77, 1996a:191, 240, 246, 311, 1996–97:882, 2002:91, 93, 113, 129, 353, 2004:51–52; Higham and Thosarat 1998:84; Higham et al. 2004:325). In his 1996–97 overview of dating pertinent to Non Nok Tha, Higham interpreted selected radiocarbon dates from several sites, including selected dates from Non Nok Tha rice temper and charcoal, as indicating a 1500–1000 B.C. range for the beginning of the Southeast Asian bronze age generally and for the Non Nok Tha burial deposition specifically. In his recent revision (Higham et al. 2014:74), he argues that the bronze age begins at Non Nok Tha in the 10th century B.C.

The adversarial tone of many of the discussions about the appearance of bronze in the region has had a prolonged dampening effect on scholarly discourse among Southeast Asian archaeologists who have witnessed it, and the acrimonious tenor has undermined the contributions of this and other important sites for Southeast Asian prehistoric archaeology. The interested reader may consult the following literature (Bayard 1972, 1987, 1996–97; Bayard and Charoenwongsa 1983; Clark 1977:345–346; Hedges et al. 1992; Higham 1984a, 1987, 1996a:191, 246, 1996–97, 2011b; Higham et al. 2011b; Loofs-Wissowa 1983a, 1983b, 1992; Smith 1979; Solheim 1983; Spriggs 1996–97; see also the final Non Nok Tha site publication Bayard and Solheim 2009, available online at http://guampedia.com/archaeological-excavations-at-non-nok-tha-thailand/.)

Irrespective of the tone of the debate, the dates from Non Nok Tha are full of contradictions. The protracted discussions did introduce and explore many important issues concerning the difficulties of chronology building for Southeast Asian metal age sites. Those issues included: (a) the quality of radiocarbon dates from different labs; (b) the reliability of dates from different carbon sources such as bone versus charcoal; (c) the accuracy of TL versus radiocarbon dates; (d) the reliability of rice temper dates; (e) the potential for old wood effects; (f) the difficulty of dating burial sequences from either burial-associated or occupation-associated charcoal; (g) the importance and the pitfalls of interpreting depositional contexts of dated samples; (h) the confusion introduced by the constantly evolving technical aspects of radiocarbon dating (e.g., changes in half-life, changes in counting methods, changes in calibrations, changes in pretreatment protocols, and introduction of AMS that enabled dating of very small samples with less than one milligram of carbon); and (i) the handicap of the absence of well-articulated regional ceramic sequences to provide a clear sequential framework and relative chronology for absolute dates. Despite recent advances in dating technology (e.g., Higham and Higham 2009b), methodological problems still call the new results into question (Webb et al. 2007; White 2013, 2015a), and chronological issues remain a major unresolved topic for the region.

Dates and Independent Inventions

When the early bronze dating for northeast Thailand was first proposed, one of the reasons that global attention was paid (as opposed to the dates being summarily dismissed or ignored) was that the dating appeared when several Anglo-American archaeologists were becoming infatuated with the idea of "independent invention" in prehistory, with copper-base metallurgy of particular interest (e.g., Renfrew 1969). Again, the story of interpretation intertwined with the story of technical changes in radiocarbon dating. One effect on world archaeological interpretations that the first applications of radiocarbon dating in the 1950s and early 1960s had on the mid-century *Kulturekreis* reconstructions was scientific evidence that the pyramids in Egypt and Mycenaean stone structures were probably not the source for monumental architecture such as megaliths like Stonehenge in western Europe. The European monuments were too old to be explained by the diffusion of ideas or the movement of people from west Asia and northeast Africa to western Europe (Renfrew 1973:16) as a *Kulturkreis* reconstruction would have predicted. Those monuments must have been the idea (invention) of the (barbarian) locals.

The appreciation for indigenous processes of development gained further ground as the implications of the first attempt at calibration (adjustments to raw radiocarbon results based on tree-ring evidence for fluctuations in atmospheric ^{14}C) was applied to dates from southeast Europe. Colin Renfrew (1969, 1973) argued that Balkan metallurgy was so early, according to the newly calibrated dates, it must have been invented independently from metallurgy in the Near East. In the late 1960s and early 1970s, the idea that not only metallurgy, but many phenomena such as agriculture, states, and other aspects of culture and civilization could have been developed independently many times, in many places, captured the imaginations of archaeologists and excited debates ensued, as they do when established paradigms are dismantled. The 3rd millennium dates for bronze from Non Nok Tha came just at a moment when the idea that metallurgy might have been *independently* invented in several areas and was not necessarily associated with high civilizations was being strenuously discussed by the discipline. The theoretical issues behind explanations of technological change are discussed more deeply in Chapter 3.

Early Debates over Ban Chiang Chronology

Discovered and excavated in the heyday of archaeological fascination with the concept of and evidence for independent invention of technologies, initial radiocarbon dates from the Penn/FAD Ban Chiang excavations at first seemed to not only confirm Non Nok Tha's early dating for bronze, but to indicate bronze might have been present in Thailand even earlier during the 4th millennium B.C. (Gorman and Charoenwongsa 1976). Within a few years, however, that conclusion changed based on a closer stratigraphic assessment of the dates from the site (White 1982, 1986).

Chet Gorman, co-director of the Penn/FAD excavations at Ban Chiang, Assistant Professor at Penn's Anthropology Department, and Assistant Curator at Penn Museum, died of cancer at a tragically young age of 43 in 1981. The Smithsonian Institution's Traveling Exhibition Service (SITES) had already begun development of an exhibit on the site and asked Joyce White to be the guest curator, since she had been involved in the Ban Chiang analysis for several years as a Penn graduate student under Gorman. Her assessment of the dating evidence, based on 33 radiocarbon determinations on charcoal from the site, pointed to early 2nd millennium or ca. 2000 B.C. as the oldest possible date for bronze uncovered by the Penn/FAD excavations at the site of Ban Chiang (White 1982). The full rationale for this conclusion was laid out in White (1986).

Moving the date for the oldest Ban Chiang bronze from 3600 B.C. to 2000 B.C. calmed some of the hubbub over the early dates for bronze in Southeast Asia, although proponents of 3rd millennium B.C. bronze or older remained (Bayard 1996–97; Rainey 1992:244; Spriggs 1996–97). Scholarly debates continued but at lower decibels for the next couple of decades, mostly focusing on a narrower time range of a roughly 500-year difference for the oldest bronze, about 2000 B.C. (e.g., J. White 1997, 2008) versus about 1500 B.C. (e.g., Higham 1984a, 1989, 1996a, 1996–97, 2002, 2004).

Over the years, many more radiocarbon dates have become available from Southeast Asian sites, but contradictions in the evidence continue (e.g., Higham et al. 2011a, 2011b, 2015; J. White 2008, 2015a). Southeast Asian archaeology and radiocarbon dating are maturing at the same time and each has had its own growing pains. Although improvement of radiocarbon dating methods as well as other absolute dating methods is only to be applauded, recognition that these techniques will be ever evolving has forced regional archaeologists to broaden their range of criteria for evaluating time/space relationships in prehistory. The point/counterpoint history for the dating debates surrounding the metal ages of Southeast Asia is not reviewed here but is planned for a future Ban Chiang monograph (White in preparation). For some recent discussions, see Higham et al. (2011a, 2011b, 2015), Rispoli et al. (2013), and J. White (2008, 2013, 2015a).

Recent Developments in Southeast Asian Chronology

Excavations in the upper Mun Valley recently led Higham to further contract his estimate for the appearance of bronze in Southeast Asia to closer to 1000 B.C. (Higham and Higham 2009a, 2009b; Higham et al. 2011a, 2011b). The younger dating initially was based on Higham's assessment of dozens of new dates run at the University of Oxford Radiocarbon Lab on freshwater bivalve shell of two genera, *Hyriopsis* and *Pseudodon*, excavated from grave contexts from the site of Ban Non Wat, and application of Bayesian statistical analyses (Higham and Higham 2009b). Subsequently, ten dates were run on bone from Ban Chiang burials that he claimed also supported his new chronology (Higham et al. 2011a). More recently, a large-scale program focusing on bone and shell dates from several sites in northeast Thailand appeared to bolster the Ban Non Wat sequence based on shell dating (Higham et al. 2015).

One of the few constants in Southeast Asian archaeology has been that new chronologies for individual sites and the region as a whole have been proposed in rapid succession, contributing to the enduring confusion over the region's chronology. New dating techniques are often championed in proposed chronologies, but each time results from a new technique are introduced, the technique has subsequently been shown to produce some problematic results. The presence of some problematic results is then used by some scholars to reject the technique overall. For example, rice temper dates championed as recently as Higham et al. (2004:325) were subsequently categorically rejected (Higham and Higham 2009b; Higham et al. 2011b). Yet despite dismissal of rice temper dates by some scholars, other regional scholars have continued to rely on rice temper dates (e.g., Glover and Bennett 2012:101) without criticism.

Higham and Higham (2009b:139) accurately stated that to adopt their chronological framework "require[s] the rejection of all previous attempts to date Southeast Asian prehistory radiometrically." In essence, their statement rejects all radiocarbon dates run prior to 2009 from Southeast Asia. However, the shell dates upon which their chronology currently relies are also subject to critique on technical grounds in that rigorous vetting protocols were not applied (White 2013, 2015a, 2015b); the possibility that diagenesis of the shell might have led to erroneously young results was not ruled out (see Webb et al. 2007). Moreover recent "collagen" dates from Thai sites can also be questioned for accuracy (White 2015a).

In short, no dating technique has been conclusively demonstrated to be a panacea for dating Southeast Asian sequences as inevitably many details and contradictions, technical and other, emerge that need to be addressed. Chronological disagreements and reassessments are in fact common. Chronological reassessments are underway in other parts of the world including in seemingly exhaustively studied areas like the Near East. The ARCANE project (http://www.arcane.uni-tuebingen.de/presentation.html), which interestingly is focusing on the 3rd and 2nd millennia B.C. of the eastern Mediterranean and Near East, is taking a gradual, multiyear, systematic, and cooperative approach to reassessment of both relative and absolute chronologies by subregion, involving numerous scholars, disciplines, and institutions (e.g., Lebeau 2011). The ARCANE program also recognizes that synchronization of the relative cultural sequences (a methodological area in which Southeast Asian archaeology lags most other regions of the world) is *foundational* to accurate interpretation of the chronometric evidence. A similarly thorough, broad-based, and systematic approach will be needed in East and Southeast Asia if current chronological discrepancies are to be resolved. In the long term, regional chronologies cannot be based only on a handful of sites or one or two scholars' views of the evidence.

Although detailed assessment of Higham's recent chronology is outside the scope of this volume, in the view of the authors it is premature to reject all radiocarbon dates from Southeast Asia that were run prior to 2009. As Spriggs (1996–97:947) commented with regard to an earlier effort to discredit conventional radiocarbon dating from selected Southeast Asian sites, "But the question then has to be asked as to why almost all of the conventional dates from NNT [Non Nok Tha] and, according to Higham, the majority from other sites such as Ban Chiang have turned out to be wrong. I know of no other equivalent suite of dates from the Asia-Pacific area where this has proved to be the case." It is anticipated that when all issues of site stratigraphy, interregional relative chronology, and technical dating protocols are resolved, the result will be a refinement of the Southeast Asian regional chronology rather than a "revolutionary" reassessment (cf. Higham and Higham 2009b).

Moreover, many aspects of Higham and Higham's chronological argument are intertwined with a political and economic argument about relationships between Thailand and specific regions and archaeological cultures in the PRC and central Asia. Yet the bases for the chronologies of these other parts of Asia have the same kinds of issues of quality and quantity of dates and site sequences that can be found in Southeast Asia; absolute chronologies in central and eastern Asia generally rely on dates that Higham would consider untenable in Southeast Asia (as noted in Higham et al. 2011a:241, 243 for dates from southern PRC). Therefore, in order to not build arguments based on comparing apples to oranges, to truly resolve the detailed macro-regional relationships in Asia two to five thousand years ago, will require not only new dates for much of the PRC—a mammoth undertaking—but also reassessment of the relative and chronometric evidence, in order to make valid arguments about synchronization at the macro-regional scale.

In the meantime, in this volume, the authors assume that the region's absolute chronology will continue to be adjusted for the foreseeable future. Therefore, emphasis is placed on clarifying technological and relative sequential evidence from a defined region in Thailand, the Chao Phraya and middle Mekong Basins. This regional scope is judged the best scale for making contributions to knowledge of Southeast Asian prehistory at this point in time. Absolute dates are de-emphasized in favor of demonstrating multisite interregional relationships whose precise chronometric placement can and will be refined as more data become available. Some may find the chronological ambiguity uncomfortable, but it will not be the first or the last time that Southeast Asian archaeologists must endure ambiguous chronologies. The volume's interim working chronology is reviewed in Chapter 2, and a regional chronological sequence is reviewed in TAM 2C (chapters 4 and 5).

Non-chronometric Issues

Aside from the perennial issue of dating, there are other key current questions and topics about prehistoric Southeast Asian metallurgy. Although independent invention of metallurgy in Southeast Asia was recently still considered a viable hypothesis by Higham (2002:116–117), most regional specialists have not entertained the possibility since the 1980s (see reviews in Pigott 2012a; J. White 1988, 2008; White and Hamilton 2009:358). There are at least four other topical areas that are critical to current debates: (1) How was metal technology transmitted to the region? (2) Under what societal contexts in the region was metal adopted and developed? (3) What impact did the adoption of metallurgy have on prehistoric societies of Southeast Asia? (4) What were the intra- and inter-regional social, political, and economic dynamics that developed and maintained metal production and metal product manufacturing?

While discussions about transmission of metallurgical technologies across the region ultimately will require the dating problems to be resolved, other contributions to the topic of how bronze metallurgy was transmitted to and within the region can be addressed. One set of questions revolves around the geography of transmission, i.e., which intermediary societies were involved, especially at the earliest stages. Two possibilities include societies along portions of the Mekong drainage basin north of Thailand (White and Hamilton 2009) or those along the Xijiang River in southeast PRC (Ciarla 2007a). These two subregions are not yet thoroughly researched by archaeologists, and regional sequences from each region along with full discussions of evidence of their extra-regional relations are eagerly awaited. This volume does not strive to resolve this debate but rather seeks to present empirical evidence about the metal technological system that was, in fact, transmitted to Thailand. The evidence for the early metallurgy reaching the region will be needed to compare with metal-related evidence to the north.

Thus, expanding beyond the topic of geography are questions of what exactly was transmitted: Was it alloy recipes, casting techniques, smelting procedures, and/or manufacturing equipment? What techniques were local innovations? By what social processes was metal technology transmitted: immigration of practitioners; copying of trade items by indigenous craftspersons; migrations of whole social units that then provided metal products to previous occupants of the region or replaced (e.g., outcompeted demographically or conquered) previous inhabitants of the region? All these questions eventually can be addressed "up the line" into inner Asia in order to decipher the likeliest transmission routes and processes.

The transmission possibilities have greatly expanded from even ten years ago as much more pertinent data from inner Asia has become available, transforming archaeological understanding of inner Asian dynamics during prehistory (e.g., Chernykh 1992; Chernykh et al. 2004; Frachetti 2009, 2012; Guedes 2011; Jones et al. 2011; Linduff 2004). Much more than metal was involved in those early interactions. Interregional interactions and transmissions of plant cultivating technologies, pastoral lifeways, and human populations within vast inner Asia are now perceived as dating earlier than metal transmission—from the 4th and 3rd millennia B.C. and perhaps as early as the 6th millennium B.C. (Jones et al. 2011). In its position as an eastern Land's End, Southeast Asia holds some key data for addressing several of these continental-scale issues.

The social contexts of bronze adoption and the impacts of the technology on societies are other important topics in early metallurgy discussions. Defining the societal contexts that were willing to adopt and adapt this complex technology is an important area of investigation for archaeometallurgists. The field has moved beyond the outdated assumption that states or cities were necessary to adopt or maintain this technology, or that adopting bronze technology would necessarily lead to elites or political hierarchy (e.g., Frachetti 2009; Roberts 2011:137; Thornton and Roberts 2009:182; White and Hamilton 2009:360). The field is wide open for examination of the full range of societies that adopted metal technologies and the variable ways technologies affected the societies and vice versa. How non-state societies adopted such complex technologies and what kinds of producer-consumer relationships developed are topics of great disciplinary interest (e.g., Doonan et al. 2014; Roberts 2014; among other chapters in Roberts and Thornton 2014).

Within the large arena of Eurasia, Southeast Asia is one example demonstrating the variable social contexts in which metallurgy took root and flourished. Although at the time Ban Chiang was excavated, early tin bronze was thought to be associated with states, hence the Ban Chiang finds were considered anomalous (Muhly 1988:16), now with the expansion of research into the bronze age across vast Eurasian regions, it is clear that throughout Eurasia many societies that adopted bronze were not states (Chernykh 1992; Frachetti 2012; Roberts 2011). Southeast Asia can no longer be viewed as anomalous but serves as an example of a non-state bronze-using society (for a variant assessment see Higham 2011a).

However, it is the societal details that intrigue. Prehistoric Thailand demonstrates an adoption of metals like bronze that was associated with low evidence for interpersonal violence in comparison with some other examples (Domett and Tayles 2006; White 1982; cf. Jantzen et al. 2011). The possibility that initial adoption was framed by community economic specialization (as opposed to a more centralized economic model [White and Pigott 1996]) may have been a factor in the distinctive pattern of metal adoption in Southeast Asia. There is much debate concerning the social particulars of bronze adoption. Was the technology appropriated by big men aggrandizers seeking new exotic prestige goods to parlay into status enhancements (Rispoli et al. 2013)? Was bronze adoption a catalyst for a "starburst" of hierarchical state-forming activity dating to around 1000 B.C. (Higham and Higham 2009b)? Or was it brought into and maintained for centuries within a regional non-state economy based on community specialization that included an exchange system of subregionally produced goods and valuables, similar to Alpine Europe (Shennan 1999)? These questions are addressed in TAM 2C (chapter 6).

The intraregional dynamics associated with the adoption and development of metallurgical technology is another fascinating area of current concern. The recent application of analytical technologies like lead isotope analysis (LIA) is enabling archaeometallurgists to indicate from which ore sources the copper from individual artifacts may have derived (Pryce 2012; Pryce et al. 2014; see also TAM 2C, chapter 3). This enables the reconstruction of specific economic and social networks of the past, providing a new window on details of social and economic relationships not previously achieved in Southeast Asia. How each subregion developed and interacted with other subregions can also now begin to be empirically examined.

Summary of Areas of Agreement and Disagreement

Today, all regional specialists agree that, contrary to Clark's 1969 statement quoted previously and summarizing the mid-20th century view, bronze production did occur in mainland Southeast Asia prior to the appearance of iron. Regional specialists also concur that metallurgy was not independently invented in Southeast Asia but came from elsewhere in Asia.

The timing for the very first appearance of copper-base technology in the region is still debated, with the disagreements based on differing interpretations of stratigraphic and chronometric evidence, but also influenced by technological and socioeconomic evidence. The dating disagreements have implications for the time-depth and duration for the presence of pre-iron bronze technology in the region, but not necessarily for the source of the technology or the geographic route of technological transmission. The dating disagreements have potential implications for models addressing the transmission of the technology from other parts of Asia, as well as interpretations of the speed, degree, and nature of impact that the acquisition of metal technology had on the prehistoric societies; but these latter two topics do not wholly hinge on dating and can proceed on their own terms.

In addition to the debates about the dating and route of the metals transmission process, there are also subtler disagreements concerning topics such as how to reconstruct technological activities, how best to envision cross-craft interaction (for instance, the relationship if any between bronze and iron technologies), and how best to view the social positions of and relationships between the metal and metal object producers and consumers. Just as the dating interpretations can lean toward earlier versus later views, the socioeconomic interpretations lean toward emphasizing more hierarchical as opposed to less hierarchical views.

As debates in modern archaeology entail many aspects beyond dating, this work emphasizes fundamental contextual, technological, and theoretical dimensions of the issues at hand that are essential for addressing the cultural meaning of metals in prehistoric societies, irrespective of how the dating specifics are eventually resolved.

Overview of Chapters in Volume 2

Volume 2 of the Ban Chiang monographs is organized into four parts. This first book, TAM 2A, provides overviews of the study sites, as well as theory, geology, and metallurgy generally needed to properly assess site-specific archaeometallurgical data. Following the general introduction to the volume in Chapter 1 where the relevance and importance of Ban Chiang archaeometallurgy to global discussions of pre-industrial metallurgy is addressed, Chapter 2 presents an overview of the excavations and sequences at Ban Chiang, Ban Tong, Ban Phak Top, and Don Klang. Chapter 3 is an overview of how ancient metallurgy has been viewed in the past, the current status of several assumptions that have long guided its study, and why the "Conventional Paradigm" for the roles of metals in ancient societies is being reassessed. Chapter 4 discusses the new paradigm emerging for the study of ancient metals. Chapter 5 discusses metals in economic contexts in pre-industrial societies. Chapter 6 describes the geological background for understanding ancient metals in northeast Thailand. Chapter 7 introduces the basics of metal production and metal product manufacturing in pre-industrial contexts. An Appendix provides a glossary of technical and non-technical terms used in this book.

The second book, TAM 2B, focuses on the metal-related empirical evidence from the four sites under study. Following an introductory chapter, chapter 2 outlines the research program undertaken with the study collection. Chapter 3 provides the classification for metal artifacts from the study corpus. Chapter 4 presents the technical analyses of the study corpus. Chapter 5 presents manufacturing evidence from crucibles (studied by William Vernon), prills, and molds. Chapter 6 examines the depositional contexts of metals and related production artifacts. Chapter 7 reviews topics highlighted by a "life history" approach to assemblage assessment, including evidence for change over time in typology and evidence for metal use in burial contexts in comparison with non-burial contexts. Appendices include the distribution plans of non-burial metals by site, level, square, and quadrant as well as supplementary data tables.

The third book, TAM 2C, puts the study corpus into a larger regional economic context. Following the introductory chapter, chapter 2 by Vincent Pigott provides an overview of copper production sites in northeast and central Thailand and Laos. Chapter 3 by Oliver Pryce presents lead isotope evidence for sources of the copper in a small sample of copper-base artifacts from the study corpus. Chapter 4 provides a regional overview of metal consumption and production in prehistoric northeast and central Thailand. Chapter 5 synthesizes the evidence for prehistoric metal technological systems of Thailand. Chapter 6 examines the economy of metal production and consumption in the prehistoric societies of central and northeast Thailand and presents a model to help understand the character and trajectory of the adoption and development of metals technology in prehistoric Thailand. A summary critique of the Higham metal age model is also offered. Contributions of this study to global studies of early metallurgy are proposed.

And lastly, the fourth book, TAM 2D, is a catalog of the excavated metal artifacts, crucibles, and prills.

NOTES:

1.1 Translations from French to English in this chapter provided by Joyce White.

1.2 For reasons explained in White (2002) as well as in Chapter 3, this volume will favor lower case references to "ages" when referring to ages in the Three Age System in general and heuristic contexts. When discussing the work of scholars who place great emphasis on the Three Age System, capitalized Ages are used to discuss their work. For Southeast Asia, as explained in Chapter 3, the preference is for avoiding the Three Age System and replacing it with pre-metal period, bronze period, and iron period as advocated.

2

Ban Chiang, Ban Tong, Ban Phak Top, and Don Klang: Summary of Excavations and Sequences

Joyce C. White

Background

The study of prehistoric metallurgy in Southeast Asia and the rest of the world since the 19th century has made it clear that two aspects are critical: (1) precise stratigraphic and contextual information for early metals must be provided; and (2) entire metal assemblages must be assessed, including fragmentary metal artifacts and related evidence from occupation contexts, not merely intact objects commonly recovered from burials. This volume, therefore, includes unusually detailed discussion concerning the early metal artifacts recovered, especially from Ban Chiang. In order to appreciate the archaeological evidence, the excavation histories and methodologies are summarized in this chapter.

Excavation Histories

The Penn Museum joined with Thailand's Fine Arts Department to excavate at Ban Chiang in the mid-1970s (Gorman and Charoenwongsa 1976; Winegrad 1993:164–180). The Penn Museum first became interested in the site because of the unexpectedly early bronze metallurgy from Non Nok Tha as well as some early thermoluminescence (TL) dates on unprovienienced Ban Chiang pottery (Lyons and Rainey 1982:6). A visit by Froelich Rainey, then director of the museum, to Ban Chiang in 1973 confirmed that ample finds of bronze and iron artifacts as well as fascinating pots had emerged from informal diggings by villagers under their houses (Rainey 1992:231). In addition, preliminary excavations at the site undertaken by Thai archaeologists (Suthiragsa 1979) showed Ban Chiang to be a deeply stratified site holding evidence of a rich and previously unknown culture.

Negotiations during 1973 led to the establishment of the "Northeast Thailand Archaeological Project" or NETAP and two excavation seasons at Ban Chiang by the joint Penn/FAD program, one in 1974 and a second in 1975, ensued. During the 1975 NETAP season, William Schauffler, a University of Pennsylvania Department of Anthropology Ph.D. candidate, conducted a reconnaissance survey in northern northeast Thailand and undertook small excavations at three sites expected to be comparable in age to Ban Chiang. Two sites in the Sakon Nakhon Basin, Ban Tong and Ban Phak Top, were thought likely to have cultural sequences cross-dateable to Ban Chiang, and the third site, Don Klang, was closer to Non Nok Tha in the upper Nam Phong Basin (Schauffler 1976). The excavation methods and recording systems employed at those three sites were very similar to those at Ban Chiang. These similarities greatly facilitate the coordinated reporting of the recovered metal-related remains from all four sites. Together the four sites were expected to shed much light on the later prehistory of northern northeast Thailand.

Following the 1975 season, all excavated materials were sent to the institutions undertaking analyses: University of Otago, New Zealand, for faunal remains; University of Hawaii for human skeletal remains from 142 Ban Chiang burials; and University of Pennsylvania for cultural materials (e.g., more than 1,300,000 sherds) and miscellaneous other remains, such as soil samples (Gorman and Charoenwongsa 1976:16). At the request of the Fine Arts Department of Thailand, most of the habitation sherds were returned to Thailand in 1983. Reconstructible pottery, small finds, and half the soil samples remained on loan to the Penn Museum for ongoing analysis.

Summary of Excavation Methods

The excavation strategy at all sites reflected the overall intent to begin definition of regional cultural sequences. Several aspects of the excavation methodology were modifications of procedures implemented at Non Nok Tha as described by Bayard (1971). Only limited exposures of each site were feasible due to constraints such as time, resources, and private property issues. Ban Chiang is a modern village, parts of which are located on top of a mound rising about 7 meters above the surrounding rice fields. Its densely spaced dwellings reflect the current village's age of over 200 years (Gorman and Charoenwongsa 1976:17); this density of occupation meant that only a small sample of the underlying mound site could be exposed. The 1974 excavation (termed the BC excavation) took place in a set of six squares and intervening baulks in the back and side yards of Mr. Pote Montripituk over the course of seven months (March through September 1974; baulks were excavated March through May 1975). The 1975 excavation (termed the BCES excavation for "Ban Chiang eastern soi") took place along a village road in a set of four squares and intervening baulks over the course of seven months (March through September 1975). These two excavation locales were situated towards the central highest portion of the mound and were separated from each other by 75–100 meters. A total of 130.8 square meters was excavated at the BC and BCES locales combined. The size of the mound is at least 8 and probably closer to 15 hectares (White 1986:29). Calculating from the most conservative estimate of the surface area overlying prehistoric deposits—8 hectares—the Penn/FAD project excavated at most about 0.16% of the mound's surface area. At the three tested sites of Ban Phak Top, Ban Tong, and Don Klang, only single squares ranging in size from 3 x 3 m to 3 x 4.2 m were excavated.

At the BC and BCES locales and similar to Non Nok Tha, excavation proceeded within squares defined by a grid system with letters indicating units along one axis and numbers along the perpendicular axis. Thus, the BC locale had six squares comprising C3, C4, C5, C6, B5, and B6. The BCES locale had squares D4, D5, D6, and D7. The intervening baulks, usually 1 meter wide, were named after the adjacent squares, e.g., D4/D5. The deposit was removed primarily by trowel, and all soils were screened with 1 cm mesh, unlike at Non Nok Tha where the hard clayey soils were not screened. Another of the differences from the Non Nok Tha excavations was that at Ban Chiang grass roofs were erected over the excavations, which slowed the drying of sediments and assisted in discrimination of soil distinctions. Cultural remains were found as deep as 4 m at BCES and 2.80 m at BC.

Initial removal of deposits during the BC excavation revealed the absence of clearly defined fine-grained soil stratigraphy. After removing the first meter or so in part of the BC locale, the excavation strategy emerged to remove distinct, usually thick, soil strata by smaller, discrete and separately numbered horizontal units usually 10 cm thick. The 10 cm units, often called "spits" in the field records, are termed excavation "layers" in this monograph. Each square had its own independently numbered excavation layers; i.e., layer 6 in one square is not necessarily stratigraphically equivalent to layer 6 in another square. Each layer was usually excavated in four quadrants. Features were defined for each layer in each square and excavated separately from the general soil matrix. Each time a new layer was exposed within a square, features were defined and feature numbers assigned for that layer beginning with "1."

Burials were also excavated as much as possible as discrete units. They were numbered sequentially within each excavation locale, beginning with the number 1. This system emerged during the BC excavation and was more consistently implemented during the BCES excavation. Similar procedures were employed at Ban Tong, Ban Phak Top, and Don Klang, except that only a single square was excavated at each of those sites as stated above. The sites varied in the amount of burial remains recovered. At Ban Chiang, at least 142 burials are recognized

(Pietrusewsky and Douglas 2002; note that BC Burial 55 was assigned its burial number during the stratigraphic analysis in Philadelphia based on the excavation records, but the skeletal elements were not recovered or analyzed). Ban Tong had two recognized burials, Ban Phak Top had no burials, and Don Klang had 17 burials which produced remains from 27 individuals (White 1978).

Thus, the excavation recording system included several concurrent numbering and naming systems. Provenience information normally includes at a minimum the excavation locale (BC or BCES for Ban Chiang, BT for Ban Tong, BPT for Ban Phak Top, and DK for Don Klang), the square or baulk (for BC and BCES), and the excavation layer. Other provenience designations could include the quadrant, a feature number and its source layer, or a burial number. In the excavation field records such as layer plans, excavation layer numbers were circled and feature numbers were put in "standing" triangles (point is up). If a feature defined in one layer continued to be removed in successive lower layers, it retained its original feature designation from the layer in which it was first defined, but the feature number was placed in a "hanging" triangle (point is down).

Bag numbers beginning at each locale with the number 1 were assigned to specific excavation tasks that removed defined deposits. For example, a bag number would be assigned for the excavation of a feature or of a 10 cm layer in a quadrant. The materials retained during those tasks, whether sherds or soil samples, are identified by those bag numbers, which tie the materials into a specific provenience. Bags represent an excavation effort within a specific provenience and context. Bag numbers at each locale were assigned consecutively as material came out of the pits irrespective of squares. The overwhelming majority of artifacts in bags were sherds from pottery vessels most probably broken in daily life.

Other cultural artifacts, including metals, were segregated during the excavation, screening, or cleaning efforts in the field and assigned a small find number. Assignment of small find numbers at BC and BCES followed slightly different procedures. In the BC records, blocks of small find numbers were set aside for types of artifacts, for example, 1–100 for clay rollers, 101–249 for pellets and beads, and so forth. In the BCES records, small finds received numbers as they came out of the ground: e.g., small find 1 could be a roller, small find 2 a bead, small find 3 a metal fragment. Additional small finds were identified after the excavation during the process of pottery reconstruction, which took place at the Penn Museum from 1976 into the early 2000s. Such small finds had been inadvertently bagged with the sherds in the field. Small finds discovered during laboratory analysis were usually assigned small find numbers beginning with 2000. The small finds retained their bag numbers, which thereby associated the small find with all other artifacts found within the bag's provenience. Small finds are conventionally designated with the locale initials, the small find number, a slash, and the bag number, e.g., BCES 263/1582.

Stratigraphic Analysis

Only a brief description of the stratigraphic analysis is given here, as a forthcoming volume will present the detailed evidence (White, in preparation). The stratigraphic analysis undertaken by White was based on her experience from 1976–1979; for most of that period she was Ban Chiang lab manager during the first few years of the site's analysis in Philadelphia. (White was not present at the excavations, which took place before and during her first year of graduate school at Penn's Department of Anthropology). Previous work by White (1982, 1986) on Ban Chiang's relative chronology drew primarily on general soil strata and burial evidence. However, as Ban Chiang is also an occupation site, a framework for temporal subdivision of the non-burial deposit is essential for a full site chronology. This chapter briefly describes the approach and rationale for the stratigraphic subdivision developed for Ban Chiang in the late 1980s and 1990s. White applied the same procedures to the other three sites discussed in this study.

Importance of Relative Chronology

The relative sequence is the backbone of any site's chronology. It is the framework that structures presentation and assessment of a site's absolute dates and provides the basis for cross-dating of cultural deposits with other sites in a region. If interpreted meticulously, a relative sequence should hold up over time irrespective of inevitable changes in absolute dating.

Figure 2.1 Example of a Ban Chiang lower Early Period supine burial. BCES Burial 51 is a male aged 40–45 years from EP IVa.

Figure 2.2 Example of a Ban Chiang lower Early Period flexed burial. BCES Burial 55 is a male aged 14–16 years from EP IVa. Bone bangles are at each elbow.

Figure 2.3 Example of an infant jar burial from Ban Chiang. BCES Burial 63 from Early Period IVb is aged 9.5 lunar months.

Relative Chronology: Previous Work

Previous work on the relative chronology is the starting point for the more refined analysis of the BC and BCES excavation locales. White (1982) outlined the Early, Middle, and Late Periods for Ban Chiang's prehistoric deposit based primarily on changes in grave styles such as placement and treatment of grave goods relative to bodies. Early Period burials have three styles: (a) most adults and subadults, including juveniles and children aged 3–15 years, were buried in a supine body position with 0–4 small to medium pots commonly placed near the feet or head (Fig. 2.1); (b) a few adults were buried in a flexed position with 0–3 vessels and occasionally other small finds (Fig. 2.2); (c) infants including one toddler aged 2–3 years old were buried inside large vessels (Fig. 2.3). The burials in the Middle Period included from 0 to as many as 24 vessels. The many large and often carinated vessels (Fig. 2.4a–d) were usually broken in association with the supine body (Fig. 2.5). During the Late Period, 0–9 intact, usually large, vessels were generally placed over the bodies of adults and subadults (Fig. 2.6).

Although pottery styles and soils influenced the definition of the three periods, not until White (1986) were Ban Chiang's ten burial phases defined

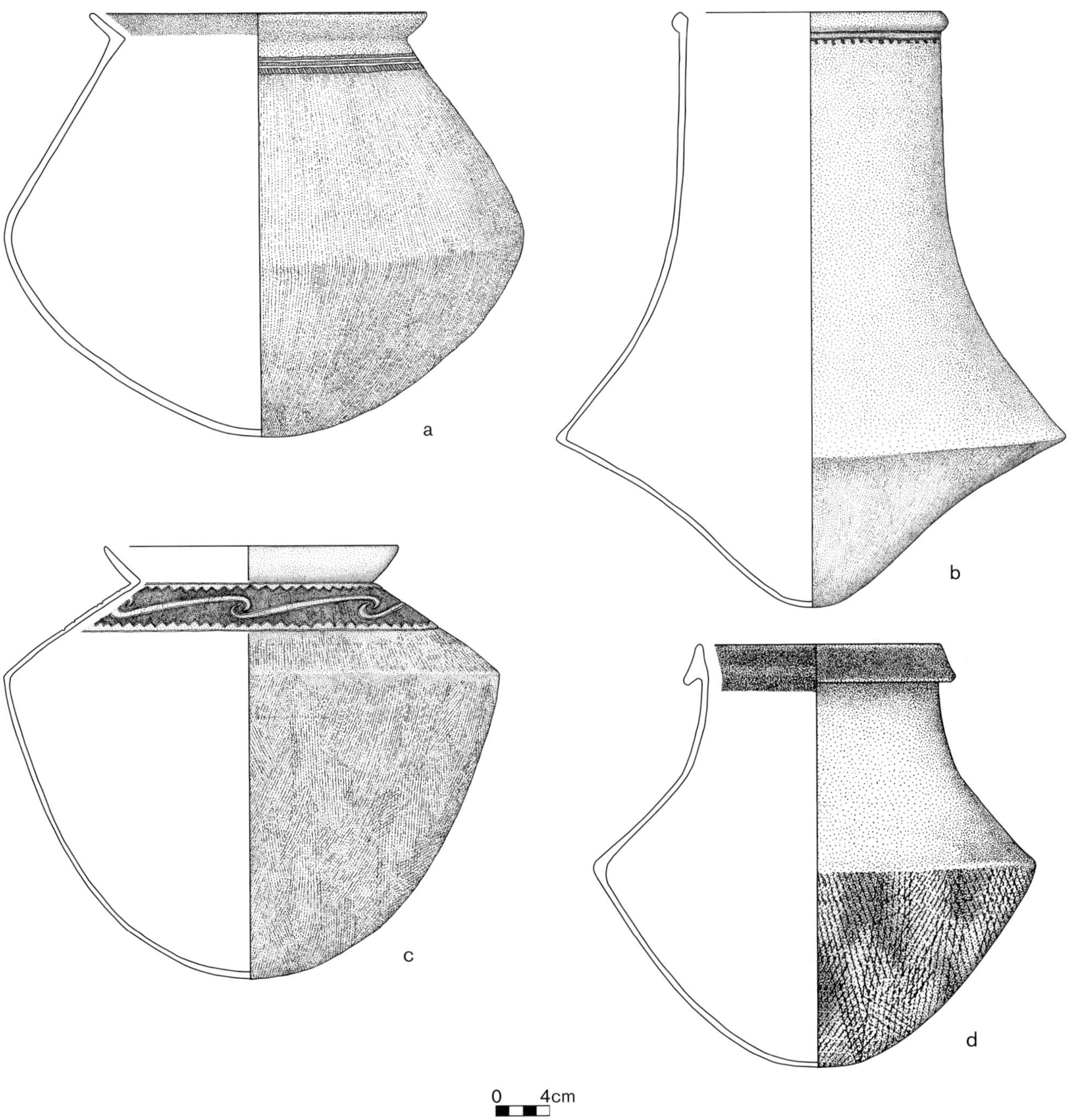

Figure 2.4 Examples of carinated pots recovered from the Ban Chiang Middle Period. (a) BC Burial 18 Pot A 712 from MP VI. (b) BCES Burial 40 Pot A 1934 from MP VIIa. (c) BCES Burial 14/15 Pot C 1047 from MP VIIa. (d) BCES Burial 12 Pot E 990 from MP VIII.

Figure 2.5 Example of a Middle Period VIIa burial with sherds broken over the body, BCES Burial 40, male aged 45–50 years. Sherds over legs had been removed for this photograph.

Figure 2.6 Example of a Late Period burial with intact vessels placed over the body but crushed and broken in situ. BCES Burial 8 from LP X is an infant/toddler aged 2–3 years.

and numbered I through X. The burial phase sequence incorporated stratigraphic burial superposition, changes in pottery styles, and wherever possible, soil evidence. Matrices of burial superpositions were graphically illustrated (White 1986:77–78) and used to demonstrate the sequence of grave styles and ceramic types in relationship to soil stratigraphy. Five Early Period (EP) burial phases, three Middle Period (MP) burial phases, and two Late Period (LP) burial phases were delineated.

Burial Phase Refinements

The same periods and burial phases have continued to be used in subsequent publications as well as in the current analysis. Phases used in Pietrusewsky and Douglas (2002) reflect some refinements in assigning burials to phases over White (1986). As can be seen in table 1.1 of Pietrusewsky and Douglas (2002:5), the presence and prominence of individual burial phases at each locale varies. For example, Late Period burial phase IX has no interments at the BC locale and five interments in the BCES locale.

In this volume, burial subphase designations are incorporated where applicable. Subphases, designated with letters such as "a," "b," etc., have been defined when burials in a phase have similar pottery, but there is stratigraphic evidence that some interments occurred earlier or later within the phase's depositional span. Such distinctions are particularly useful in phases such as Ban Chiang's Middle Period Phase VII. Burials from Middle Period Phase VII usually have several carinated pots broken over the supine bodies, but these burials can be divided into two subsets. The higher subset of Middle Period Phase VII burials, Middle Period Phase VIIb, has many vessel forms not found in the lower subset of burials termed Middle Period Phase VIIa. Ban Chiang's Early Period Phase V, another example, has burials with globular cordmarked pottery of a very similar shape and size deriving from a range of depths that could represent a considerable time span. The higher Early Period Phase V burials can include more than one pot, more size variation, and some vessels may have painted and incised decoration, while the lowest Early Period Phase V burials, such as BCES Burial 59, have one single undecorated globular cordmarked vessel. Three Phase V subphases are recognized: Va, Vb, and Vc.

Occupation Deposits: Issues of Relative Chronology

No matter how refined the definition of the burial phases, these phases proved inadequate for temporal subdivision of the non-burial deposit due to characteristics of the site formation processes. For example, in some instances the non-burial deposit seems to have been formed during a hiatus in mortuary activities at a locale. In some parts of the site, burials from more than one phase were placed closely together during a time when depositional buildup was slow. Multiple burial phases could originate from a single, relatively thin soil deposit, but the superposition, changes in grave goods, and dating evidence indicate interments occurred over a considerable time span. Thus, the thin non-burial deposit likely also had occupation materials from a substantial time period.

The problem of temporally subdividing occupation deposits into the same phases used for burials is compounded by a variety of other issues including the unclear tropical soil stratigraphy and taphonomic considerations. Ultimately, establishing two separate although related relative chronologies, one for the burials and one for the non-burial deposits, proved not merely a pragmatic solution to a descriptive problem, but also a better reflection of the site formation processes postulated for these sites.

Site Formation Processes

Viewed from the point of view of site formation processes, mortuary and occupation deposition at these sites contribute two distinct types of accumulation. As a deliberate interment within a moment of archaeological time, a human burial contains a set of indisputably contemporaneously interred items (corpse with accompanying grave goods) in a constructed space designed to hold the burial assemblage in one place within the earth in perpetuity. In most cases, archaeologists assume that the grave goods were manufactured close to the lifetime of the interred individual. The possibility that some grave goods were heirlooms is hard to rule out, although no clear example has been found in prehistoric Thailand. The burial's in situ position as recovered by the archaeologist was chosen by the ancient inhabitants. Such deliberately deposited "time capsules" provide a totally distinct interpretive challenge from occupation midden deposits.

Accumulations from occupation activities at these sites appear to be largely incidental and incremental byproducts from a variety of activities of daily life that are not necessarily contemporaneous or intrinsically related. The prevalence of postholes and rarity of living surfaces at Ban Chiang and similar sites indicate that structures were made of organic materials and were usually raised over the ground surface on stilts, as is common in traditional mainland Southeast Asian dwellings within recorded memory.

Inferring from traditional villages in the region today, many daily activities, such as cooking, eating, and sleeping, likely took place off the ground in the raised dwellings. Debris from daily life accumulated incrementally under and around village structures and was most likely subject to ongoing processes of redeposition, such as water flows and human and animal traffic. Formation of surfaces containing discrete cultural features with meaningful associations to each other (e.g., a defined house with identifiable activity areas) would have been rare. Reoccupation of previously abandoned localities along with processes of redeposition from the digging of holes for mortuary and occupation activities and surface natural and cultural processes meant that materials of vastly different ages and functions may ultimately be recovered together by the archaeologist's trowel.

Definition of Levels

Nevertheless, by looking at evidence from superposition in deep deposits, the lateral pattern of features such as postholes and artifact clusters that are visibly intact and coherent, and evidence from pot seriation, it is possible to discern horizontal "levels" of deposition at each locale. Similarities among features across horizontal spans at each locale can provide a means upon which to base a relative sequence of the non-burial deposit. These horizontal units, termed levels, are not to be confused with discrete strata defined by clear horizontal soil changes. Rather the levels are means of aligning deposits across an excavation locale with similar materials that approximate a temporal period.

Deposits at each locale discussed in this study therefore have been subdivided into levels, which are numbered from the base of the locale (see Bayard 1971:13 for a similar approach). Unlike the excavation *layers* described above which are numbered separately within each excavation square starting at the top, levels laterally crosscut all squares within a locale. Criteria for assigning deposits to a level are multifaceted and include depositional and cultural attributes. Included are position within the larger soil units, relative and absolute depth, superposition, and vertical and horizontal changes in cultural materials and types of features. Tables of correlations of layers in neighboring squares have been constructed based on similarities among these criteria and will be presented in White (in preparation). Pottery similarities across the locales from both burial and non-burial contexts have been particularly important aides to correlating excavation layers among each locale's squares.

Levels are integrated with Early, Middle, and Late Periods first defined in White (1982) and the burial phases defined in White (1986), with subsequent refinements discussed above providing a comprehensive relative sequence at each site. The levels have proven useful for grouping both burial and non-burial deposits at all sites considered here. Both levels and burial phases can be assigned to periods on the basis of depositional, chronological, and/or cultural evidence. The period framework provides the basis for regional comparisons. Moreover, the Early Period has been usefully divided into an upper Early Period and a lower Early Period at a point in the Ban Chiang sequence where there are changes in ceramics in concert with changes in burial orientations, even though grave styles (White 1982) continue. This comprehensive stratigraphic analysis enabled White to assign each individual metal and metal-related artifact, in fact virtually all Ban Chiang deposits, to locale-specific levels, as well as to periods. Grave goods can additionally be assigned to burial phases (see Fig. 2.7).

White has applied the same analytical approach to Ban Tong, Ban Phak Top, and Don Klang. Levels have been defined for the locales excavated at each of these sites, and site-specific burial phases have also been named where applicable. All levels and phases have been assigned to periods, which provide the framework for comparisons among sites (see Table 2.1).

Relationship of Mortuary and Occupation Activities

For these four sites, assignment of deposits into levels forms the basis for organizing metals and related artifacts from non-burial contexts, as distinct from metals from burial contexts. The question arises, however, concerning what is the general taphonomic relationship between burial and non-burial deposits? Was burial and occupation deposition occurring simultaneously? Or do they reflect alternating uses of the locale? The answer to this question has cultural as well as analytic implications.

The view taken here, more fully explicated in White and Eyre (2011), is that corpses were interred under and/or between residences and other structures while the sites were continuously occupied. This practice, termed "residential burial" (Adams and King 2011), is common in a wide range of societies globally from neolithic Çatalhöyük (Hodder and Cessford 2004) to classic Maya urban sites (McAnany 1995). The physical integration of burials and occupation features means that the most economical, elegant, and obvious interpretation of their deposition is that daily activities of life and death were closely connected in space and time.

One pragmatic objective for defining levels is to be able to provide a useful relative chronology for the non-burial occupation deposit despite the rarity of clear occupation floors. The conclusion that burials were deposited within the space also occupied during ongoing habitation at the site also implies that burials should be attributable to levels. But in practical terms, attribution of burials to levels must be considered approximate because it is often difficult to determine the precise depth and source from which an interment was cut.

The source of a grave cut was often difficult to identify during the excavation. Frequently the first indication that the excavators were coming upon a burial was the appearance of part of a human skeleton. Grave cuts for many (but not all) skeletons were not observed, or not observed above the appearance of the skeleton. Other evidence, such as patterns of disturbance, might demonstrate that the bodies were, indeed, interred, as opposed to deposited within a mound built over the body. There appears to be some variation in grave depth for interments. Some corpses during Middle Period Phase VII may have been placed on the surface and materials mounded over the body, and some interred corpses appear to have had grave goods placed on the grave after interment, as has been observed in ethnographic contexts (Lewis and Lewis 1984:234–235). In general, however, the most common practice seems to have been interment of corpses in shallow graves, usually less than 0.75 m, and often less than 0.5 m in depth. (Although such depths may seem implausibly shallow, they compare favorably with Condominas's

Period		BCES Level	BCES Burial Phases and Burials	BC Level	BC Burial Phases and Burials
Recent		4D		13	
Historic		4C			
Historic		4B (no plan)			
Historic		4A			
LP–Proto				12	
LP		3I			
LP		3H	LP X (1, 2, 3) LP IX (4*, 9)	11	LP X (1, 3, 14, 51, 55)
LP		3G	LP X (7, 8, 13) LP IX (4*, 5, 6)		
MP–LP		3F	MP/LP (39) MP VIII (12)	10	MP–LP (2)
MP–LP		3E	MP VIIb (11, 19*, 20*, 22*, 23*, 26*)		MP VIIb (22?)
MP		3D	MP VIIb (19*, 20*, 22*, 23*, 24, 26*, 80)		
MP		3C	MP VIIa (10, 14, 15, 16, 17, 40, 41, 71, 73, 75, 78)		MP VIIa (11)
MP		3B		9	
EP–MP	upper Early Period	3A	MP VI (21, 25?)		MP VI+ (27) MP VI (4, 9?, 15, 17?, 18, 20?)
			EPVc (28, 29, 30, 32, 53)		EPVc–MP (49) EP Vc (5, 6, 7, 8, 10, 12, 13, 16, 19, 28, 35, 53)
EP		2H (no plan)			
EP		2G	EP Vb (27, 36, 56)	8	
EP		2F	EP Va (33, 42)		EP Va (21, 23, 29, 30, 36, 37, 39, 50)
EP		2E			
EP	lower Early Period	2D	EP IVc (31, 34, 35, 37, 38, 48, 64, 82)		
EP		2C	EP Va (59)** EP IVb (46, 61, 63, 66, 77)	7	
EP		2B	EP IVa (43, 50, 51, 54, 55, 69) EP III–IV (45, 65) EP IIIb (44, 47, 49, 70) EP II–IV? (67, 68) EP II–III (81)		EP IV (26) EP III–IV (31, 54) EP IIc (24*, 43*, 43A*, 46*)
EP		2A	EP IIIa (49A, 52, 62?, 76) EP II–III (58, 72, 79?, 79A?, 79B?) EP IIc (57, 60?, 74)		
EP				6	EP IIc (24*, 43*, 43A*, 46*, 33, 33A, 34, 34A, 38, 40, 42, 45)
EP				5	EP IIb (32, 41, 47, 52)
EP				4	EP IIa (25, 48)
EP				3	
EP				2	EP I (44)
		1A		1	

* Burial is assigned to more than one Level

** BCES B.59 is similar to other EP V burials in orientation, ceramics, and grave style. Its depth suggests that the source of the grave cut was not identified during excavation, and that it likely derived from a higher level.

? Indicates best estimated phase placement at time of writing

Figure 2.7 Stratigraphic framework for BC and BCES Periods, levels, burial phases, and burials.

Note: BC = Ban Chiang 1974; BCES = Ban Chiang Eastern Soi 1975; EP = Early Period; EP–MP = Early Period to Middle Period; MP = Middle Period; MP–LP = Middle Period to Late Period; LP= Late Period; LP–Proto = Late Period to Protohistoric Period; B. = burial.

Table 2.1 Correlation of Ban Tong, Ban Phak Top, and Don Klang Sequences with Ban Chiang Periods

BC/BCES		BT	BPT	DK	DK Burials
Modern/recent		Level 9	Level 12	Level 5	
Historic		Level 8B	Level 11		
Protohistoric				Level 4B	
LP–Protohistoric transition		Level 8A		Level 4A	
Late Period	LP X	(small bit of LP)	Level 10	Levels 3C, 3B, 3A, 2/3, 2C, Phases 4, 3	Burials 1, 2, 3, 4, 5, 6, 7, 8, 9, 17
		Level 7B			
	LP IX	MP–LP?			
Middle Period	MP VIII				
	MP VIIb				
	MP VIIa	Level 7A (Burial 1)	cultural hiatus	cultural hiatus	
	MP VI				
upper Early Period				Level 2B, Phase 2b	
	EP Vc				
	EP Vb	Levels 6, 5, 4	Levels 9, 8B, 8A, 7		
				Level 2A, Phase 2a	
	EP Va				
lower Early Period	EP IV			Levels 1/2, 1C, Phase 1b	Burials 10, 11, 12, 13, 14
	EP III	Levels 3, 2/3, 2B (Burial 2), 2A			
	EP IIc				
	EP IIb		Levels 6, 5, 4, 3, 2, 1		
	EP IIa			Level 1B, Phase 1a	Burials 15, 16
	EP I				

Note: BC/BCES = Ban Chiang 1974 and Ban Chiang Eastern Soi 1975; BT = Ban Tong; BPT = Ban Phak Top; DK= Don Klang; LP = Late Period; MP = Middle Period; EP = Early Period.

[1977:291] observation of a grave cut only 65 cm deep for a highly ranked Mnong Gar man interred inside a log coffin.) Post-excavation analysis of the burial sequences in matrices (schematics showing burial superpositions and intercutting similar to a Harris Matrix approach) allowed the approximate source and depth of most grave cuts to be positioned at least in relative terms (White 1986:77–78). That burials also are often in superposition relationships with habitation features assisted the assignment of burials to the sequence of defined levels.

Once the basic positioning of burials on empirical grounds within the levels was undertaken, analysis showed that no simple correlation could be demonstrated between the phase units of the burial sequence and level units of the occupation deposition. In a few cases, burials assignable to a single level were all from a single phase. Similarly, in a few cases, all burials from a single phase at a locale come from the same level. However, in other cases, such as if there was a period of a slow rate of accumulation of occupation debris, burials from more than one phase could be attributable to a single level. On the other hand, levels could usually be assigned to periods, although some levels contained materials from contiguous periods, possibly indicating transitional and/or commingled deposition. Figure 2.7 provides the breakdown for BC and BCES by periods, levels, and phases.

From an analytic point of view, the lack of a simple relationship between burial phases and levels is handled by the ability to collapse and expand scales of analysis depending on the question being addressed and the degree of temporal discrimination allowed by the pertinent data. The variety of chronological distinctions employed in this analysis, including both depositionally based distinctions such as levels and culturally based distinctions such as burial phases, enables sequences to be subdivided (i.e., "lumped" or "split") in a number of different ways as the data allow. Fine- or coarse-grained temporal discrimination is dictated by both the questions being addressed and the degree of discrimination possible within the data set. Larger-scale questions may be best addressed based on comparisons by periods, such as when undertaking regional comparisons. Questions involving temporal comparisons that combine mortuary and occupation evidence at Ban Chiang are also best addressed by periods, which provide the framework for "lumping" several levels. Smaller-scale questions may call for finer-grained temporal examination and be best addressed by comparisons based on levels within individual locales. For example, seeking evidence for differential spatial use of a site over time is based on comparisons of levels from separate locales. Detailed questions of change in mortuary behaviors rely on the phase sequence. The differential but interlocking units of analysis allow a wide range of intra- and inter-site questions to be addressed in this study. In short, both kinds of classifiers—lumpers and splitters—can be accommodated by the stratigraphic system developed for these sites.

Depositional Contexts

In addition to sequential units of analysis noted above, another analytical unit developed for Ban Chiang as well as for the other three sites is implemented in this volume, namely depositional context categories. As discussed above, the two basic types of deposition, mortuary and occupational, are intermingled within excavated locales. The occupation deposit was not definable by bounded residential units, but features, such as postholes and groupings of artifacts in horizontal scatters or clusters as well as the general accretion of debris, demonstrated occupation use and provided some structure for sequences of occupational use. As noted above, provenience details recorded in bag descriptions usually provide information on the type of deposit being removed, such as a spit, feature, or burial. The deposit type noted for the bags is, by implication, a context, as perceived by the excavators. Although a highly elaborated system for context/deposit categories has been developed, for the purposes of this volume, four basic context categories have been delineated, to which most artifacts can be attributed: (1) grave good (gg); (2) burial-associated material (ba); (3) feature material (fea); and (4) material from the general soil matrix (gsm).

The first general context is *grave good.* Grave goods were in most instances described as such by the excavators and the designation indicates artifacts whose condition and obvious association and placement with a skeleton implies deliberate inclusion with the burial. They are part of the burial's "time capsule."

The second general context category, termed *burial-associated*, is an ambiguous context for artifacts that are not clearly grave goods but were recovered in close association with a skeleton. For example, an occasion for assigning a "burial-associated" context would be one described as "removing grave fill." Metal-related (and other) artifacts from such a context might in theory be displaced grave goods, for example, a bangle fragment close to a disturbed burial but not on a wrist or ankle. On the other hand, the bangle fragment near a skeleton might have been disinterred from the deposit into which the burial was interred, and then reinterred with the grave fill and hence predate the grave (J. White 2008:94). Others have made the same argument that non-metal, non-grave good items (e.g., charcoal) recovered close to skeletons predate the burial (e.g., Higham 1984a:231, 1989:12, 126). At other times, arguments have been put forth that objects categorized as "burial-associated" could be displaced from higher and hence later contexts (e.g., Higham 2011a). Here the authors assume that, in the case of Ban Chiang, redeposition of burial-associated metal objects into intrusive features such as graves from preceding deposits (including of the same time period as the burial) is more likely than downward movement into graves from significantly later deposits through post-depositional mechanisms such as bioturbation. Artifacts from burial-associated contexts are examined in TAM 2B (chapter 6) to see if better understanding can be achieved on their depositional process, i.e., whether they are more likely displaced grave goods, or redeposited materials from non-burial contexts.

The third general context is *feature*. Most features are deposits resulting from occupation use of the site. Several feature types can be defined ranging from deliberately intrusive features such as postholes and pits to features that represent debris deposited on ancient surfaces, such as artifact scatters. It is assumed that materials from most features were recovered from in situ contexts as they were deposited by ancient inhabitants, many by small but specific events. Some features were intentionally placed such as a posthole, others created haphazardly by, for example, tossing a broken pot over the side of the porch. Given that the feature outlines and contents are still intact, it is assumed that most materials recovered from features have not significantly moved since deposition of that feature in prehistory. Given their likely depositional stability over time, features provide important contexts from which the non-burial deposit has been delineated into levels.

The fourth general context is the *general soil matrix*. Materials recovered from excavating spits, i.e., excavation layers, would be assigned to this context. As noted above, most materials recovered from the general soil matrix are presumably accumulations of ongoing incremental and undifferentiated deposits of debris of daily life. Materials recovered from the general soil matrix have the highest probability among all contexts to have had repeated displacement from a variety of natural and cultural taphonomic processes.

The four context categories outlined here provide structure for comparing metal and related finds from burial and occupation contexts in the forthcoming, TAM 2B.

Absolute Chronology

Ban Chiang is well known for having produced dates from a variety of materials with several techniques (Carriveau and Harbottle 1983; Hedges et al. 1997; Higham et al. 2011a, 2011b, 2015; Hurst and Lawn 1984; J. White 1997, 2008). There have been a considerable number of chronologies proposed and the site's dating has been debated for more than 40 years (Gorman and Charoenwongsa 1976; Higham 1984a; Higham et al. 2011a, 2011b, 2015; J. White 1982, 1986, 1997, 2008, 2015a). The chronology remains under debate, although details of the arguments have changed over the years, particularly as new dates have become available. There remain serious contradictions in the technical evidence, along with different points of view on taphonomic and stratigraphic interpretation, on how to interpret evidence from ambiguous deposits, and on how to construct regional sequences with cross-dating. It is the point of view of the authors that, as of this writing, there is no final detailed absolute chronology for the Ban Chiang sequence specifically or for prehistoric Thailand in general.

Detailed discussion of the evidence for the absolute chronology of Ban Chiang will be presented in a forthcoming volume on Ban Chiang stratigraphy and chronology (White, in preparation). For the purposes of this volume, the approach is to acknowledge different points of view and to briefly review some key technical issues underlying the debates to help explain to interested readers why no final chronology is offered here. An interim, general, working chronology is proposed for the periods described above using a "paired dates" approach to temporally define key transitions. The discussion in subsequent chapters of the monograph emphasizes the relative chronology. Thus, while the absolute chronology is likely to change going forward, any changes can be applied to a stable relative chronology.

Summary of Recent Disagreements in Ban Chiang Absolute Dating

> *A key assumption in radiocarbon dating, that a sample only contains carbon that was present in the living organism, is only valid if all exogenous sources of carbon are removed prior to dating. (Wood 2015:63)*

Currently, absolute dating of Holocene archaeological sites in Thailand relies primarily upon the radiocarbon dating technique. Although the technique of dating ^{14}C from excavated organic materials has been around since the 1950s, the technique has continuously evolved from its initial appearance to this day (Wood 2015). The evolution and improvement of the technique over the years has been prompted by challenges, such as discrepancies among dates expected to be close based on archaeological considerations and the need to date small samples. The issue of removing all exogenous carbon and being confident that only the carbon from the original living organism is being dated is ongoing and of key importance, as very small samples currently being dated with accelerator mass spectrometry (AMS) are highly susceptible to contamination (Wood 2015:65). Another ongoing issue is the problem of samples excavated from particular contexts by the archaeologist having been displaced, though to an unknown degree, from an original depositional context by disturbances of various types, such as roots, burrowing animals, and digging of pits and graves.

Different organic materials have become associated with distinct types of potential problems, resulting in radiocarbon determinations that can be older or younger than the deposit the archaeologist is seeking to date. Dates on charred wood may be suspected of being too old due to "old wood effect," i.e., the possibility that the charcoal came from heart wood of an old tree or came from a piece of wood that had been used for generations (e.g., as timber in a house) before becoming charred and deposited into an archaeological context. Different organic materials can be susceptible to different sources of contamination. Post-depositional chemical changes (diagenesis) can introduce older and/or younger carbon; recent carbon such as carbon from humic acids can alter the buried object's carbon chemistry (e.g., Webb et al. 2007). Living organisms can incorporate old carbon during their lives through particular diets or simply geochemical context (reservoir effect [see Jull et al. 2013]). Another variable adding to the complexity of interpreting radiocarbon results is that pretreatment protocols designed to remove contamination can vary from laboratory to laboratory and produce different final results (Fiedel et al. 2013; Kuzmin et al. 2014).

The dating evidence available from Ban Chiang, as well as Thailand and Southeast Asia as a whole, is replete with issues such as these. Unfortunately, dates from the most desirable samples, such as charred short-lived plant foods like cereal grains excavated from impeccable contexts like from grave good vessels or hearths, are very rare in Southeast Asia (Thomas and McLauchlan 2006). At the time of writing, the radiocarbon dates under consideration for Ban Chiang and key comparative sites include dates on charcoal, freshwater shell, human and animal bone, rice temper in pottery, charred plant macrofossils, charred food residues on pots, tooth enamel, and phytoliths from the contents of a grave good pot that contained rice. The field is dealing not only with apples and oranges, but also peaches, cantaloupes, and dried apricots.

Not surprisingly, there are discrepancies among these dates and there are various approaches to interpreting the dates and the discrepancies among them. Setting aside for the moment stratigraphic issues, one can summarize two main concerns that have been raised about technical biases in the dates themselves. Older than accurate dates may be suspected from charcoal, rice temper in pottery, and phytoliths, and hence Higham et al. (2011a:230, 239) have argued these should generally be considered *termini post quem*—or dates after which the target cultural event occurred. On the other hand, shell and bone dates from northeast Thai sites may in contrast be young and hence *termini ante quem*, or dates before which the target cultural event occurred (see White 2015a).

Shell Dates

Recent dates on shell from sites in the upper Mun basin, particularly Ban Non Wat, are argued as likely contaminated with young carbon (White 2013, 2015a). Webb et al. (2007) have shown that freshwater shell recovered from geochemical conditions of high salinity and high magnesium:calcium ratios can become contaminated with young carbon. Such soil chemistry is associated with sediments from old sea beds. Much of northeast Thailand overlies old sea beds of such richness that salt extraction is an important industry region wide, thus, freshwater shells excavated from these depositional contexts seem likely to have been affected. Moreover, geochemical studies undertaken at Ban Non Wat by Boyd and colleagues and referenced by King et al. (2011:2224) have apparently identified high magnesium and sodium: "Soils surrounding the burials are rich in organic matter, and contain high concentrations of minerals such as magnesium, and manganese (Boyd et al. 1999). The area's soils are also rich in evaporate, giving them high concentrations of sodium (ibid)." The elements that signal problematic geochemistry for shell diagenesis thus appear to have been found in sediments contiguous with Ban Non Wat burials.

In a recently published reference book on radiocarbon dating, Taylor and Bar-Yosef stated (2014:74) that the "use of non-marine shells for dating should probably be restricted to situations where...detailed studies of the geochemistry of the freshwater environment have been done." Webb et al. (2007:803) also stated, "groundwater chemistry should be considered before accepting bivalve-based radiocarbon dates uncritically." Yet Higham et al. (2015; see also Higham and Higham 2009b) have not provided evidence that they have undertaken the geochemical studies other specialists advocate should be done when evaluating freshwater shell dates.

Moreover, routine protocols for assessing shell for contamination can miss the diagenetic alterations caused by geochemical conditions related to sediments from sea beds (Webb et al. 2007). Raman spectroscopy that can identify these diagenetic changes was not employed to assess the Ban Non Wat shell (Thomas Higham, pers. comm. 2015). In their effort to identify recrystallization that can signal contaminated specimens, Higham et al. (2015:4/20) only mentioned that, "A shell fragment was...tested for recrystallization using the Feigl staining method; when no recrystallization was observed the fragment was crushed and prepared for dating." The Feigl protocol tests for the presence of secondary calcite on aragonitic carbonates, according to Higham (2015c:1235). However, this argument misses the point raised by Webb et al. (2007) on vetting protocols for freshwater shell for recrystallization. Tests only for calcite will not detect the recrystallization of aragonite that can occur under geochemical conditions associated with old sea beds. Webb et al. (2007:805) noted that up until recently "vetting techniques for freshwater shells have been based on the assumption that diagenetic alteration would yield calcite." By using Raman spectroscopy, they found *calcite-free diagenetic alteration*, whereby aragonite precipitated instead of the more common calcite under certain geochemical conditions, specifically high magnesium:calcium ratios and salinity. Given these issues of incomplete vetting protocols used for freshwater bivalve shell dates from the upper Mun Valley, the recent shell dates from upper Mun sites are best considered *termini ante quem*.

Bone Dates

In the history of radiocarbon dating, bone has been a challenging material with a performance record of often dating too young (Denaire 2009; Zazzo and Saliège 2011). In a study comparing bone dates, charcoal dates, and dendrochronology pertinent to the European neolithic, Denaire (2009) found that the charcoal dates matched the dendrochronology and that the bone was consistently young, often by a few hundred years. He cautioned against building chronologies based primarily on bone dates. Scholars have found that bone dates can produce unreliable results due to soil chemistry, pretreatment protocols, diet, and other factors. Although pretreatment protocols employing ultrafiltration have increased the efficacy of bone dating in many cases, particularly for Pleistocene samples, problems persist and some of these are reviewed below.

BONE DATES FROM BAN CHIANG

> *Most bone derived from the typical...archaeological site in most temperate and tropical environments contains little* unaltered *collagen... it is probably inappropriate to label an organic extract from an archaeological bone as "collagen" unless a biochemical analysis on the organic fraction...has been obtained. (Taylor and Bar-Yosef 2014:81; emphasis in original)*

Higham et al. (2011a, 2015; see also Higham and Higham 2012) recently have proposed a chronology for Ban Chiang based solely on bone dates. There are several reasons not to accept the proposed chronology *pro forma* or at face value. Although this monograph is not the place to do a complete review of the evidence pertinent to the Ban Chiang chronology, a few comments on some key technical and chronometric issues can illustrate the need to keep an open mind about the Ban Chiang chronology and why dates from charcoal and rice temper should not be arbitrarily excluded in discussions.

The dates for Ban Chiang used by Higham et al. (2011a, 2015) rest solely on what they term bone "collagen." They argue that the Oxford protocols for ultrafiltration of bone essentially ensures that the dates are accurate. However, a recent inter-laboratory assessment of bone dating compared protocols and results from four radiocarbon labs—Oxford, Groningen, Kiel, and the University of California, Irvine—for a bone sample of well-established age (Fiedel et al. 2013). The study found that "collagen" samples, including some that underwent ultrafiltration at Oxford, were significantly younger than expected for a sample dated in the four different laboratories. Gillespie et al. (2014) also reported erroneously young ages from the application of Oxford's ultrafiltration method. Recent problematic "collagen" dates have been noted for late Pleistocene-aged bone recovered from contexts with obvious humic contamination (Fiedel et al. 2013) and also from collections that have been curated and conserved over a number of years (Nalawade-Chavan et al. 2014). Taylor and Bar-Yosef (2014:82) noted that problematic "collagen" dates have also been found in Holocene contexts; bones with less than 1% "collagen" yield often have seriously anomalous dates and, though less common, "bones with collagen yields between 1% and 5% have yielded problematic ^{14}C ages, typically at the level of a few hundred years" (Taylor and Bar-Yosef 2014:82).

A probable reason for bone dates that are too young is that contaminants remain in ultrafiltered samples (e.g., Nalawade-Chavan et al. 2014). As Boudin et al. (2013:2039) recently noted: "Ultrafiltration of bone collagen…is an effective method for the removal of low molecular weight contaminants from bone collagen *but it does not remove high molecular weight contaminants, such as cross-linked humic-collagen complexes* [emphasis added]." Furthermore, on the subject of collagen dating, Marom et al. (2012:6878) stated: "the extracted bulk gelatin can be heterogeneous and include, or be cross-linked to, potential contaminants from the depositional environment, such as humic and fulvic acids, rootlets, cellulose, sediments, and other plant and animal remains including amino acids from bacteria and micro-organisms."

In "Supporting Information," Higham et al. (2015) noted that the bone samples they dated met criteria for suitability for dating, stating that the C:N (carbon to nitrogen) ratios in bone samples, the routine way ultrafiltration samples are vetted, "were well within acceptable ranges." However, Marom et al. (2013:705) commented that merely determining that the C:N ratio falls between 2.9–3.6, a routine approach for identifying contamination, is not precise enough to rule out contamination.

The Oxford University Accelerator Unit (ORAU) has recently undertaken analyses of "collagen" extracts from their ultrafiltration protocol in order to investigate the heterogeneity of the product. The study (Brock et al. 2013:445) noted: "the final product of 'collagen' extraction at ORAU appears to be an aggregate consisting of a range of proteins of different molecular weights, including collagen, as well as some other organic matter and inorganic species. Ultrafiltration is removing some, but not all, of the <30kDa fraction from the samples." They went on to state (Brock et al. 2013:461): "Reaching definitive conclusions regarding what ultrafilters remove from bone 'collagen' is rendered more challenging because of the wide spectrum of diagenetic processes and contamination influences on archaeological bone."

Taylor and Bar-Yosef (2014:77) have noted that the term "collagen" is being inaccurately used in many publications presenting bone dates. Collagen "should be used only when there is some independent data presented that the fraction contains only or primarily the protein collagen. One standard method to demonstrate this analytically is to obtain a profile of the amino acid components of such a fraction....In the absence of this or other appropriate data, the term 'total acid insoluble' can be used to correctly characterize this fraction" (Taylor and Bar-Yosef 2014:77). No amino acid profile has been presented by Higham et al. (2011a, 2015) to support their claim that only collagen was being dated from Ban Chiang bone.

In summary, there is a growing appreciation that the phrase "collagen date" is a misnomer implying a false precision to the ultrafiltration protocol. The Higham et al. (2015) terminology in calling their bone dates "collagen" likely misrepresents the material actually dated in their bone samples and the accuracy of their results. Dating publications are beginning to put "collagen" in quotations (e.g., Brock et al. 2013), and other phrases such as "high-molecular weight" (HMW) fraction (Minami et al. 2013) are appearing in the literature. Taylor and Bar-Yosef (2014:81) suggest "ultrafiltered gelatin," "base-insoluble," and other terminology. In this volume the phrase "bone date" is used.

RED FLAG BONE DATES

These issues are pertinent to the ultrafiltered bone dates from Ban Chiang. This can be illustrated with the following example. Although the Higham team constructs their argument emphasizing correlation of temporal units at the scale of the "Neolithic," "Bronze Age," and "Iron Age," archaeologists usually cross-date and build relative sequences on the basis of more refined, more temporally restricted, smaller-scale units such as levels or phases defined in terms of highly specific and time-restricted cultural traits, e.g., specific types of ceramics. The underlying idea is that levels at two or more sites with clearly similar material evidence shared items of material culture and/or the specific stylistic and/or technological system used to make those items. In other words, societies in contemporaneous phases were either exchanging object types that they have in common, and/or existed in related production networks that were manufacturing objects with then current styles and techniques. The most convincing cross-dating relative frameworks rely on the identification of material items in production during short temporal spans and seek to differentiate the short-term items and styles from products that for whatever reason were enduring and long lived. Then similar dates are expected from phases with very similar material culture. In general, for a variety of reasons, including highly variable prehistoric pottery and a late start in modern archaeological research, Thai archaeology is far behind other regions in the world in developing these detailed phase correlations based on regional ceramic sequences.

A "red flag" signaling that the Ban Chiang bone dates reported by Higham et al. (2011a, 2015) are problematically young is a clear example where bone dates from Ban Chiang do not cross-date harmoniously with shell dates (which as noted above have their own potential problems) from Ban Non Wat, although the bone and shell are associated with grave pottery with sufficient similarities that regional specialists would expect them to be close in time. The pertinent Ban Non Wat shell date is associated with a vessel (Fig. 2.8a) from Burial 28 from phase Neolithic 1 (hereafter N1) (OxA-18133, 3170±27 BP, 1500–1405 cal. B.C.). The pertinent bone dates from Ban Chiang come from the double burial of BC Burials 43 and 45, which were interred with a large pot with i&i (incised and impressed) decoration (Fig. 2.8b). Both BC Burials 43 and 45 have rib bone dates, respectively OxA-25014 (2984± 26 BP, 1284–1117 cal. B.C.) and OxA-x-2438-16 (2958±29 BP, 1260–1055 cal. B.C.). The plotted calibrated ranges for these three dates are illustrated in Figure 2.9. The Ban Chiang bone date ranges are far younger than the Ban Non Wat shell date. Moreover, the Ban Chiang bone dates[1] not only do not overlap the Ban Non Wat shell date even at the 95.4% level of confidence, they differ on the order of a couple hundred years.

Why would archaeologists expect the specific vessels associated with these dates to be close in age? Beginning with the bigger picture, this grave pottery is relevant to discussions of the i&i ceramic style. The i&i ceramic style has been argued by many regional specialists (e.g., Bellwood et al. 2013:160; Higham 2004; Higham and Rispoli 2014; Higham et al. 2011a; Rispoli 2007; Rispoli et al. 2013) to reflect a multiregional ceramic horizon. The style has also been thought to be associated with the expansion of neolithic, Austroasiatic-speaking rice farmers (Higham 1996b). As research has moved forward,

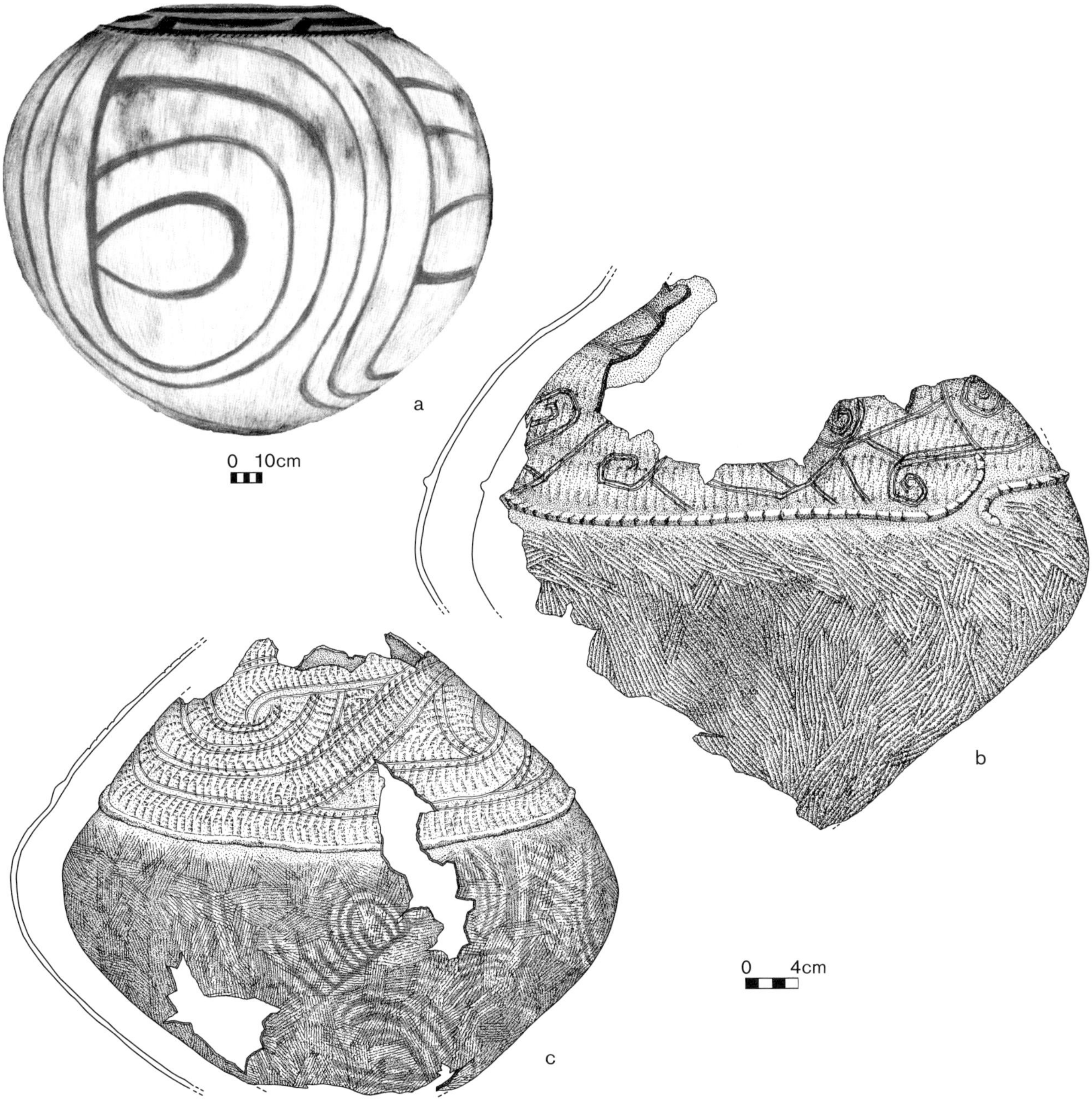

Figure 2.8 Possible comparanda from early phases at Ban Chiang and Ban Non Wat. (a) Example of a BNW Neolithic 1 burial jar that contained an adult, Burial 28 cat. 1039. (b) The i&i burial jar for BC Burial 43 Pot A 1374. (c) Example of a lower Early Period Phase IIc i&i burial jar, BCES Burial 46 Pot A 1608.

this decorative style has been found on a variety of forms and shown variation in technique and motifs within mainland Southeast Asia (Sarjeant 2014), as well as within Thailand (Rispoli 1997; White 2017; Wiriyaromp 2010). Sometimes paint decoration accompanies the incised and impressed decoration and, in some cases, paint displaces the decorative impressing within incised zoned motifs (Wiriyaromp 2010). This variation suggests that local potters copied a regionally shared style rather than a scenario where pottery centers produced vessels that were widely traded (White 2017).

There are no identical i&i vessels from Ban Chiang and Ban Non Wat. However, comparison

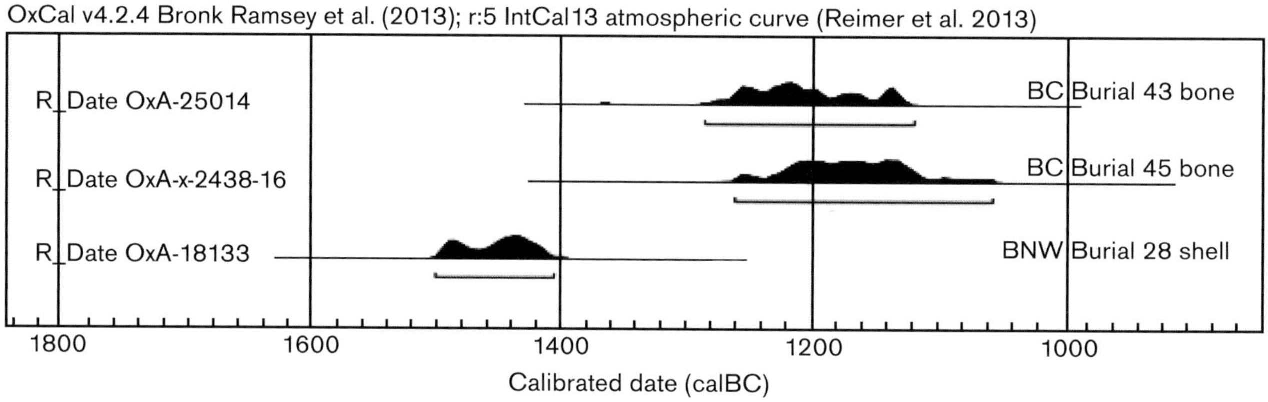

Figure 2.9 Problematic calibrated dates associated with comparable early "neolithic" i&i burial pottery including bone dates from Ban Chiang (BC Burials 43 and 45, Figure 2.8b) and a shell date from Ban Non Wat (BNW Burial 28, Figure 2.8a).

of the overall burial ceramic assemblages from Ban Non Wat's earlier neolithic and Ban Chiang's lower Early Period reveals that Ban Non Wat's pot form 3 (example in Fig. 2.8a) and large ovoid jars from Ban Chiang Early Period Phase IIc (examples in Fig. 2.8b and c) share distinctive commonalities in manufacture and use.

A quick glance at the drawings in Figure 2.8 at first might lead the viewer to think the Ban Non Wat and Ban Chiang vessels are too dissimilar to be meaningful comparanda. The different proportions and prominence of the decorative shoulder registers, the bold, large-scale painted design below the decorative shoulder zone on the Ban Non Wat vessel, the conspicuous cordmarked surface treatment on the bodies below the decorative registers in Ban Chiang examples, together could create a first impression that the pottery types are unrelated.

However, a closer examination of the photographic, descriptive, and contextual evidence reveals notable commonalities. It becomes apparent that some of the seeming dissimilarities between the Ban Chiang and Ban Non Wat vessels in Figure 2.8 stem from the markedly different illustration styles used in the Ban Non Wat and Ban Chiang drawings. The illustrator of the Ban Non Wat Burial 28 vessel has enhanced the clarity of the painted design below the appliqué, whereas the photograph of the vessel (Higham and Wiriyaromp 2010a:19) shows the painted design to be so faint as to be barely perceptible. Faint white paint employing a similar motif (but at a smaller scale) is also found on some of the Early Period Phase IIc Ban Chiang vessels as can be seen in Figure 2.8c. The photograph and the description of the Ban Non Wat Burial 28 vessel also clearly indicate that the vessel's body surface below the decorative register is cordmarked, as are the bodies below the decorative registers in the Ban Chiang vessels, but the cordmarking is only ambiguously hinted at in the illustration of the Ban Non Wat vessel. The illustration differences, of course, are not errors, but instead reflect different priorities and approaches to archaeological illustration of these vessels.

When these vessels are examined from the perspective of each site's early ceramic assemblages, they are meaningful comparanda in terms of their large size, their globular to ovoid body form, the placement and components of their surface treatments and decoration, their design elements and positioning, and their ritual use. Breakage patterns indicate that both forms usually originally had rims, although often they were not recovered from excavated examples. The overall composition of the body's surface treatment typically includes a shoulder band that is usually burnished with incised complex motifs with impressed infill. The horizontal design zone is separated from a cordmarked body by an appliqué cord.

The decorated shoulder zones above the appliqué cords are also somewhat difficult to compare just from the illustrations. From the descriptions of the Ban Non Wat examples and from Ban Chiang data one can note a difference between the Ban Non Wat and Ban Chiang examples is that the incised complex motifs at Ban Chiang usually have impressed rockerstamping infill, whereas the motif on the Ban Non Wat Burial 28 vessel has painted infill. It is noteworthy, however, that other Ban Non Wat phase N1 vessels of different forms do have impressed infill of incised motifs (Higham and Wiriyaromp 2010b). Furthermore, Rispoli et al. (2013) included the Ban Non Wat Burial 28 vessel as an example of ceramics from the initial "neolithic" occupation of Thailand, inferred to be broadly contemporaneous with i&i pottery from other sites in Thailand including Ban Chiang (see also Sarjeant 2014).

One particularly striking comparison between some of the Ban Chiang large globular i&i vessels and the Ban Non Wat Burial 28 vessel is the faint painted motifs on the cordmarked surface below the appliqué cord. On BC Burial 46 Pot A (Fig. 2.8c), a motif of concentric arcs whose bases intersect with lines in very faint white paint is found. Although executed at different scales, this highly idiosyncratic motif is comparable to the painted motif on the Ban Non Wat Burial 28 vessel also below the appliqué cord that also includes concentric arcs intersecting with a line, but at a much larger scale. Such similar idiosyncratic designs positioned on the same part of vessel bodies of quite similar forms and overall composition of surface treatments is an additional argument for expecting Ban Non Wat pot form 3 (including Ban Non Wat Burial 28) and Ban Chiang Phase IIc to be very close in time.

Adding to the formal similarities among these vessels is their identical ritual use. These vessels were used for jar burials at both sites. Examples from Ban Chiang usually contained skeletal remains of infants including a toddler aged 2–3. Four examples from Ban Non Wat contained infants, and there is one case where the vessel form contained an adult. The similarities in the formal attributes described above and the ritual use of Ban Non Wat pot form 3 and Ban Chiang Early Period Phase IIc large globular i&i pots imply that their corresponding societies existed in a regional network that shared related technological, stylistic, and symbolic behavioral systems. The producers of these vessels would need to have existed very close in archaeological time, such as at the scale of the same or adjacent generations, for the living memory embodied in these vessels and their associated ritual usage to have been shared and transmitted. For this suite of correlated ritual and stylistic behaviors to materialize seemingly briefly and autonomously six generations apart in different parts of northeast Thailand is highly implausible.

Another way of evaluating the Ban Chiang bone dates is to compare what Ban Non Wat pottery *would be* contemporaneous with BC Burials 43 and 45 with its associated i&i pot (Fig. 2.8b) if the Ban Chiang bone dates and Ban Non Wat dates were accurate. In the Bayesian analysis of Ban Non Wat dates, the boundary transition between Ban Non Wat's early neolithic N1 burials and the "late Neolithic occupation" is 1442–1340 B.C., and the boundary transition between the late "Neolithic" and the "Bronze Age" is 1126–844 cal. B.C. (Higham and Higham 2009a:22–23). The Ban Non Wat Bayesian analysis therefore would place its later neolithic roughly in the range of ca. 1400–950 B.C.

If the Ban Chiang bone dates for BC Burials 43 and 45 are accurate (and if the Ban Non Wat shell dates and their Bayesian analysis is correct), BC Burials 43 and 45 dating roughly to 1200 B.C. correlate squarely with Ban Non Wat's later neolithic and N2 burials. However, the Ban Non Wat N2 burials characteristically have small to medium globular cordmarked vessels with everted rims and no elaborate incising or impressing. In fact, the BNW later neolithic N2 pottery and grave ritual appears strikingly similar to the Ban Chiang *upper* Early Period Phase V pottery. When Ban Chiang's bone dates for Early Period Phase IIc create a correlation with Ban Non Wat's later neolithic, but the obvious cultural comparanda are with Ban Non Wat's early neolithic, the resultant cross-dating based solely on the radiocarbon determinations, ignoring material evidence, is highly likely incorrect. This evidence strongly suggests that significant technical issues in the dates are present, particularly with the Ban Chiang bone dates.

Another broader though still informative comparison is with a range recently provided for early "neolithic" i&i pottery in southern Vietnam and central Thailand (ca. 2400–1800 B.C.; Bellwood et al. 2013:155). The bone dates from Ban Chiang in particular fall far outside the range that many regional specialists expect for the i&i decorative style. However, the Ban Non Wat shell dates associated with i&i pottery are also non-overlapping with the Vietnamese evidence; hence, they can also be suggested to be implausibly young (White 2015a).

This example of clear inconsistencies among dates for cultural comparanda at Ban Chiang, Ban Non Wat, and southern Vietnam provides evidence that both the bone and shell dates recently published for Thailand (Higham et al. 2015) are best considered *termini ante quem*. Even though Higham and Douka (2015) have argued that the Ban Chiang bone samples are unlikely to have been contaminated and that their results should be accurate, the evidence presented here for incompatible chronologies among relative site sequences based on cultural evidence including ceramic and ritual comparanda demonstrates that chronological problems for prehistoric Thailand remain unresolved.

The cultural evidence indicates that Ban Chiang's lower Early Period approximately correlates with Ban Non Wat's early neolithic, and Ban Chiang's upper Early Period approximately correlates with Ban Non Wat's later neolithic. At a minimum conclusion, the roughly 200-year difference between the non-overlapping Ban Non Wat shell and Ban Chiang bone dates associated with the clearest examples of comparable early pottery from the two sites does not support the claim by Higham et al. (2011a:263) that the bone dates from Ban Chiang "present a near perfect correspondence with the dated sequence for Ban Non Wat."

NOTE ON BAYESIAN MODELLING

> *Models are affected by the quality of data included. The neat graphics and mathematical terminology produced by programs such as OxCal can give the illusion of a robust analysis, even where radiocarbon dates are suspected to be systematically incorrect... (Wood 2015:63)*

Bayesian modeling of date sequences is a trend in the field and such modelling appears to tighten many site chronologies and helps to eliminate outlier dates. But as noted in Wood's (2015) quote above: Bayesian modeling cannot correct inherently flawed data. Bellwood et al. (2013:154), although appreciating the potential contributions of Bayesian modeling in general terms, wished to avoid the possibility "that modelling would produce results that are importantly wrong" when technical, stratigraphic, and sample issues are less than robustly reliable. In the view of the authors of this volume, the bone dates from Ban Chiang are systematically young, and, therefore, their Bayesian modelling (Higham et al. 2015) has produced significantly flawed results.

Higham et al. (2015) argued that they have dozens of dates from five sites, whose Bayesian analyses all produce the same results. However, the appearance of robust Bayesian analyses masks the absence in their argument of rigorous cross-dating based on ceramic comparanda. As shown in the above example comparing the dates of Ban Non Wat Burial 28 and Ban Chiang BC Burials 43 and 45, the dates cannot be meaningfully evaluated with Bayesian modeling without carefully argued comparanda.

A WAY FORWARD

In summary, the authors of this volume view the absolute dating of Ban Chiang and other late Holocene sites in Thailand as unresolved and a work in progress. This is hardly surprising when even regions such as the Near East, which have many times the number of dates, scholars, excavated sites, and a much longer history of archaeological scholarship, have contentious debates on chronology (Finkbeiner et al. 2015:1). The ARCANE Project (Associated Regional Chronologies for the Ancient Near East and the Eastern Mediterranean) provides a suitable model for what Southeast Asian archaeologists must eventually undertake, namely a collaborative, multi-institutional, and multidisciplinary assessment of all chronological data. The relative and the absolute chronologies *must* be compatible for a regional chronology to be broadly useful to the field. The goal should be detailed regional chronology first and foremost, not the date for an "Age."

In the meantime, for this volume a working chronology is defined with the recognition that it will need revising as more evidence becomes available and as the technical contradictions and stratigraphic and cross-dating issues are addressed and resolved. The working chronology is reviewed in the discussion below of evidence for dating key transitions in the Ban Chiang sequence, with particular emphasis on paired radiocarbon determinations (two or more determinations from the same "event" such as the same burial) where paired dates pertinent to the transitions, and ideally bracketing the transitions, are available. In addition, when ceramic and grave style comparanda from other sites can be suggested that also provide meaningful insight to dating the Ban Chiang transitions, such comparanda are noted.

Key dating evidence is reviewed below for four boundaries or transitions at Ban Chiang: (1) the appearance of the first mortuary phase (pre-metal), namely the lower Early Period Phase I; (2) the lower Early Period to upper Early Period transition; (3) the upper Early Period to Middle Period transition; and (4) the Middle Period to Late Period transition. The calibrated date ranges are given at the 95.4% confidence level according to OxCal 4.2 applying the Intcal13 calibration curve (Table 2.2).

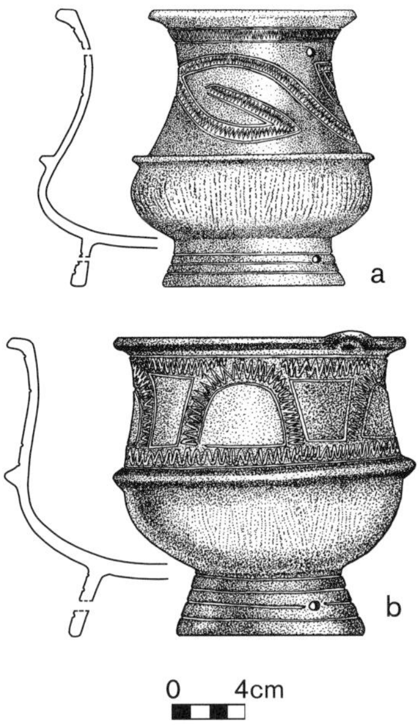

Figure 2.10 Two examples of i&i pottery from Ban Chiang lower Early Period Phase I. (a) BC 1241 Pot A. (b) BC Burial 44 Pot C 1339.

Towards a Working Chronology

The Beginning of the Lower Early Period

Although there are radiocarbon dates on charcoal from the 4th and even the 6th millennium B.C. from basal deposits at Ban Chiang (Hurst and Lawn 1984), their cultural contexts are not currently clear. Furthermore, these early dates are not directly relevant to the chronology of the metal age, which is the main period of concern in this volume. To delineate the appearance of metal in the Ban Chiang sequence, it is important to define the evidence for the start of the mortuary/occupation settlement system characteristic of the Ban Chiang cultural tradition. The clearest evidence for the beginning of the Ban Chiang mortuary/occupation settlement system at the Penn/FAD excavations is BC Burial 44, the deepest burial in the 1974 excavation, which has an associated pot with decoration of the recognized regional i&i (incised and impressed) pottery style (Rispoli 2007). This burial, along with a BC feature of similar depth with similar pottery to BC Burial 44 (Fig. 2.10), is regarded as Phase I of the lower Early Period at Ban Chiang (White 1986:223).

There are discrepant paired dates from BC Burial 44: (1) a date on phytoliths recovered from inside a BC Burial 44 grave good pot that held rice (CAMS 41264, 3730±50 BP, 2287–1977 cal. B.C.); and (2) a bone date from a human fibula from BC Burial 44 (OxA-25015, 3242±26 BP, 1608.5–1442.5 cal. B.C.). The several-century discrepancy between the two non-overlapping dates is noteworthy. Considering the contamination concerns noted above for both phytolith and bone dates, the conservative view can regard the phytolith date as a *terminus post quem* and the bone date as a *terminus ante quem* for BC Burial 44. One can reasonably conclude that BC Burial 44 and consequently Ban Chiang Phase I of the lower Early Period falls somewhere between the late 3rd millennium B.C. and the mid-2nd millennium B.C.

However, there are grounds to consider the phytolith determination to be closer to an accurate date. Phytolith dating appeared promising in the 1990s, but the reliability of the methodology has since been questioned on a variety of grounds (reviewed in Taylor and Bar-Yosef 2014:89–90) as contamination is difficult to detect and remove. On the other hand, the late 3rd to early 2nd millennium B.C. phytolith date and its association with i&i pottery are consistent with other dates in mainland Southeast Asia associated with similar pottery (e.g., Bellwood et al. 2013; White 2015a). As noted above, Bellwood et al. (2013) recently discussed the basal pottery assemblage that included i&i ceramics at An Son in southern Vietnam, which regional archaeologists would consider stylistically related to Ban Chiang Early Period Phase I (as well as Early Period Phase IIc). These excavators date the An Son basal horizon from which i&i pottery was excavated to the 3rd and early 2nd millennium B.C. on the basis of several charcoal and charred food residue dates. Bellwood et al. (2013:153) rejected the "old wood" interpretation for the charcoal dates, and they further draw

Table 2.2 Radiocarbon Dates Presented in the Order Discussed in Chapter 2

Lab Number	Date (bp)	Uncertainty	Conventional ^{14}C date	Calibrated Date at 95.4%	Phase	Context	Material	Note
OxA-18133[a]	3170	27	3170±27	-1499.5–1404.5 B.C.	BNW N1	BNW B.28	shell	freshwater
OxA-25014[b]	2984	26	2984±26	-1283.5–1117 B.C.	EP IIc	BC B.43	bone	human rib
OxA-x-2438-16[b]	2958	29	2958±29	-1259.5–1054.5 B.C.	EP IIc	BC B.45	bone	human rib
CAMS-41264[c]	3730	50	3730±50	-2287–1977 B.C.	EP I	BC B.44	phytoliths	rice in grave vessel
OxA-25015[b]	3242	26	3242±26	-1608.5–1442.5 B.C.	EP I	BC B.44	bone	human fibula
AA-15577[c]	3495	105	3495±105	-2131–1533 B.C.	EP IVc	BCES B.34	rice temper	
OxA-30654[b]	2690	29	2690±29	-897–803 B.C.	EP IVc	BCES B.34	bone	human rib
AA-15582[c]	3320	50	3320±50	-1739.5–1495.5 B.C.	EP Va	BCES B.59	rice temper	
P-2454[d]	3270	180	3270±180	-2027–1086 B.C.	EP Va	BCES B.59	charcoal	
OxA-30666[b]	2568	29	2568±29	-806–560 B.C.	EP Va	BCES B.59	bone	human rib
AA-15581[c]	2970	60	2970±60	-1388–1010.5 B.C.	EP Vb	BCES B.56	rice temper	
OxA-x-2590-19[b]	2513	30	2513±30	-790–539 B.C.	EP Vb	BCES B.56	bone	human misc.
OxA-x-2559-13[b]	2584	23	2584±23	-805–765.5 B.C.	EP Vb	BCES B.56	bone	pig
OxA-22380[b]	2516	26	2516±26	-789.5–541 B.C.	EP Vb	BCES B.56	bone	pig
P-2634[d]	2780	50	2780±50	-1047–818 B.C.	MP start	base of red-brown soil	charcoal	hearth
OxA-30648[b]	2498	29	2498±29	-784.5–534.5 B.C.	MP VIII	BCES B.25	bone	human rib
AA-23284[e]	2440	45	2440±45	-754.5–405 B.C.	MP VIII	BCES B.12	rice temper	same pot as 23285
AA-23285[e]	2380	55	2380±55	-751.5–371 B.C.	MP VIII	BCES B.12	rice temper	same pot as 23284
OxA-30433[b]	1886	26	1886±26	64.5–214.5 B.C.	LP IX	BCES B.6	bone	human rib
OxA-30434[b]	1864	27	1864±27	79.5–225.5 B.C.	LP IX	BCES B.9	bone	human rib

Note: BNW = Ban Non Wat; BC = Ban Chiang 1974; BCES = Ban Chiang Eastern Soi 1975; EP = Early Period; MP = Middle Period; LP = Late Period; B. = burial.

[a]Higham and Higham 2009a. [b]Higham et al. 2015. [c]White 2008. [d]Hurst and Lawn 1984. [e]unpublished.

comparisons with Nong Nor phase 1 and Khok Phanom Di Ceramic Period 1, noting compatible dating in the late 3rd and early 2nd millennium B.C. (ca. 2400–1800 B.C. [Bellwood et al. 2013:155]).

Unless An Son's basal horizon current dates are rejected by the excavators and they revise the dating for that horizon to ca. 1500 B.C. on the basis of different dating, this monograph's authors will not arbitrarily disregard the phytolith date from BC Burial 44 on the basis that it "might" be flawed. Instead the authors choose to favor the phytolith date on the basis that it is consistent with a larger regional picture advocated by other regional specialists for "a 'Greater Mekong' cultural network dating from about 2500 B.C. onward" (Bellwood et al. 2013:160). Therefore, the beginning of the Ban Chiang cultural tradition and Early Period Phase I is considered by the authors as likely to date from the late 3rd millennium B.C., and a provisional date of ca. 2100 B.C. is used for the start of lower Early Period Phase I in the working chronology employed in this volume.

Dating the Transition from the Lower Early Period to the Upper Early Period

The start of the upper Early Period is marked by changes in burial orientations and pottery styles that occur at the boundary between Early Period Phase IVc and Early Period Phase Va. At BCES, there is also a change from tightly spaced graves in rows to spatially dispersed interments. There are charcoal, rice temper, and bone dates pertinent to this transition, including paired dates from graves. These dates, however, have internal contradictions.

From Phase IVc, flexed burial BCES Burial 34 is key in the discussion as it is the highest and probably the most recent burial in the lower Early Period from the Penn/FAD excavations. BCES Burial 34 has paired dates on rice temper and bone. The rice temper date is AA-15577, 3495±105 BP, 2131–1533 cal. B.C. A human rib bone date from the same burial is OxA-30654, 2690±29 BP, 897–803 cal. B.C.

From the upper Early Period Phase V, the lowest and quite possibly the oldest burial from Phase Va is BCES Burial 59 with three associated dates on charcoal, rice temper, and bone. The charcoal from burial BCES B.59 is P-2454, 3270±180 BP, 2027–1086 cal. B.C. The rice temper date from burial BCES Burial 59 is AA-15582 3320±50 BP, 1739.5–1495.5 cal. B.C. The human rib bone from burial BCES Burial 59 is OxA-30666, 2568±29 BP, 806–560 cal. B.C.

As with dates from BC Burial 44, the bone dates from BCES Burial 34 and BCES Burial 59 are significantly younger than other dates (from rice temper and charcoal) within the same burial contexts and non-overlapping by even greater margins.

Although for many years advocating the desirability of rice temper dates, (e.g., Higham 1988b:75, 77, 1996a:191, 240, 246, 311, 1996–1997:882, 2002:91, 93, 113, 129, 353, 2004:51–52; Higham and Thosarat 1998:84; Higham et al. 2004:325), Higham more recently has critiqued them, arguing that they are likely older than the vessel providing the sample (e.g., Higham et al. 2011b) because of old carbon in the clay. However, one issue affecting the potential contribution of old carbon to the dated sample that has been clarified recently has to do with the combustion temperature used in preparing the sample. And it has also become evident that individual radiocarbon labs combusted the pretreated organic temper samples at different temperatures.

Referring to the set of determinations run at Oxford (Hedges et al. 1997) that have *not* been utilized by White because of their internal high inconsistency (White 2015a), Higham et al. (2011b:588) have stated: "The Oxford Ban Chiang [rice temper] samples were all combusted at 900°C." This high a temperature is now thought to release geological carbon from the clay, hence producing dates that are too old. However, the rice temper dates from Ban Chiang published in J. White (1997, 2008), including those from BCES Burial 34 and Burial 59, were run at the Arizona AMS Facility. There the Ban Chiang rice temper samples were combusted at 400°C in accordance with Arizona's protocol (correspondence on file in the Ban Chiang office). The rationale for the lower temperature is that "temper-derived carbon tends to be preferentially removed with a lower temperature combustion than the carbon bonded within the clay fraction" (Higham et al. 2011b:588). A likely reason for the differences between the Arizona and Oxford sets of Ban Chiang rice temper dates is differences between the labs in their 1990s combustion protocols (White 2015a), with Oxford's combustion protocol now recognized as erroneous and Arizona's combustion protocol consistent with current practices.

Currently, direct dating of organic-tempered ceramics using the low temperature combustion protocol is being successfully undertaken in other areas such as north Asia and Africa (e.g., Janz et al. 2015; Messili et al. 2013). Some Southeast Asian archaeologists continue to rely on rice temper dates (e.g., Glover and Bennett 2012:101) without being challenged. And four rice temper dates undertaken for Ban Non Wat's phase IA2 (Higham and Higham 2009a:20, 21, 24) and run at Arizona did produce one in accord with and two "a little earlier" than the date range expected by the Higham team (Higham et al. 2010:1024), hence *not* dramatically older as they may have anticipated. In summary, although rice temper samples may not be ideal candidates for radiocarbon dating, the Arizona Ban Chiang rice temper dates cannot be arbitrarily dismissed on the basis of combustion temperature (in contrast to the Oxford Ban Chiang rice temper dates which can be rejected on the basis of their use of high combustion temperatures). As the Arizona rice temper samples were combusted in accordance with protocols in place today, it seems likely that the dates reflect the ages of the rice temper and not old carbon in the clay that is released at higher combustion temperatures. In sum, the Arizona rice temper dates seem likely to be close to the age of the vessel. It is also interesting to observe that the rice temper and charcoal dates from BCES Burial 59 are neatly compatible. Nevertheless, applying the most conservative argument that the bone dates represent *termini ante quem* and rice temper and charcoal dates *termini post quem* for their respective burials, the lower Early Period/upper Early Period transition seems likely to fall somewhere in the 2nd millennium B.C.

Comparanda for this transition with other sites are not as clear-cut as basal An Son and Ban Non Wat's early neolithic are for Ban Chiang's Early Period Phases 1 and IIc. But ceramic style and grave style comparanda for Ban Chiang's upper Early Period Phase V appear quite plausible with Ban Non Wat's later neolithic and Neolithic 2 burial phase (hereafter N2) as noted above. Medium to small globular cordmarked pots with simple everted rims and simple grave styles characterize both Ban Chiang upper Early Period Phase V (Fig. 2.11a–c) and Ban Non Wat N2 burials. The Ban Non Wat N2 burial deposition is given a lower boundary of 1377–1133 B.C. on the basis of a Bayesian analysis (Higham and Higham 2009a:23). The transition boundary for the end of the phase N1 burials and the beginning of the later neolithic occupation provided by the Bayesian analysis is 1442–1340 B.C. Considering the reliance of the Ban Non Wat chronology on shell dates that are likely *termini ante quem*, these two Ban Non Wat boundaries are plausible minimum dates for the transition to later neolithic deposition at Ban Non Wat.

These ceramic and burial style comparanda between Ban Chiang upper Early Period Phase V and Ban Non Wat later neolithic support provisional placement of the lower Early Period/upper Early Period transition at Ban Chiang in the mid-2nd millennium B.C. This boundary is certainly one that will need greater temporal definition going forward, but since the shell dates likely bias the Ban Non Wat chronology to the young side, and the Ban Chiang bone dates are also considered here *termini ante quem*, a provisional boundary date for Ban Chiang's lower Early Period/Upper Early Period transition of ca. 1500 B.C. is used in this monograph.

Upper Early Period/Middle Period Transition

The transition from the upper Early Period to the Middle Period at Ban Chiang has been defined on several stratigraphic and cultural grounds. At BCES, there was a distinct soil change associated with a charcoal date from a hearth (P-2634, 2780±50 BP, 1047–818 cal. B.C.). Associated with this soil change were changes in grave ritual and burial ceramics. While small- to medium-sized globular cordmarked pots similar to those of Early Period Phase V in the lower grey stratum continued at least initially in Middle Period Phase VI burials, new styles of larger vessels were added to grave good suites (see the example in Fig. 2.4a). The new larger vessels were commonly buff and/or whitish in exterior color, and these large vessels were usually deliberately broken in association with bodies; in Phase Middle Period Phase VII the broken vessels occur as sheets over the bodies (Fig. 2.5).

There are no paired dates directly associated with the final Early Period subphase Vc or the initial Middle Period Phase VI burials. However, there is an Early Period Phase Vb grave, BCES Burial 56, that has paired dates, an Arizona rice temper date plus one human and two pig bone dates from Oxford

Figure 2.11 Example of an upper Early Period Phase V grave at Ban Chiang. (a) BCES Burial 33 illustrating the supine interment with a medium-sized globular cordmarked vessel. (b) BCES Burial 33 Pot A 1722 in situ. (c) BCES Burial 33 Pot A 1722 drawing. Comparanda for Ban Chiang's upper Early Period may be found at Ban Non Wat upper neolithic, N2 phase.

(the latter taken from the same bone). Like other burials with paired bone and rice temper dates, they are non-overlapping and separated by hundreds of years. The rice temper date is AA-15581, 2970±60 BP, 1388–1010.5 cal. B.C. The three bone dates include one human (OxA-x-2590-19, 2513±30 BP, 790–539 cal. B.C.) and two pig (OxA-x-2559-13, 2584±23 BP, 805–765.5 cal. B.C.; and OxA-22380, 2516±26 BP, 789.5–541 cal. B.C.).

There is one Middle Period Phase VI human rib bone date from BCES Burial 25 (OxA-30648, 2498±29 BP, 784.5–534.5 cal. B.C.). There are no rice temper dates from this or other Middle Period Phase VI burials.

The charcoal date P-2634 with a fairly small standard deviation, situated just above the soil transition at the very beginning of the deposition of the lower red sediments where the Middle Period burials begin, is the kind of cultural and stratigraphic context that is normally highly desirable for archaeologists. This date should not be arbitrarily dismissed. That the hearth date lies above younger bone dates from the stratigraphically lower Phase Vb BCES Burial 56 (and the bone dates do not even overlap the overlying hearth date), provides additional evidence that the Ban Chiang bone dates are consistently too young for their stratigraphic context and are most accurately considered *termini ante quem.* On the other hand, the rice temper date from BCES Burial 56 is compatible with the hearth date.

It is the opinion of the authors that the hearth date at the stratigraphic interface defining the Early/Middle Period boundary provides the best available estimate that the Early Period/Middle Period transition occurred during the early 1st millennium B.C. The provisional date of ca. 900 B.C., therefore, is used in the working chronology for the transition from the Early Period to the Middle Period.

Middle Period/Late Period Transition

In the cultural sequence at Ban Chiang, the transition from the Middle to the Late Period is defined by the end of the burial practice of deliberately breaking the large whitish and peach-colored, often carinated, vessels in the burials and the beginning of placing intact vessels (Fig. 2.6), often with painted red swirling designs, over the body. It is likely that the change in ritual and ceramics between the Middle and Late Periods was not abrupt, based on observations of the graves in the open-air museum at Ban Chiang's Wat Phosrinai, where ceramics overlying burials appear to show a mix of Middle and Late Period characteristics. However, at BCES the pertinent excavated burials fall clearly into Middle or Late period style groups, and potentially can help position the Middle/Late Period transition. Other differences between the two periods are the appearance of glass beads and rollers in Late Period contexts; these items were not found in the Ban Chiang Middle Period. The rollers and glass argue that the beginning of the Ban Chiang Late Period roughly correlates with Ban Non Wat's Phase IA1, where glass and a roller have been identified.

The most recent Middle Period burial is clearly defined with BCES Burial12, which is the only certain Middle Period Phase VIII grave (an example of a Middle Period Phase VIII carinated vessel is seen in Fig. 2.4d). BCES Burial 12 has no bone dates but has ceramic temper dates run at Arizona (AA-23284, 2440±45 BP, 754.5–405 cal. B.C. and from the same pot AA-23285, 2380±55 BP, 751.5–371 cal. B.C.). No Late Period burials have ceramic temper dates, but there are a couple of human bone dates (BCES Burial 6 rib bone, OxA-30433, 1886±26 BP, cal. A.D. 64.5–214.5; BCES B.9 rib bone, OxA-30434, 1864±27 BP, cal. A.D. 79.5–225.5). There are thus no good examples of paired dates from different samples pertinent to this transition. However, applying the conservative interpretation of the available pertinent dates, the transition appears to fall between the mid-1st millennium B.C. and the early 1st millennium A.D. Consideration of the comparanda (such as rollers) of the Ban Chiang Late Period with Ban Non Wat's IA1 (dated 420–100 B.C.; Higham and Higham 2009a:25) suggests that 300 B.C. is a reasonable provisional date for the transition from the Middle Period to the Late Period.

The Working Chronology

The previous discussion provides the basis for the working chronology employed in this volume and shown in Table 2.3. The relative sequence is emphasized throughout the remainder of this volume, so that future refinements to the absolute chronology can be easily applied to this volume's data.

Table 2.3 Working Ban Chiang Chronology

<table>
<tr><th colspan="2">Period</th><th>Burial Phase</th><th>Working Date Range</th></tr>
<tr><td colspan="2" rowspan="2">Late Period (LP)</td><td>X</td><td rowspan="2">ca. 300 B.C.–A.D. 200</td></tr>
<tr><td>IX</td></tr>
<tr><td colspan="2" rowspan="3">Middle Period (MP)</td><td>VIII</td><td rowspan="3">ca. 900–300 B.C.</td></tr>
<tr><td>VII</td></tr>
<tr><td>VI</td></tr>
<tr><td rowspan="5">Early Period (EP)</td><td>upper</td><td>V</td><td>ca. 1500–900 B.C.</td></tr>
<tr><td rowspan="4">lower</td><td>IV</td><td rowspan="4">ca. 2100–1500 B.C.</td></tr>
<tr><td>III</td></tr>
<tr><td>II</td></tr>
<tr><td>I</td></tr>
<tr><td colspan="3">Initial Period (occupation before Burial Phase I)</td><td>?–ca. 2100 B.C.</td></tr>
</table>

Summary

Although the relative chronology for Ban Chiang is well developed and fairly straightforward, the absolute chronology remains a work in progress. The recently proposed chronology for Ban Chiang based on bone dates (Higham et al. 2015) is not synchronized with other regional evidence, once dates for valid comparanda based on pottery are examined (White 2015a). Cross-correlations among sites based on "Ages" (e.g., the "Neolithic," the "Bronze Age"), which form the structure of the relative chronology in Higham et al. (2015) are too crude to be meaningful for tight regional archaeological sequences. The field of Southeast Asian archaeology needs to work toward refined regional ceramic sequences that can provide the best bases for cross-dating between sites and the interpretation of absolute dates in regional context. Dating of substances such as amino acids like hydroxyproline (Nalawade-Chavan et al. 2014) could be undertaken and compared to "collagen" and shell dates from the same burials to see if the incongruities of dates from different materials from the same cultural context can be resolved.

The discussions referencing the Ban Chiang chronological sequence in the remainder of this volume emphasize the relative sequence. In this way, the discussions can be adapted to inevitable changes in absolute chronology as dating methodologies advance.

NOTES:

2.1 "Non-overlapping" dates in this study refers to calibrated date ranges at 95.4% confidence levels unless otherwise stated.

3

Debunking the Conventional Paradigm: Metals and Prehistoric Society

Joyce C. White and Elizabeth G. Hamilton

The early practitioners of archaeology and anthropology used technology as a direct measure of the evolution of societies. (Chapman 2003:33)

The study of ancient metallurgy, often termed archaeometallurgy, is a field that has roots in two distinct disciplines: archaeology and metallurgy. Scholars from each discipline bring different knowledge, sets of skills, and often different questions and answers to the study. Archaeologists may stress the social and economic ramifications of metal technology in their investigations, and materials scientists are usually intrigued by the technological knowledge and capabilities of a past society. Many important advances in the understanding of ancient metallurgy have emerged from partnerships between archaeologists and metallurgists, such as the thirty-year collaboration of James Muhly, Robert Maddin, and Tamara Stech (Ferrence 2011). Though archaeologists and metallurgists are the primary collaborators, some projects also involve historians, numismatists, and philologists, along with "geologists, materials scientists, chemists, physicists, limnologists, botanists, toxicologists, mining engineers, blacksmiths, goldsmiths, and conservation scientists" (Killick and Fenn 2012:560). In short, the study of ancient metals is an exemplar of collaborative scholarship (though not always an easy partnership [Muhly 2003]) that integrates the hard and social sciences and frequently the humanities as well, in order to best address the meaning and role of metals in past societies.

Both metallurgists and archaeologists have studied prehistoric metals in Southeast Asia as they have investigated ancient metals the world over. This chapter and Chapters 4, 5, and 7 seek to provide background from two perspectives needed in the study of ancient metals—societal/behavioral and technological/material. Both approaches are essential for understanding metals in the past, and both are essential for the conceptual and methodological frameworks applied to the analysis of the prehistoric metallurgy from northern northeast Thailand that is the main subject of this volume.

The great interest of both social and materials scientists in the advent and early development of metallurgy in large part stems from the supposition that the emergence of metal technology represents an important threshold in both human technological and social development. From a technological perspective, to smelt, cast, and work metal artifacts is a process qualitatively different from making stone tools, pots, or cloth. These latter crafts are based primarily on mechanical alterations of natural raw materials. Making tools or ornaments from stone is a process of *subtraction* of material until a desired shape is attained, be it merely a sharp edge or a beautifully configured point. Making pots is an *additive* and *transformative* process. Clay is dug, sieved, mixed with water and temper, shaped, and heated. A soft substance is transformed into one that is hard, but its origin in clay is still apparent, especially for low-fired earthenwares and terracottas; rub the broken edge of a potsherd with a rasp, and rough clay, the original material, will come off. Textiles are produced by rearrangement of fibers; long threads are spun from short strands of wool, cotton, or flax, and those threads are intertwined to form fabric. Yet if one inspects the cloth, one can see the original threads. Similarly, the raw materials used to plait baskets are minimally processed fibrous plants.

Metal, in contrast, is a thermochemically *derived* material (Franklin 1983) extracted from its natural parent material, ores. With the exception of rare and limited deposits of what are called "native" metals, such as naturally occurring elemental gold and copper, most metals must be extracted from ore rocks by *smelting*. Smelting metal is a process of transformation in which one material, in the presence of high temperature and specific atmospheric conditions, is irreversibly transformed into another material with entirely different physical and mechanical properties. From hard, brittle, green rock comes shining red-orange, malleable copper. From dull black or red stone comes bright, reflective, sharp-edged iron. Except for native metals and meteoritic iron, the origin of the metal in the natural world, that is, its original ore, is not readily apparent. The ore cannot be reconstituted from the smelted metal. Extracting copper from its ores is a process so complex that some scholars have argued that it was developed only once and subsequently spread to all other areas of the world, including the New World (Wertime 1964, 1973). A less extreme version has smelting invented in Southwest Asia and from there spreading to the rest of the Old World, but with metallurgy in the Americas resulting from separate invention in the Andes (Roberts et al. 2009).

From a social perspective, the procedures involved in *producing* metal (i.e., extracting metal from ores by smelting) mean that a society undertaking smelting and then manufacturing of metal products requires a certain baseline level of social complexity beyond the simplest societies, such as isolated hunter-gatherers organized into small mobile bands who maintain a minimal material culture. To acquire and process the raw materials not only for one or more metals (in the case of alloys), but also for the technologies, materials, and facilities used to derive and transform metals (e.g., to make the crucibles, molds, furnaces or other installations, bellows or other air delivery equipment, etc., and manipulate them under conditions of very high temperature and some minimum of atmospheric control), the society must have the capability for planned and coordinated actions over space and time among several individuals, and probably more than one community. To maintain such a technological tradition within a society, complex knowledge and skills involving several interrelated steps and processes, and several materials (not only metal, but clay, stone, and other resources used to create the requisite manufacturing equipment and facilities) must be transmitted among persons and across generations. In comparison with raw materials that are comparatively common and more evenly and widely distributed on the globe, such as wood or clay, the geographic distribution of ores is patchy, and often rare or absent within a region. The society that produces metal must have mechanisms to acquire raw materials from a distance, possibly from more than one distant region, as well as to distribute manufactured products among various social units and locations; the few individuals in the overall society and region likely to undertake the prospecting and production activities are unlikely to be the sole consumers or users of the metal products. The presence of metal in an ancient society therefore implies some kind of craft specialization and some kind of exchange system. In short, the existence of a system of metal production inherently has wider implications for the societies undertaking it.

However, as is shown in this chapter, archaeological evidence now demonstrates that production of bronze did not necessarily require centralized or hierarchical complex societies for its manufacture or consumption. The wide range of pre-industrial social contexts where metal production and metal product manufacturing have occurred are used to critique many of the traditional assumptions about the social implications of metal production. These reviews in turn introduce many of the theoretical issues involved in studying metallurgy in pre-industrial societies, especially the proposal that metallurgy played a catalytic role in the emergence of social stratification and political and economic centralization.

By debunking the Conventional Paradigm for interpreting ancient metallurgy, this chapter provides the intellectual rationale for why the archaeometallurgy of Thailand needs to be studied using the New Paradigm discussed in Chapter 4. The argument is made in five parts. First is a detailed overview of the assumptions underlying the Conventional Paradigm for early metallurgy and society and how they stand in light of current archaeological evidence. Second is a summary overview of social contexts of early metallurgy world-wide, which helps to put the Thai evidence into a global perspective. Third is a discussion of how the logical underpinnings for the Conventional Paradigm are flawed, and why they should be replaced. Fourth is a brief discussion of selected paradigmatic shifts that have occurred generally in the discipline of archaeology that have implications for how archaeometallurgy can be studied with a New Paradigm. And finally, the periodization used in this volume is presented, which seeks to move beyond the flaws of the Conventional Paradigm.

Background to Metals and Social Theory

> *Like so many myths, once they get repeated and written down, they tend to become accepted as fact, even if the basis for this has been forgotten or was non-existent in the first place....Why should myths like this matter? They matter a great deal because they act as a kind of smoke-screen to distract us from what really happened. Even worse they also distract us from the need to question and undertake the research necessary to determine what really happened and lead us to accept background ideas or a "given" framework into which to fit the results... (Gilmour 2009:138–139)*

In reviewing past scholarship on early metallurgy, it is important to keep in mind that perceptions and interests of disciplines, and hence interpretive frameworks, change over time. Some questions are answered by new data; some previously accepted answers are found to be inadequate; theories are rejected, but some rejected theories are reborn; new questions are asked using new methodologies (Trigger 2006:534). All scholarship exists in a historical context, and this volume is no exception.

Although the contributors to this volume do not consider themselves proponents of any single school of archaeological thought, they do think that Southeast Asian archaeological scholarship could benefit from theoretical and methodological advances that have arisen in the discipline of archaeology over the last three decades. At the beginning of modern archaeological investigation of Southeast Asia in the 1960s and 1970s, the dominant paradigm for understanding prehistory was evolutionary; prehistory was commonly defined in terms of universal stages of progressive technological and social development—from small simple groups to states, and from stone-tool-using to bronze- and then iron-using societies. This volume attempts to bring in a range of different perspectives in its presentation of the prehistoric metallurgy of Southeast Asia. However, a review is needed of the older conventional approaches to discussing past social and technological change and their limitations for understanding archaeological data that have emerged in recent decades in Southeast Asia and elsewhere on the globe.

Universalist and Contingent Approaches

> *misleading and incorrect ideas about technological change...pervade modern Western societies. (Schiffer 2011:10)*

Most scholarship on metals in antiquity until recently has explicitly or implicitly assumed the existence of progressive "universal sequences" of technological change (e.g., Tylecote 1976, 1992). Even though Muhly (1988:2) declared that, "The idea of a universal sequence of steps or stages in technological progress...seems to be gone forever," terminology and residual assumptions from a progressive, universalist mindset linger tenaciously in the literature. As noted by Pfaffenberger (1988) and Schiffer (2011:14), these lingering tacit assumptions impede a better understanding of technology and technological change and in turn, a better understanding of prehistoric metallurgy generally and in Southeast Asia in particular.

In this study, the older approaches and their associated assumptions regarding metal technology are grouped under an umbrella phrase: the Conventional Paradigm. Signature attributes of the Conventional Paradigm are reliance on: (1) universalist approaches (assuming that what holds true for one society will hold true for others); (2) linear progressive sequences (technological developments in all cultures follow the same sequence of advance); (3) normative analyses (generalizing a standard from a small, subjectively selected sample, thereby ignoring variability);

(4) essentialist reasoning (asserting something is true because it is assumed to be inherent in, i.e., essential to, the category); (5) technology-determinism (that technological change is an independent variable and a primary mover of culture change).

From a Conventional Paradigm perspective, technological change is expected to develop in a predictable trajectory with an inherent technological logic, irrespective of cultural context. Many critiques of these and related positions have appeared over the past four decades, for example, Pfaffenberger's (1992) critical assessment of "The Standard View of Technology" and Schiffer's (2011) discussion of "crap detection." Lechtman (1980) and Ehrhardt (2005) have critiqued assumptions encompassed in the Conventional Paradigm (i.e., "the Standard View") with specific reference to early metallurgy.

Nevertheless, despite the critiques of assumptions and beliefs that are encompassed in the Conventional Paradigm, they remain common in both scholarly and popular literature. This chapter reviews several of the assumptions and beliefs pertinent to the study of early metals in the context of current empirical data in order to heighten awareness of biases, both overt and implied, in previous as well as current scholarship on this topic.

Metal and the Ages of Man

> *But how much use must be made of a material for it to have an age called after it? (Tylecote 1992:7)*

A universalist perspective on temporal change in metal technology has a deep pedigree in Old World cosmology. In Western literature, the idea that tools made of specific materials, including metals, characterized different periods of time that also had distinctive social qualities dates back at least to classical times. Hesiod's Golden, Silver, Bronze, and Iron Ages corresponded to periods of varying moral and political characteristics. For example, strife and metal weaponry were considered by Hesiod to characterize his Bronze and Iron Ages, perhaps planting the indelible seed that has endured in Western cosmology associating early development of those metal technologies with societal violence.

References to technological sequences continued into later times, and the 1st century Roman poet Lucretius mentions a sequence of materials used by humans to manufacture tools, from stone to bronze to iron (Renfrew and Bahn 2005:265). Interestingly the same sequence was mentioned in ancient Chinese literature by a philosopher of the Eastern Zhou period, 770–221 B.C. (Renfrew and Bahn 2005:265) and by a Han dynasty writer named Yüan Kang (Wertime 1964:1257). In the west, 16th century Renaissance scholar Mercati revisited the idea that stone tool finds, for some centuries considered to be implements used by elves or to be of some other fantastical origin, instead represented tools used by humans prior to availability of metal (Renfrew and Bahn 2005:265).

Although there may have been an intellectual awareness that technologies available to societies could have changed over the course of human existence, a concrete practical application of a temporal sequence of material technologies known as the Three Age System was developed only in the 1820s by Christian Jürgensen Thomsen of Denmark. Thomsen adapted the earlier general concepts to the organization of a large collection of ancient artifacts excavated from the barrows and middens of Denmark that ultimately became the basis for the National Museum of Denmark. To him, the sequence represented evidence for technological progress during human existence, and this technological sequence could be used to organize the temporal sequence of the societies that produced or used the tools. Thomsen organized the museum collections based on what he considered to be the most advanced material technology used to make tools present in the source society: the Stone Age (including both the "old stone age" with flaked stone and the "new stone age," or Neolithic, with polished stone tools); the Bronze Age; and the Iron Age. Thomsen's formulation looked not only at tool materials but also at the assemblages of all other artifacts *associated* with the tools. This appreciation of the specific associations of cultural materials in the ground and the implications of these associations for the age and social context of those associated materials became a fundamental principal underlying the emerging discipline of archaeology.

It is hard not to see a historical relationship between the intellectual traditions of Mediterranean antiquity's Ages of Man and the subsequent universalist approaches to history and technology that flourished from the 19th century to the latter half of the 20th century in European and American archaeology. Many social evolutionary sequences have been proposed over the last 150 years in addition to the Three Age System. In the 19th century, Lewis Henry Morgan's Savagery, Barbarism, and Civilization stages (2000 [1877]) mirrored Herbert Spencer's ideas that both social and biological evolution were directional from lower to higher forms. In the 20th century, Fried's (1967) simple egalitarian, rank, and stratified societies and Elman R. Service's (1962; see also Sahlins and Service 1960) politically framed sequence of band/tribe/chiefdom/state continued the tradition of viewing social evolution in directional and progressive terms conceptualized as stages. In Service's sequence, in brief, bands are egalitarian mobile hunter-gatherers; tribes are egalitarian sedentary agriculturalists, chiefdoms are characterized by social inequalities including political leadership; and states have several layers of hierarchy with central administering governments. Although strong criticism of the utility of such evolutionary schema over the past three decades (e.g., Chapman 2003; Ehrenreich et al. 1995; Feinman and Neitzel 1984; Upham 1987) has reduced greatly their prominence in archaeological literature (Earle 2008), they remain a starting point for many broad overviews of the archaeological past and human prehistory (e.g., Fagan 2008). Their terminology and point of view are still commonly referenced in the archaeological literature (e.g., Redmond and Spencer 2012), including that of Southeast Asia (e.g., O'Reilly 2007a).

Can Technological Stages Be Equated to Archaeological Ages?

> *The ordering of history according to dominant material has become the most popular and useful form of chronological classification in our own day. (Wertime 1964:1257)*

The empirical linking of technology and social progress was expounded upon in an article entitled "Archaeological Ages as Technological Stages" by V. Gordon Childe (1944). Here Childe explicitly set out the argument that progressive technological stages were foundational to progressive archaeological Ages based on specific archaeological evidence. He also said that the Bronze Age "approximates more closely to a true Age than any earlier Stage" (Childe 1944:8). While his discussion was quite nuanced and acknowledged the exceptions provided by the prehistory of Europe itself, he asserted that the earliest civilizations in the Near East emerged in the Bronze Age and the "Bronze and Iron Stages are...associated with distinct types of civilization" (Childe 1944:23). The early "Bronze Age" came to be equated by European scholars with the rise of stratified urban and state societies, writing, and other socioeconomic traits often tied to the concept of "civilization" (e.g., Childe 1950). Distinguishing civilized from uncivilized/non-civilized/barbaric/savage/primitive societies was of great interest to mid-20th century European archaeologists (see review in Tringham 1974).

One crucial distinction for the development of "civilization," according to some scholars, was between metalworking cultures and stone-using cultures. As Franklin summarized (1983:280), "Bronze objects...were seen as signaling the transition into adulthood of a civilization, thus becoming an essential element in the definition of civilization itself." The distinction rested not so much on the mere practice of metallurgy, for that could and did spread to "barbarian" lands, but on which lands first developed the practice along with its attendant traits. As Sir Leonard Woolley wrote, "the vast gulf that separates the cultural status of Neolithic man from that of the developed Bronze Age was due not only to the use of metal as such but to the many other arts and activities, such as sculpture and carpentry, which accompanied and were encouraged by metallurgical knowledge, so that the expression 'Bronze Age' is in fact but a label denoting something far more than the words imply" (Woolley 1963:548). He went on to write that the non-urbanized cultures "remained...in the primitive state of barbarism and, at best, borrowed from their more inventive neighbours such techniques and practices as might ultimately equip them to contribute in their turn to general progress" (Woolley 1963:821).

Childe's view of technology developed over the course of his lifetime, from viewing technological change as a prime mover of social and economic change to realizing how technological changes were bound up with social and ideological factors (Trigger 1986). His work set many agendas for the archaeology of the 20th century even as new lines of thought and new data emerged that changed or challenged specific syntheses that he put forward (Wailes 1996a). His proposals that craft specialization must have had a critical role in the development of differentiated socioeconomic systems initiated an enduring research agenda in archaeology that continues to be explored (e.g., Costin 2001; Costin and Wright 1998; Hirth 2009a; Hruby and Flad 2007; Wailes 1996a). Childe asserted that metal technology would have been the first technology to require specialized producers. In his view, the early metalworkers must have been full-time, itinerant specialists who were detached from the kinship structures of tribal society. Woolley (1963:554) adopted this view and wrote, echoing the mid-century *Kulturekreis* thinking, that technical advances spread from "civilized" societies to "barbarian" societies largely because of the explorations of metal prospectors seeking raw materials in lands where metal exports had already found a market.

During the second half of the 20th century, much effort was expended by archaeologists attempting to refine universalist schema in order to articulate an all-inclusive progressive framework. Similarly, specifically with reference to prehistoric metal technology, archaeometallurgists sought to varying degrees to define specific advances that would characterize, for example, the Early Bronze Age as distinct from the Late Bronze Age, irrespective of geographic location (e.g., Charles 1980; Tylecote 1992).

As archaeological research expanded in the New World and into previously less-researched parts of the Old World, the universality of even the basic Three Age System was cast into doubt. New World archaeologists found no use for the system because metal objects had little utilitarian role in pre-Columbian societies; their early state-level complex societies generally used stone tools. Moreover, the rare examples of early New World metal technologies did not fit the technological sequence envisioned by those advocating models of technological progression. Aesthetic qualities such as sound and color more than utilitarian properties such as hardness appeared to be the focus of metallurgical development in many New World cases (e.g., Hosler 1994; Lechtman 1980). Not knowing quite what to do with the New World metal technological sequence, Wertime (1973:876) termed it "peculiar," presumably because it did not fit the perceived "norm" established by Near Eastern and European metallurgy.

As more data on the human past came to light in the latter half of the 20th century, much variability emerged that did not conform neatly to universalist schema. Mere refinements of universalist concepts, such as proposing subtypes of stages, like simple versus complex chiefdoms (Cordy 1974; Steponaitis 1978; Wright 1984), or more than one line of social or technological development, as in "multilinear evolution" (Childe 1951; Earle 1997:208; Steward 1955), ultimately did not provide satisfactory cross-cultural frameworks to account for important findings in the archaeological record (Trigger 2006:436, 440). Repeated discovery of exceptions or interesting variants that did not fit proposed social types or developmental norms, including the findings of bronze in non-state societies in Thailand (Bayard 1979; Muhly 1988:16; Solheim 1968; Wertime 1973:876), eventually stimulated many scholars of antiquity to reexamine their expectations for the universality of linear progressive change and the analytical usefulness of successive social or technological ages and stages (see review in Chapman 2003).

In recent decades, archaeologists have increasingly explored a wide range of intellectual positions that move beyond universalist approaches and grand narratives for studying cultural and technological change in human history (Gnecco and Langebaek 2014; Hodder 2012a, 2012b; Trigger 2006:444–483). Universalist approaches, which are rooted in large-scale "origins of" intellectual frameworks—like "the origins of civilized society" (e.g., Service 1975; Woolley 1963)—are now recognized as too crude and oversimplified for the in-depth and fine-scale studies typical of contemporary archaeology (e.g., Fowler 2013; Roberts 2011:137; White 2017). An "origins of" perspective on humanity's past is like looking at history through a rear-view mirror, only perceiving a narrow path that led to the present and not perceiving the options available and choices made along the way, or the other avenues pursued by other actors (Gould 2001; Pinch and Bijker 1987; Schiffer 2001c; White 2017). Contemporary studies of past societies demonstrate that outcomes of societal change are affected by a range of dynamics, processes, relationships, and factors outside the purview of conventional, universalist, rear-view mirror approaches (Chapman 2003; Crumley 1995; Jones 2002; see also Hodder 2012b).

Among factors usually minimized by universalist approaches are case-specific conditions, such as the specific historical background, local environmental variables, or prevalent ideology in a society. Archaeologists increasingly appreciate that the trajectory of any particular region is largely *contingent* upon the historical and environmental context specific to that particular region. One of the contingent factors not sufficiently represented in universalist approaches is "culture" in the anthropological sense of the term, namely a specific society's customary ways of believing, thinking, and behaving, which can shape trajectories of societal change in profound and idiosyncratic ways. Striving to acquire a deep understanding of cultural contexts is increasingly recognized as essential to understanding technologies and their changes in the past or present (Basalla 1988; Jones 2002; Killick 2004; Kuhn 2004; Pfaffenberger 1992; Schiffer 2001c, 2004, 2011).

Although no single paradigm has replaced universalist approaches to the past (Hodder 2012c), approaches that actively seek to incorporate local social, geographic, and historical variables, as well as individual and community actions and strategies (agency) have become prominent in current archaeological scholarship (Andrews and Doonan 2003; Dobres 2000; Hodder 2012a, Jones 2002; Killick 2004; Trigger 2006:491). Many of the new approaches are termed "bottom up" perspectives, in contrast to "top down" approaches common in the 1970s and 1980s that emphasized determining who or what had "control over" people, resources, regions, etc. Hodder (2012b:163; see also Fowler 2013) discusses an "emergent" viewpoint that is likely to become prominent as new paradigms replace conventional paradigms. These recent, more forward-looking, theoretical approaches to the past that incorporate agency perspectives (Barrett 2001, 2012), and/or behavioral approaches (LaMotta and Schiffer 2001) have yet to be widely applied to Southeast Asian archaeological data (White 2017). Application of contingent approaches to the study of the technology and economy of early East Asian and Southeast Asian metallurgy is also in its infancy.

While overt reliance on universalist approaches as the overall intellectual basis for studying the past has receded in many realms of the New World (e.g., Gnecco and Langebaek 2014), unfastening the perception of metal-based technological development from theories of sociopolitical development is generally not far along in the Old World where the Conventional Paradigm was born and tenaciously lingers (e.g., Earle 2002, Higham 2002, 2004; Higham et al. 2011a; Rispoli et al. 2013; Thorp 2006:8). Universalist frameworks are also still common in the archaeometallurgy literature (e.g., Charles 1980; Bourgarit 2007; Strahm and Hauptmann 2009; Zhang 2012), and universalist terminology including the Three Age System is common in descriptions of the human past in introductory textbooks and popular syntheses (Darvill 2002; Fagan 2008; Higham 1996a, 2002, 2014; Higham and Thosarat 2012; Johnson and Earle 2000; Milisauskas 2011). The terminology and many of the intellectual assumptions underlying progressive stage sequences are still prevalent in Southeast Asia (e.g., Higham 2004, 2014; Higham et al. 2011a; O'Reilly 2007a, 2007b:584–587).

Yet metallurgical sequences in many parts of Europe, Africa, as well as East, South, and Southeast Asia are now known to have their own distinctive characteristics and relationships to regional social, political, and economic development (Hoffman and Miller 2009; Killick 2009; Linduff and Mei 2009; Roberts 2009a, 2009b, 2011, 2014; White and Pigott 1996; White and Hamilton 2009). Despite increasing evidence that cultural sequences in many parts of the Old World do not fit comfortably into the Three Age System (e.g., Janzon 1988:104; Nelson 1993; White 2002:xvi–xvii, 2017), the nomenclature has generally not been replaced. Unfortunately, along with the residual nomenclature, lingering assumptions regarding how metal technology interacts with societies distort perceptions and interpretations of archaeological data. Therefore, some deeper appreciation of how archaeologists came to use and still often use universalist terms and assumptions, especially in relationship to metal technology, is important for the critique presented in this volume.

Debunking Presumptions of the Conventional Paradigm

The Initial Adoption of Metal

> *Metallurgy really begins with smelting. (Killick and Fenn 2012:563)*

When metallurgy "began" is not only a question of dates and sites, but also of what constitutes "metallurgy." Does it include the manipulation of any metallic substance, including working of native metals? Or can metallurgy only be discussed when the practice of smelting is demonstrated? Or is "true" metallurgy best reserved for contexts where production of alloys is the norm? The distinctions that were clear to earlier scholars have become more blurred in recent archaeometallurgy literature (e.g., Ehrhardt 2009, 2014).

The Conventional Paradigm has situated the beginnings of the use of metals in a sometimes lengthy period during which native metals, particularly copper, were worked. Scholars like Childe, Woolley, and Tylecote considered that these earliest uses of copper were small scale and did not seem to have had any profound impact on society. They drew a sharp distinction between the knowledge of native metals such as copper and the developed practice of metallurgy. The Copper Age, or Chalcolithic, was, according to Woolley (1963:550), merely a late phase of the Neolithic, because the scant supply of copper (assumed to be native copper) was used only for small objects and ornaments and was treated as a form of pretty stone. A Copper Age was never accorded membership in the Three Age System.

The Copper Age was contrasted with the "full" or "true" Bronze Age (Tylecote 1992:35) which was characterized by extractions of metal from various ores and alloying of tin and copper in controlled proportions for "true" tin bronzes (Tylecote 1992:25) that were stronger and harder than cast copper. There was also an expectation that bronze should become a "dominant" and "useful" material (Wertime 1964:1257), and hence produced on a significant scale. Thus, it was perceived that a true Bronze Age had a significant impact on society.

Much of the literature written from the perspective of the Conventional Paradigm highlighted the usefulness of metal materials and implied an intentionality of prehistoric craftspersons striving to apply materials in useful ways, which in turn provided the reasons behind their adoption (e.g., Wertime 1964; Tylecote 1992). This focus on utility and intentionality falls into the logical trap of "presentism," according to Killick and Fenn (2012), i.e., because metals are such centrally useful materials in current industrial society, it was assumed that past societies would have recognized and capitalized on their potential usefulness. Transformations to society resulted both from the availability of better tools and weapons and from the development of elaborate production infrastructure and extensive networks required to bring the materials together to produce the bronze products. The Conventional Paradigm presumes bronze metallurgy to be an important socioeconomic catalyst.

Although Conventional Paradigm proponents focused on the advent and impact of tin bronze metallurgy, as archaeological evidence has accumulated over the decades it has become clear that early metal adoption followed various paths, occurred in a wide range of contexts, and had varying kinds and amounts of societal impact. The distinctions so important to the Conventional Paradigm are no longer compelling in light of current evidence (Roberts and Thornton 2014). For example, the prolonged and often elaborate working of native copper in Archaic North America never led to copper smelting or even melting, but did entail far-flung trading networks and the production and exchange of sophisticated prestige objects (Ehrhardt 2009). Yet Ehrhardt also noted that spear points, knives, and awls, along with bracelets, rings and beads made of native copper were produced by mobile hunter-gatherers of the Old Copper Complex of North America, and studies show the many utilitarian-looking items do have physical evidence of use. Copper was not just trinket material in these early New World societies. Given the expert manipulation of native copper's properties by heating and hammering and the sophisticated symbolic social roles of the metal objects produced, Ehrhardt (2009) questioned the meaning and value of excluding the prehistoric copper working in North America from the ranks of "true metallurgy" based on the absence of smelting. Killick and Fenn (2012:563) on the other hand have continued to regard working of native copper as "pre-metallurgical."

Another aspect of the Conventional Paradigm sequence challenged by current data is the primacy of working native copper. Gold may have been the first worked metal in South America (Lechtman 1980:275), and there is little evidence that in the Andes working native copper preceded smelting of copper ores, which began with the Moche civilization (Lechtman 1980:287–288, 2014).

Archaeometallurgists now accept that prehistoric New World metallurgy illustrates a variety of technologically and culturally distinctive variations and trajectories. Nonetheless, this recognition has had limited impact on several preconceptions entrenched in Old World archaeometallurgy that tend to look to properties seen as intrinsic to metal itself and to cast metal and metal artifacts as independent variables of singular influence on sociocultural trajectories. Although recent compendiums on early metallurgy emphasize a holistic approach (Thornton and Roberts 2009), ideas and assumptions from the Conventional Paradigm, which often decontextualize metal evidence, remain common, as seen in assertions, for example, that mere presence of iron tools "would have encouraged and enabled the construction of banks and rice field boundaries…in order to control water" (Higham 2014:268).

Bronze and War

Given the utilitarian importance of metal in the modern Western world, many scholars have tended to align with Childe (1944) in assuming that the reason for the adoption and spread of metalworking was too obvious to be discussed—it was for implements, particularly weapons (Wailes 1996a:8). That metal "lent itself directly to war" fostered its dissemination, according to Wertime (1973:876). The advantages of metal in warfare would appear to be decisive for any kind of conflict or competition with non-metal using groups (e.g., Chernykh 1992:5; Goody 1971:46–47; Renfrew 1972). Early in his career, Renfrew (1972) believed that the main value of bronze was that it could produce superior weapons. He amplified the resulting causative chain, bringing in the role of elite control over economies. Superior bronze weapons led to greater power for the weapon-wielders, which led to: (1) the domination of one group over others; (2) the intensification of the trade and exchange networks necessary to supply the copper and tin, the sources of which were seldom nearby; and (3) an increase in the production of surplus subsistence products to exchange for the new goods. Elites' control of limited and valuable resources necessitated their control of the production and distribution of the metal.

The role of warfare in stimulating the spread of bronze metallurgy and weaponry along the lines that Renfrew suggested may have been important in some regions such as northern Europe (Earle 1997; Sherratt 1994a, 1994b), but weaponry is not prominent in several early copper-base and bronze assemblages, either of state or non-state societies. When so-called "weapons" were present, they do not necessarily indicate the occurrence of warfare. The Indus civilization metal evidence is one particularly notable case since the sociopolitical context suggests a "state-like" civilization (but see Possehl 1998) that in theory could have had the capacity for organized warfare. While spear and arrow points are present, they are not particularly prominent among the wide range of more common items like tools, ornaments, vessels, and other items made of copper-base materials (Hoffman and Miller 2009).

Production of weaponry also does not appear to have been the focus for the initial adoption of bronze technology in central plains China. The few copper-base pieces from Longshan sites include a bell, awls, miscellaneous small pieces, and one flat axe (Yan 2000:106). During the subsequent Erlitou period Phase 1, small objects like awls, fishhooks, blades, and arrowheads and larger items including bells and plaques were recovered (Yan 2000). The earliest spectacular bronze vessels appear in a subsequent phase, and they seem to be the main focus of local innovation in the production of bronze artifacts.

Even when artifacts that could be termed a type of "weapon" (e.g., decorated daggers in Korea [Kim 2001], arrow points in the Ganeshwar-Jodhpura culture [Rizvi 2007]) were present, without other supporting evidence of violent activities, their role was more likely for ceremony (in the case of Korea) and/or hunting (in the case of the Ganeshwar-Jodhpura arrow points). In this volume, the more neutral term "implement" is used rather than "weapon" for such artifacts.

Metal and Functionalism

Even if metal's usefulness in warfare was not the primary reason for the dispersal of metal technology, might other utilitarian purposes have been the primary reason for adoption, such as use in agriculture or woodworking? This view would be in line with Childe's (1930) argument that the transformative power of metal rested in useful implements that would materially alter the economy, reflecting his Marxist emphasis on production and the relations of production.

This functionalist view of technology, a particular target in Pfaffenberger's (1992) critique of the Standard View of Technology, still recently has been taken for granted by many important historians of metallurgy (e.g., Tylecote 1992). An unstated assumption in the functionalist view is that technological innovation and adoption is driven by human practical needs. In Tylecote's magisterial *History of Metallurgy* (1992), for example, the question of *why* this new material was adopted is never addressed.

Yet over the course of the 20th century as archaeological research has expanded into regions outside of the source areas for Western civilization (i.e., beyond Europe and the Near East), the picture of early metallurgical development has become increasingly diverse, and tools and weapons are not invariably present in early metals-bearing assemblages. Both archaeologists (e.g., Renfrew 1986) and metallurgists (e.g., Smith 1981), in assessing the ever-expanding archaeological and historical record, began to point out the essential problem with utilitarian and functionalist points of view for early adoptions of metal, as utilitarian purposes may not have been considered important for some centuries or even millennia after the earliest appearances of the material.

> Indeed, the explanatory problem…is not to explain why people did not at once use this great technology. It is, on the contrary, to understand why they bothered at all. For until the technology develops a great deal further, as it can only do through intensive use, early copper metallurgy does not produce anything decisively useful at all. (Renfrew 1986:146)

As more of the world's prehistory has become known and each region's sequence has been reconstructed in greater detail, it has become clear that in much of the world (including Southwest Asia) the earliest use of metals and ores was often for ornaments (Killick 2009; Lechtman 2014:369; Roberts et al. 2009; Schoop 1995). The functionalist view that matters of style, decoration, and ornament are at most frills on a materialist and utilitarian base (Basalla 1988; Ingold 1990; Pfaffenbeger 1992) has been contradicted time and time again as the archaeological and historical record pertinent to early metallurgy became more fully known. Smith, more than three decades ago (1981), argued that this was true of almost all new materials and techniques, i.e., that most technical achievements of humanity—pottery, metalwork, printing, weaving—had their origin in aesthetic curiosity, the desire to create beautiful and ingenious objects (Killick and Fenn 2012). Although a particular material or technique may later become a part of a particular culture's way of satisfying what it defines as needs (the definition of which varies with each culture), often the initial use of the material did not occur in utilitarian artifacts.

Over time, reusability may have become metal's main utilitarian advantage over stone and bone (Castro et al. 1998, as cited by Chapman [2003]). Though the initial production of the metal was more laborious than obtaining stone and bone, its mutability may have been highly valued. The same piece of metal can be in turn an awl, a bracelet, and a dagger; the form of the artifact it composes at any one time does not limit its potential for subsequent uses.

Metals and Progress

> *The idea of directional change was common in the 19th century when it was linked to ideas of progress. (Hodder 2012b:167)*

> *Our crap detector should sound the alarm every time we encounter the word* progress *in an account of technological change. (Schiffer 2011:14; emphasis in original)*

Another assumption conventionally underlying discussions of technological change is that change is progressive and linear, not just in the social realm, but in the physical realm as well (e.g., Strahm and Hauptmann 2009). In this view, as technological change occurs, more knowledge and more difficulty are incorporated into the system, the knowledge builds upon preceding steps in a logical, often "acultural" manner (Hosler 1994:4; Pfaffenberger 1992). The progressive, linear, acultural sequence of copper-base metallurgical development stated in Wertime's (1964, 1973, 1980) seminal articles has influenced thinking on metallurgical technologies (Killick 2001). The progressive sequence begins with hammering and annealing native copper, followed by (in general terms) melting and open mold casting of native copper, smelting copper from carbonate ores and later sulfide ores, smelting and casting arsenical coppers, closed mold casting, and deliberate alloying of smelted copper with tin, lead, or other metals to form bronze alloys. The utilitarian capabilities increased with each stage, so that qualities like hardness were progressively enhanced. From this sequence of technological development, graded universal, cross-cultural, metal technological stages were defined to which individual societies were assigned (e.g., Tylecote 1992).

This progressive linear "universalist" view has been at times a useful point of reference in a general manner by providing a model that can be tested. By their very basis in systematically exploiting knowledge of aspects of the physical world, technologies tend to build upon prior knowledge and do not randomly appear and disappear like fads. However, as archaeometallurgical data have accumulated, and as archaeologists have achieved a more detailed understanding of specific regions and a broader understanding of technological development, it is now clear that the progressive model does not map well onto real archaeological data, even in the Near East (e.g., Thornton 2009a, 2014; Yener 2000).

Excessive attachment to the model can lead to false or distorting assessments when specific cases are evaluated only as they fit into the model. For example, in many parts of the world, exploitation of arsenical copper co-occurs with other copper-base materials and alloys and may have been a choice rather than an accidental alloy used at a particular "stage" (e.g., Iran [Thornton 2014:677]). Hence, arsenical copper is best viewed as one variant copper-base material that may be selected, if available, for smelting for a variety of reasons; its presence cannot be inferred as indicating a metallurgical "stage" laying the foundation for a next step toward tin bronzes.

A collateral assumption to the expectation of progressive technological change is an assumption that societies will normally employ the newest, most "advanced" technologies available (Schiffer 2001c:216). However, an example from pottery technology illustrates that the most "advanced" technologies are not necessarily favored. Thus, if the potter's wheel becomes known to potters, it is commonly assumed that wheel-made pottery will rapidly replace hand building. However, ethnographic and archaeological evidence has now shown that potters do not necessarily adopt the wheel when exposed to the technique (Berg 2007). Whether the wheel is even retained in a production system has much to do with the scale and organization of the learning network employing the technique (Roux 2008).

Moreover, hand-building and wheel-building technologies can co-exist, as can be seen recently in northeast Thailand where some villages make hand-built vessels and others wheel-fashion vessels. As both kinds of vessels are widely exchanged and used in the region, the different practices are not due to practitioner isolation. The wheel-built vessels are made from different clays, undertaken in a workshop labor organization, fired in kilns, and used for different purposes than the hand-built vessels, which are made with household labor organization and are open-air fired (Joyce White, field observations 1978–1981; see also Boileau 2005). Villages, including the production villages, consume both product types but use them for differing functions. The adoption, implementation, and sustainment of these two potting technologies is embedded in particular cultural and economic systems.

Detailed analyses of regional sequences of metallurgical development are now uncovering many variant paths and metallic sequences. Regional metallic sequences depended on variables ranging from local ore resources to cultural choices and priorities (Lechtman 1980). In South America, aesthetics and ritual trumped usefulness in the early development of metallurgy and, while metallurgically complex, the products were rare. Lechtman (2014) presented sequences in the Andes that focused on gold, silver, and copper to produce metals manipulated for qualities of iconographic form, color, sound, and sheen, and not for qualities of utilitarian advancement. The earliest alloys there were of gold and silver. Tin bronze did not become a "dominant material" as predicted from the utilitarian Childean framework, but the complexity and sophistication of metallurgical technology cannot be denied (Lechtman 1988, 2014). The South American metallurgical technology did not develop in the context of a large pan-regional production system. Instead, subregional innovation and manipulation of rare and aesthetically refined materials emphasized ores and metals available nearby to produce a variety of alloys that brought out variation in color and sound. Lechtman also found some areas preferred to cast shapes and other areas preferred "plastic deformation" or direct shaping of the malleable material by hammering.

In the Old World, Thornton found on the Iranian Plateau many diverse metallic sequences with various forays into polymetallic ore selection and mixing completely outside of the "mantric narrative" of the conventional metallic sequence (Thornton 2009a:302). When the archaeometallurgy of other less-thoroughly studied areas including Southeast Asia is investigated to a degree comparable to the Andes and Iran, variability in metallurgy and its sequential development within local regions is likely to be uncovered.

Furthermore, the progressive expectations of the universalist technological sequence offer nothing to help explain the marked decline in copper production in the Carpatho-Balkan area after flourishing for several hundred years in the 4th millennium B.C. (Chernykh 1992:52). Local raw materials, regional population histories, and cultural factors slowly have come to be recognized not as ancillary but as foundational criteria in the delineation of any particular metallurgical sequence (Chernykh 1992:53). Even if as a general principle technological change tends to build upon preexisting technological knowledge, nonetheless there is no a priori technological sequence, no inherent direction or pace for change, for the development of metallurgy universally applicable to societies irrespective of context. Each metallurgical trajectory must be examined and explained on the basis of its own evidence and context (Thornton and Roberts 2009).

Another collateral assumption to progressive views of technological change that has not held up with recent data is that new technologies are usually *rapidly* adopted and in turn catalyze *rapid* "progressive" sociocultural change (e.g., see Higham and Thosarat 2012:107). Instead, evidence supports gradual adoptive processes and eventual relationships to social change. Holl (2009) for example identified a long period of low intensity of production in western Africa before more systematic routinized metal artifact production constellated. Roberts (2009a), synthesizing the data for early metals in Western Europe, showed a slow, community-based adoption process occurring well after access to the technology was present; "there was no evidence for any metallurgical revolution." (Roberts 2009a:461). Thornton (2009a) also reported a lag time of as much as a millennium from first appearance to true adoption of metallurgy on the Iranian Plateau. Although in some regions copper metallurgy spread widely after its first appearance, taking only a couple of centuries to become thriving (e.g., Levant [Golden 2009]; Balkans [Jovanović 1988]), rapid adoption of metal technology appears to be the exception rather than the rule based on current archaeological evidence from many regions around the world (Thornton and Roberts 2009:182). Sporadic local adoption that commonly characterizes a region's adoption of metal technology (e.g., Hosler 2014:333; Roberts 2014) may mean that communities could choose to be "late adopters" or "laggards" while others were "early adopters," resulting in temporally and geographically uneven archaeological evidence for the appearance of the technology within regions (e.g., see Dolfini [2014] for punctuated spread of early copper metallurgy in Italy).

Metals and Social Change

> *most regions [adopting early metallurgy] demonstrated a significant lag (often of millennia) between the initial adoption of metals and the dependence of elites upon them. (Thornton and Roberts 2009:182)*

> *Metallurgy is one of the great innovations of the prehistoric past, and in virtually all instances, it contributed significantly to the rapid development of elites in society, those who could control the acquisition and ownership of bronzes. (Higham and Thosarat 2012:107)*

Identifying crucial explanatory variables to account for social change has been a goal for much of Anglo-American archaeology in the 20th century. One attraction of the Three Age System among some Old World archaeologists and archaeometallurgists relates to its implicit (and often explicit) theory of socio-technological causality, a kind of technological "determinism" with metals playing a starring role. As Leslie White, an influential neo-evolutionist, expressed it, "social systems are...determined by technological systems, and philosophies and the arts express experience as it is defined by technology and refracted by social systems" (White 1949:391).

Metal technology has long been perceived as having a central role as a catalyst for past social, political, and/or economic change. Metal technology was seen by Childe (1951) as an "independent" and critical variable in relation to several social manifestations such as the development of specialized economies, cities, and social classes. Wertime (1973:876) viewed metallurgy as a "crucial ingredient or subsystem in the process of early urbanization." In a recent example of this line of thinking, Higham (2012:534) argued for northeast Thailand "that the novelty, probably the rarity and certainly the desirability of owning and displaying bronzes contributed significantly to the rise of social elites."

Childe's writings have shaped much 20th century thinking on the relations of metallurgy to society. His early theory that metal craft—involving abstruse knowledge of ores, mining, smelting, casting, and forging—required from its beginning that the practitioners be *full-time* specialists (Childe 1930) has endured several decades in the literature (Charles 1980; Knauth 1974; Whitehouse 1974; see Chernykh [1992] for a recent archaeological expression of this view), though even by 1965 it was recognized that there was no archaeological evidence for this hypothesis (Piggott 1965:126).

As envisioned by Childe (1930), full-time specialist metalworkers would have reconfigured tribal economic relations such that the metalworkers would have been dependent on trading the products they manufactured for food produced by non-kinsmen. To obtain metal tools obliged farmers to produce a surplus for exchange. Independence from the rigors of food production allowed smiths to travel about in a way that farmers could not and, given the long distances over which metals such as tin must be traded, even required them to travel. Detachment from food production meant that smiths were the first men (and Childe assumed that they were men) to be "detribalized," i.e., separated from the demands and constraints of tribal society. Thus, the invention of metal tools, according to Childe specifically of bronze, provided the first impetus to the production of surpluses, the growth of trade and exchange networks, and the creation of an economy separate from "Neolithic" tribalism. Later Childe (1944) presented a more nuanced view but, while he rejected a simplistic equation of bronze and iron with particular levels of societal complexity, he kept technology at the core of his scheme and devised a series of "modes" that were defined almost entirely by how and what metals were used.

As early bronze metallurgy has been investigated by archaeologists throughout Eurasia over the past several decades, the causal chain envisioned by Childe has not been supported as the typical sequence of events in societies adopting bronze (Thornton and Roberts 2009). Gibson (1996) argued that in prehistoric Europe, metalworkers not only were not itinerant, they were dependent on localities and kin relations, and probably practiced only part-time. White and Pigott (1996) argued for seasonal part-time specialization for early metalworkers in Thailand that lasted for more than a millennium with metal making taking place outside of the growing season. Ethnographic studies have disproved the idea that metalsmiths must necessarily be full-time specialists (e.g., de Barros 2000; Rowlands 1971) as well as debunked other common archaeological assumptions surrounding pre-industrial metalworking. In short, the evidence shows no necessary relation between the emergence of metalworkers and the development of a social class of full-time craftspersons reliant on surpluses from non-kin farmers.

METAL AS SOCIAL CATALYST

Nevertheless, manufacture of this new material and of goods from the new material has long been thought to have some catalytic relationship with social inequality, such as by fostering an elite segment invested in controlling metal production or by stimulating and supporting an elite consumer base for metal products that could then store and signal elite wealth. However, as archaeologists have investigated early metal-producing societies world-wide, these ideas have not been supported either, at least in terms of a rapid response to the appearance of metal technology.

For example, in China, Underhill (1996:145) observed that metals were initially present in the central plain in relatively small quantities and minor roles for some centuries before developing a significant economic presence. She stated that only after states had already emerged was bronze production "an important causal factor in sociopolitical change." Franklin (1983:288) went even further, arguing that the remarkable bronze technology of the Shang "was more a consequence and indication of a new social order than its cause."

Several compendia substantially devoted to archaeological evidence pertinent to this topic have generally repudiated the Childean view that metals were a prime mover of social change in most known cases throughout the globe (e.g., *Journal of World Prehistory* volume 22 issues 3 and 4, 2009; Roberts and Thornton 2014; Wailes 1996a, 1996b). Well before these published compendia, archaeometallurgists Stech and Maddin, who spent decades studying both southeast and west Asian ancient metals, noted that contexts shaped metallurgy more than the reverse:

> Technological systems, such as metallurgy, should be interpreted as integral parts of the cultural context in which they occur rather than as entities possessing their own dynamic, independent of context...*Metallurgy did not change the cultural dynamics in either end of Asia but was a dependent subsystem in both areas, as it was elsewhere.* (Stech and Maddin 1988:173; emphasis added)

Yet despite the recognition of the complex and contingent relationship of metal craft production and social, economic, and political aspects of prehistoric societies, simplistic assertions continue to be published. Catalytic roles for metal technology—its mere introduction directly stimulating social change—continue to be emphasized by some archaeologists (e.g., Higham 2014:15). In Thailand, for example, the early presence of copper-base metal has been argued to be the medium by which aggrandizing personalities in consumer villages could gain control over supply and ownership of prestige goods including metal and thereby attain wealth and elite status (Higham and Thosarat 2012:266; Higham et al. 2011a:264). Rispoli et al. (2013:104) asserted a related view, noting that "copper-base metallurgy...served as a catalyst to socio-political changes," and it was the skilled metalworkers who "may represent a group of...'aggrandizers' seeking a higher social status through mastering the new craft." The evidence for these ideas in Thailand is reviewed in TAM 2C where occurrence of copper-base artifacts in ordinary metal age graves and "founder burials" without outstanding wealth suggest a different, more complex scenario.

BRONZE AND PRESTIGE GOODS ECONOMIES

Even if metallurgy was not a deterministic player in past sociopolitical systems, when present, it may have had key roles in the economies of the elite. Archaeologists have called "prestige goods economies" (Brumfiel and Earle 1987) economic systems where elite goods are produced and distributed as a means to signal elite status and/or to store wealth. Prestige goods economies may be controlled to varying degrees by elites through means such as direct control over the producers, patronage, or simply cultural fiat of the society's dominant players (Hayden 1998). Prestige goods are commonly non-utilitarian items, often notable for their large size and/or elaboration (e.g., Dian drums, Sipán headdresses) and are usually costly to produce or acquire. They may be made of materials acquired from great distances and/or require a long and involved production process such as metals or silk. Prestige goods frequently show evidence of highly skilled and laborious craftsmanship. Archaeological context is also important in defining an object as a prestige good, including association with other indicators of wealth and/or status such as being found only in "rich" graves or in hoards of similarly elaborate artifacts.

Golden (2009) provided a particularly plausible example in the Levant of copper-base metal employed in the manufacture of prestige goods (*sensu* Earle 1991b:10) that could signal high status and store wealth. These goods included elegant and elaborate lost wax castings in arsenical copper of standards, vessels, "crowns," mace heads, and other ornamental goods that were recovered from Chalcolithic deposits, especially hoards. Interestingly, when tin bronze appears in that area, the elaborate castings ceased, illustrating the lack of any intrinsic connection of bronze with the manufacture of prestige products.

Because of their rarity, complexity, and laborious fabrication processes, metals, especially bronze, have been considered automatic candidates for treatment as prestige goods (e.g., Strahm and Hauptmann 2009:116). However, archaeological evidence now shows that copper-base metals, including bronze, were not automatically incorporated into a prestige economy, even when a prestige-good economy existed prior to the appearance of metallurgy. Rare did not necessarily or automatically equate to prestigious.

A particularly clear example of the non-incorporation of bronze into a prestige economy is the

Yellow River valley and other parts of northern People's Republic of China (PRC) in cultures such as Majiayao, Qijia, and Longshan from the early 3rd millennium B.C. (Jaang 2015; Linduff 2000; Liu 2003; Shao Wangping 2000:213; Underhill 1996). In that time, ceramics and/or jade were the favored prestige goods, judging from grave assemblages of elite burials (Liu 2003:22; Underhill 1996). The distribution and nature of the 3rd millennium bronze artifacts in northern China suggest that this was a minor and somewhat rare technology, used for small items including ornaments and simple utilitarian artifacts, with no particular association to elites (Linduff 2000:11; Shao Wangping 2000:213). For example, at the large Longshan site of Taosi, burials show marked social stratification with elaborate pottery, jade, and wood artifacts as grave goods in elite burials; but only one grave had a copper-base artifact, a small bell, and that grave had no other grave goods (Liu 2003:22). Copper-base metallurgy including copper production and bronze was present in northern PRC from the early 3rd millennium B.C. (Linduff 2000), thus, for perhaps 800–1000 years before the Erlitou period when elite/state-sponsored bronze metallurgy took off.

According to Liu (2003:22; see also Jaang 2015:204–205), bronze first became incorporated into a prestige good system in China when multipiece mold casting of ritual vessels was developed at Erlitou in Phase 2 (1800–1700 B.C.). Prior to that time in the region, bronze may have been novel and rare, but there is little evidence that copper-base artifacts functioned as prestige goods that stored wealth and/or signaled membership in aristocratic circles. The early state structures of Erlitou, beginning in Phase 2, controlled production of the metal and manufacture of metal goods, principally the large bronze ritual vessels. The use of bronze ritual vessels to signify state prestige can be interpreted as a continuation of the use of fine, labor-intensive ceramic vessels that were a symbol of status in the previous Longshan period (Lui 2003; Underhill 1996). As Underhill (1996:145) observed, only *after* states had already emerged was bronze production an important factor in sociopolitical life. Furthermore, as Franklin (1983) noted, only when a social hierarchy was sufficiently entrenched to be able to compel forced labor for large-scale mining and smelting in China did production of bronze objects in large quantities begin.

In other parts of the world, there was also a substantial lag time, in terms of hundreds of years, between the introduction of extractive metallurgy and the use of metal artifacts as a true prestige good (e.g., west Mesoamerica [Hosler 2014]; Iberia [Rovira 2002]). In parts of Europe, Roberts (2014:434) argued that for more than a millennium after introduction, from the late 4th through 3rd millennium B.C., copper production was small scale with occasional manufacture of only small numbers of objects. Their rare role as grave goods suggested a striking and valued material; however, "metal probably did not quite carry the prestige for prehistoric communities as is frequently imagined." Rather the disposition of copper-base artifacts in the archaeological record seems consistent with appreciation for an intriguing material, occasional use for small tools like awls and hooks, and/or applications for personal expression like trinkets or jewelry. Although some of the striking larger copper artifacts like flat axes were used (i.e., the Iceman's axe), the prevalence of unused copper flat axes in some regions such as parts of northern Europe argues that copper axes may often have functioned more for show or other symbolic purposes in many areas (Sørensen 2014:175). Overall the initial appearance of copper-base objects in a material culture repertoire did not seem to mark a technological or social revolution.

Although most archaeologists now know that the mere introduction of production of copper or even bronze was not usually a rapid or major catalyst of socioeconomic change, nor were metal objects necessarily rapidly used to signal high status or store wealth, in many regions metals were eventually incorporated into those roles, usually some hundreds of years after the technology was first introduced (Thornton and Roberts 2009:182). China was not unique in having social stratification precede the development of use of metal in prestige roles.

Cases when prestige roles for metals clearly did eventually emerge within a metallurgical tradition can provide a useful frame of reference to examine the range of roles that metal may have had in the development of social complexity. In addition to the ritual vessels appearing during the Erlitou period, swords and regalia in Denmark (Vandkilde 1996), mirrors and bells in Korea (Kim 2001), and Dian cowry containers (Minghua 2006) are examples of obvious development of prestige roles, but in different manifestations. Materialist approaches to evaluating the prestige value of such metal objects emphasize their ability to encode statements about the control of scarce and desirable resources, and how this form of storable wealth could become a focus for competition among some groups (e.g., Ehrenreich 1995; Kim 2001; Kristiansen 1987; Liu and Chen 2003; Renfrew 1986; Shennan 1986). In these scenarios, an increase in metal use was both cause and result of social changes; to obtain and control the supplies of metal gave prestige, a prestige stored and announced in the artifacts. Possession of the artifacts themselves lent prestige as well as reflected it. Durable and rare prestige goods (Clark and Parry 1990) could store value (Renfrew 1986) that was transferable and could be donated. Plourde (2009) provides a variant model that could be useful for metal age societies, emphasizing prestige goods as "costly signals" not so much of resource control, but of skill and expertise, including leadership in contexts where collective action is advantageous.

Metals can have ritual, ceremonial, and even potentially prestige value in non-state societies as they can have in state societies. Copper and bronze metallurgy emerged in west Mexico well before the development of local states. Although some tools and implements were made, the primary use of metal was to mark, through sound and color, elite and especially sacred domains (Hosler 1988, 1994, 2009). The Dian castings in Yunnan (Murowchick 1989) and the Dongson drums are other examples of artifacts in non-state but hierarchical contexts that demonstrate a ritual and expressive value for bronze in corporate contexts rather than for individual members of an elite. In contrast, the chiefdom societies of Transalpine Europe in the early 2nd millennium valued bronze primarily for weaponry—daggers, swords, arrowheads—and armor, whether for use or display by individual warriors and chiefs (Sherratt 1994a, 1994b).

Though its "statehood" might be debated (Possehl 1998), the Indus civilization demonstrates that even in the context of a major urban civilization capable of creating monumental architecture, the typology and distribution of copper-base objects argues that they were *not* valued primarily as prestige goods (Hoffman and Miller 2009, 2014). Utilitarian implements like knives, points, rods, hooks, chisels, and adze/axes predominate. Ornaments include bangles and beads. There is a low frequency of mirrors, figurines, and token/tablets. Copper-base artifacts are most commonly found in ordinary household deposits, rarely in burials or public locations.

Lechtman (2014) found in Andean metallurgy that technological elaboration did occur in the elite copper-base products but not in the utilitarian sector during Moche times. The evidence suggested that it was the presence of elite needs that stimulated technological innovation in selected portions of metal product manufacturing and not metal product manufacturing that stimulated socioeconomic development.

There are examples of bronze-producing contexts where social differentiation was not pronounced and the prestige value of the metal was unsurprisingly not marked. Bronze age Agaric southeast Spain produced copper alloy tools and ornaments, which could be placed in graves, but the small numbers of bronze grave goods and lack of marked social differentiation in grave assemblages suggest that metal did not function as a means to store wealth or express power (Rovira 2002). In one study, roughly 20% of more than 1000 graves had metal grave goods and only a few of those held more than three metal objects (Rovira 2002:11).

Thus, the details of the prestige role (if any) of bronze and other metal artifacts varied from society to society, as the societies themselves varied in social, economic, and political configurations. Materialist assumptions are not always apt, and in some societies, a ritual value for the use of bronze appears primary, but the precise form of expression depends on the cultural context. In that the values attributed to metal by any given society clearly reflect cultural meanings specific to the place, time, and social group irrespective of societal scale, it is clear that there is no a priori prestige value for bronze or any other metal, even in stratified societies. That metal-using, stratified societies such as Longshan and Harappa never incorporated metal into prestige good economies illustrates the necessity of *examining* cultural values and not *assuming or imposing* cultural values in understanding the role of metals in any specific context. It is likely that metal will be incorporated gradually into the preexisting pre-metal material and value system of the culture before in some, but not all cases, it gains a specialized place. The concept of prestige goods is further elaborated in Chapter 5.

BRONZE AND CENTRALIZED MANAGEMENT

Related to past assumptions that metals, particularly bronze, had inherent prestige value, is another frequently unexamined assumption: that early societies sought to control metal production and distribution in a top-down manner. In part, this assumption derived from some of the best-known evidence for how early metals were produced that came from sites like Timna and from Egyptian tomb paintings. That evidence fostered the impression that early copper production took place on a large scale and in managed workshop settings (Knauth 1974:51; Raymond 1984:16–18).

However, increasing evidence of small-scale metal production during the early as well as later stages of metallurgy in many regions of the world suggests that large or industrial-scale prehistoric mining and manufacturing under the control of elites was exceptional, developing under special circumstances after long periods of simpler, smaller-scale production (Budd and Taylor 1995). Decentralized small-scale metal production has been identified for a wide range of societies, from domestic societies of 2nd millennium B.C. Spain (Rovira 2002) to the greater Indus civilization (Hoffman and Miller 2009; Kenoyer and Miller 1999), where no clear evidence for centralized extraction, alloying, casting, or fabrication has been identified. The distribution, plentitude, and accessibility of ore sources can contribute to the pattern of managerial development. For example, abundant small outcrops of copper were widely distributed in Spain, providing little occasion for centralized management of a resource that simply was not scarce (Rovira 2002). In the case of the Indus, they appear to have "outsourced" metal extraction to neighboring groups and done much independent small-enterprise melting and recycling of imported materials (Hoffman and Miller 2014).

Even when highly stratified societies had unequivocal metal prestige goods, exclusive control over metal by elites is not necessarily in evidence. Mesopotamia, for example, was found to have public sector independent metal product manufacturing co-occurring with elite sector attached production (Zettler 1996). In the Andes, the Moche metalworkers produced utilitarian copper-base items for urban residents at the same time that sophisticated and elaborate products were manufactured for religious elites (Lechtman 2014). These examples show that metal was not exclusively controlled by and for the upper echelon even in early states.

Large production-site size, furthermore, may not be an indication of centralized control; small-scale production (in terms of labor units and scale of activities) may have been the practice at many sites with large amounts of metal production evidence. This has been demonstrated by excavations at Kargaly and Samara near the Ural Mountains, where, despite the large quantities of production debris, there was no evidence for centralized large-scale operations (Doonan et al. 2014). Archaeometallurgists increasingly appreciate that episodic and small-scale mining, and decentralized smelting and artifact production undertaken by households in specialized independent communities can, over the course of hundreds of years, produce quite large and concentrated quantities of debris at production sites (e.g., Hanks and Doonan 2009; Pigott 1998).

Nevertheless, in those cases when metal production and metal artifact manufacturing were brought under centralized management, the scale of production and genres of manufactured products could be quantitatively and qualitatively distinctive (e.g., Shang vessels) in comparison with less managed contexts. The capacity to enhance metal products on such scales with investment in management could have endowed the technology with potential ancillary socioeconomic impact distinct from other prestige items such as jade, shell, or amber in specific socioeconomic contexts.

In summary, the processes of adoption and incorporation of bronze technology into ancient societies and economies, what those societies initially decided to do with bronze, and how people revised and expanded the social and economic place of bronze in their society, are complex, vary with each society, and require detailed archaeological and technological investigation.

METALS AND SOCIAL COMPLEXITY

The relationship between early metallurgy and early complex societies is perhaps the most knotty of entanglements in the Conventional Paradigm. Appreciation of the complexity of activities needed to produce bronze artifacts contributed to an expectation that only "complex societies" with social and economic hierarchy could be the producers. This expectation echoed Childe's (1930, 1951) idea that metalsmiths had to be full-time economic specialists, reliant on producing metal artifacts for consumers in exchange for other life necessities, and that there had to be consumers who wanted the metal products. In other words, metals could only be produced in societies in which economic specialization existed.

The impressive bronzes uncovered in the late 1920s in the royal cemetery at Ur in southern Mesopotamia, later dated to the mid-3rd millennium B.C., fostered the sense that the "Bronze Age" implied complex societies with not only economic specialization, but also overt, marked social stratification, states, and ruling elites—who were the ultimate consumers. The images of slave-based metalworking in Egyptian tombs also supported this view. Such highly stratified societies have long been presumed to have emerged at the later end of sequences of social evolution. As discussed below, investigations of bronze-producing societies in recent decades have shown that the relationship between bronze production and social contexts is much more diverse than was expected when archaeological cultures were first tagged as "Bronze Age." Nevertheless, archaeologists have continued to employ outdated connotations of Bronze Age social complexity (e.g., Earle 2002; Earle and Kristiansen 2010:19–21; Higham et al. 2011a; Kristiansen and Earle 2015; Kristiansen and Larsson 2005;). To appreciate the impediments of a misleading linkage between bronze and complexity, understanding is needed of what archaeologists have meant and currently mean by "social complexity."

What Is a "Complex Society"?

Review of the use of the word "complex" by archaeologists in discussions of past societies reveals that connotations have evolved from the initial use of the phrase in archaeological and anthropological literature. Study of complexity in these disciplines has long had an association with a social evolutionary framework. At first, the phrase "complex society" was used by social anthropologists (see Banton 1966) as a very general term to distinguish "larger-scale societies" that have differentiation in roles, particularly in economic and technical spheres (i.e., a division of labor) from purportedly undifferentiated and egalitarian "simple" or "small-scale societies." In the latter societies, relatively few distinct roles exist, individuals perform multiple roles, and virtually no social relationships are impersonal (Benedict 1966:25; Fried 1967).

Some archaeologists in the early 1970s began to use "early complex societies" to refer to what previously might have been termed "early Civilizations" (e.g., Renfrew 1973:9). Initially such societies were equated with the earliest states. The phrase became more common in archaeological literature after the publication of conference papers from a colloquium "Reconstructing Complex Societies" at the Massachusetts Institute of Technology (Moore 1974). Which archaeological societies can be called "complex" was clearly a topic of discussion at the colloquium (e.g., Longacre 1974:29; Willey 1974:145; Wright 1974:123), and the study of complexity in archaeological contexts was recognized as involving a range of phenomena, which were variably present in specific examples. (For example, an assumption by some scholars that complex societies must have a written language was undermined by the case of the Inca, a state that had no written language but did have *quipu*—a record-keeping system using knotted string.) Willey's (1974:145) summary foreshadowed the later trend among archaeologists (e.g., Earle 1991a, 1991b) where the study of social complexity became synonymous with the study of centralized societies that could be classified as either of the last two rungs in Service's (1962) social evolutionary schema, i.e., as either chiefdoms or states.

During the 1980s, when the study of complex societies was a major focus of the discipline, archaeologists came to assume that social complexity implied centralization in political and economic spheres, social stratification, and hierarchical power differentiation. Over the next two decades, much attention was paid at conference symposia and in publications to archaeological evidence for the genesis of social inequality and elites (e.g., Price and Feinman 1995a), concentration of political power (e.g., Earle 1997), control over resources (e.g., Earle 2002), the presence of settlement hierarchies (e.g., Wright 1977, 1986), warriors and warfare (e.g., Carneiro 1981; Earle 1991b), elaboration of ceremonial expressions of elite legitimacy (e.g., Earle 1991c, 1997; Price and Feinman 1995b), and centralization of various kinds (e.g., Earle 1994, 1997:208). From the 1990s, the concept of "political economy" received much attention, as economies in the service of political agenda were considered a key attribute of so-called complex societies. Perhaps the most comprehensive statement concerning the role of political economies in the development of social complexity can be found in Earle's *Bronze Age Economics*. The following quote illustrates some of the vocabulary common in publications written from a political economy point of view, particularly an emphasis on top-down "control."

> political economy…involves the ways that surpluses are mobilized and allocated to support political activities, lifestyles, and operations of social institutions and their leaders.…The mobilization of a surplus requires a productive economy and its practical *control*. The control derives from *command* over quite specific activities involving production, distribution and even consumption. (Earle 2002:9; emphasis added)

At the same time that so much attention was focused on finding these phenomena in the archaeological record, archaeologists also discovered that many societies did not fit neatly into the types, models, and expectations drawn from study of the "archetypal" examples of chiefdoms like Hawaii and states such as early Mesopotamian states (e.g., Feinman and Marcus 1998; Upham 1987). During the 1990s, the perceptions of what constitutes "social complexity" shifted as the complexities of some hunter-gatherer societies were better appreciated (for instance, the sedentary and socially stratified northwest coast Amerindians of North America [e.g., Kelly 1991; Maschner 1991]) and as differentiated and integrated societies without clear social or political elites or conventional measures of centralization were identified (see for example articles in Ehrenreich et al. 1995; the Indus civilization as discussed in Possehl 1998).

The archaeological study of complexity now is moving away from the past emphasis on evolutionary stages for the development of hierarchy and top-down power relations and is going back to basics. The study of complexity increasingly is focusing on the study of its fundamental elements—evidence of *differentiation* (number and types of parts), and *integration* (number and types of connections among parts)—in the contexts of specific societies (Chapman 2003:82–84). Yet assumptions, terminology, and structures of reasoning from earlier views of complexity linger on in some archaeological and archaeometallurgical literature (e.g., Earle 2002; Higham 2014; Higham and Higham 2009b; Krause 2009; Zhang 2012). However, Chapman's use of complexity as related to differentiation and integration is very pertinent to interpreting Thailand's early metallurgy, as demonstrated in TAM 2C.

The Origins of Metallurgical Technology

> *Any discussion of origins must first face the fact that such investigations are no longer popular. (Muhly 1988:3)*

Because archaeologists study cultural change over time and space, they address issues of when, where, and how something—a style, a ritual, a technology—"originated." Was the source internal or external to the society being studied? If external, how did it spread? If internal, what were the conditions that led to its emergence? Such questions, which are perennial topics in the discipline, are easier to ask than to answer, especially in prehistoric contexts. Discovery of another site, another culture, another date, another technique to recover formerly invisible evidence, will set off another round of debates. Origin debates are also heavily influenced by subjective factors such as the theoretical trends of the times (see Chapter 1), the educational backgrounds of the debaters, and even nationalism (Trigger 2006:482). Discussions of origin and adoption of technologies have become more sophisticated with changes in the discipline of archaeology (see Chapter 4), but review of older frameworks is necessary as they still recur in much of the archaeological literature.

Conventionally, three basic mechanisms have lain at the core of origin discussions: migration, diffusion, and independent invention. These three mechanisms as important explanatory frameworks weave in and out of fashion within the discipline of archaeology depending on both available archaeological data as well as general trends in scholarship, as shown in Chapter 1 with the changing perceptions of prehistoric bronze in Southeast Asia.

MIGRATION

Migration is the mechanism inferred when new technologies appear in a particular area because groups of people who already use that technology have moved into the area from elsewhere. An unequivocal historical example would be appearance of ironworking in North America after migration of Europeans in the 16th century. Migration as the primary explanation for culture change over time was in vogue among 19th century antiquarians, culminating in the early 20th century with Grafton Eliot Smith's position that settlers from Egypt brought all advances in culture to all other parts of the world. The subsequent scholarly reaction against such extreme hyperdiffusion plus implications of incorporation of radiocarbon dating into interpretation of prehistory (reviewed in Chapter 1) sent the role of migrations in prehistory to the back burner or even the trash bin until recent times, when significant population movements have been brought back into discussions of technological expansions, for example for the spread of early cereal agriculture (Bellwood 2004).

DIFFUSION

Diffusion is inferred when knowledge and ideas behind a technology or other cultural trait spread from one area to another without a migration of people; one society borrows the technology or trait from another society. A good example would be the spread of the horse-drawn chariot throughout many cultures of Eurasia from its probable center of invention in the steppes (Anthony 2007:403). That diffusion without colonizing populations, also known as stimulus diffusion (Kroeber 1940), was the primary mechanism of culture and technological change was accepted by archaeologists for much of the 20th century and was the primary assumption behind the *Kulturkreis* school. Ideas spread from one culture coming into contact with another and via trade. Combining *Kulturkreis* diffusionary explanations with the concurrent interest in the origins of civilization laid the groundwork for a fundamental assumption of the Conventional Paradigm, namely that "higher" cultures, such as Mesopotamia and China, provided "lower" more peripheral cultures like Europe and Southeast Asia with technological and cultural advances (see review in Tringham 1974). This early version of a "core/periphery" viewpoint is also reflected in what Holl (2009) has called the "colonial template" for technological change.

INDEPENDENT INVENTION

Independent invention is, as it sounds, the invention in situ of a new process or technology, with no influence from other areas. The development of agriculture and metallurgy in the New World was, based on all current knowledge, completely independent from the initial development of those technologies in the Old World, for example. Current evidence indicates that there were no contacts between the Old and New Worlds in time ranges that could explain the transference of agricultural or metal technology from one continental system to the other.

When interest in independent invention reached its peak among western archaeologists during the 1970s, the discipline of archaeology was undergoing a period of far-reaching changes in both method and theory. The impact of not just the discovery of radiocarbon dating in the late 1940s, but subsequent application of tree-ring calibration to correcting radiocarbon dates beginning in the late 1960s, dramatically changed the traditional interpretive framework for Old World prehistory (Renfrew 1973). Debates about other possible instances of independent invention have emerged with the application of radiocarbon dating to the world's region's beginning in the 1950s.

The realization that culture did not simply move from one advanced spot on the globe to all other parts of the globe resulted in some prehistorians being receptive to other regions having autochthonous development of metallurgy, including in Southeast Asia. Although the Southeast Asian possibility was dropped by most regional prehistorians in the 1980s, the discussions surrounding the possibility of invention of metallurgy in Southeast Asia did contribute to much "rethinking" of the history of metals in the 1970s and 1980s (e.g., Muhly 1980:31–32, 1988; Needham 1980:512–514; Wertime 1973:876).

The key issue in current debates regarding polycentric invention concerns the beginnings not generally of metallurgy or of bronze per se, but of the smelting of copper. The extraction of copper from ore has been argued as the critical leap in knowledge and expertise fundamental to the eventual production of the highly useful alloy tin bronze and many subsequent metallurgical developments (Killick and Fenn 2012). As global prehistory has become better known, it does appear that the independent development of copper smelting as a consistently performed activity has occurred at least twice as noted above, once in the New World and once in the Old World.

Although some scholars of ancient metallurgy have remained passionately interested in the debate between mono- versus polycentric origins of metallurgy (e.g., Roberts et al. 2009; Ruiz-Taboada and Montero-Ruiz 1999), most are currently much less preoccupied with evolutionary sequences and independent invention generally than in the 1970s. The history of the development and dispersal of technologies including metallurgy is currently undergoing a new round of rethinking, Newer perspectives emphasize nuanced and intricate assessments beyond referencing migrants or spreads of ideas. Discussions focus more on "transmission" of technologies including investigation of how and why societies adopt innovations from other societies.

The Social Contexts of Early Metallurgy

A brief overview of what is now known about early metallurgy and its social contexts globally and outside of Southeast Asia will assist in setting the Ban Chiang metal remains and early metallurgy in Thailand into a broad perspective.

Early Native Metal-using Societies in the Americas

In the Eastern Woodlands of North America, copper working began more than 7000 years ago (Ehrhardt 2014:304). Among the most prolific native copper exploiters known are the hunter-gatherers of the Old Copper Complex in the Great Lakes area of North America from 4000 to 1000 B.C. (Ehrhardt 2009:221). These groups made a wide range of utilitarian and ornamental objects such as awls, beads, knives, and points from working naturally occurring metallic copper. Local native copper resources are abundant and include both geological deposits of copper and "float" copper (nodules in sedimentary contexts). The distribution of the artifacts in mortuary contexts suggests modest social differentiation in the exploiting societies, such as achieved status (Ehrhardt 2009:223). While extracting native copper from some types of deposits could be labor intensive, generally exploitation of accessible deposits like surface nodules is such that it is conceivable that one individual could undertake all the steps necessary to make simple tools and ornaments, from prospection to fabrication by hammering.

Subsequent North American cultural complexes, such as the Hopewell and Mississippian, also exploited native copper. Those societies produced elaborate refined ceremonial artworks from native copper that were widely traded and were used as prestige goods in copper-rich elite graves. Ritual regalia and personal ornaments were made including hammered and embossed headdresses, plates, masks, rattles, maces, and hair and ear ornaments (Ehrhardt 2014:319; Sampson and Esarey 1993). No evidence has been found that copper smelting emerged from these North American native copper-using societies.

In the Andes, small pieces of hammered native gold and copper beads and foils are occasionally found in Initial Period deposits of hunter-gatherers and unstratified mixed subsistence cultivators from the 2nd millennium B.C. The foils were associated with ceremonial centers and graves (Lechtman 2014). This sparse metallurgical evidence is considered an experimental phase before the Chavín period ca. 800–0 B.C., during which large hammered gold objects were made, rarely combined with silver and small amounts of copper, for sacred ceremonial purposes. Some items are considered intentional alloys, and welding and soldering were employed.

Early New World Copper Smelting Societies

Subsequently, the Moche (ca. A.D. 50–700), a state society, created metal and other material objects to be buried with the "ritual class." "The male lord-priests or priest-warriors buried at Sipán are literally covered with metal objects" (Lechtman 2014:375). Copper smelting began in this period, and ternary alloys of gold, silver, and copper were common. Utilitarian copper objects are also found in this time period, including needles, small knives, chisels, fishhooks, and tweezers recovered from burial and habitation contexts. Bronze was used during the subsequent Middle Horizon (A.D. 600–1000), with arsenical bronze or tin bronze used depending on the location and its proximity to local mineral resources. It is during this time frame that copper-base metallurgy appears in west Mexico, an area with rich copper deposits; both cold-worked and lost wax cast copper artifacts are found in non-state contexts. Sporadic interaction over hundreds of years by maritime peoples from Ecuador with developed indigenous village societies brought metal objects and metalworking knowledge from Andean metalworking traditions to Central America (Hosler 2014:333). Prior to the introduction of metalworking, the villagers polished copper ores for ornaments, and, presumably, that familiarity helped in the transmission of knowledge of how to extract copper from the ores. Lost wax cast bells and cold-worked needles, tweezers, and rings—items very similar to Andean counterparts—were subsequently made with local resources, strongly suggesting that Andean metalworkers trained west Mexicans in the required skill sets.

Early Old World Societies with Native Metals

In Old World contexts, worked native copper has been found comparatively rarely, initially in small early agrarian villages and agro-pastoral communities in 11th to 9th millennium B.C. in Iraq and Turkey (Roberts 2011:138–139; see also Craddock 1995:98; Stech 1999:60–61). The earliest evidence for annealed copper comes from eastern Turkey in the 9th millennium B.C. (Roberts 2011:139). From the 8th to 7th millennium, native copper is found in Mesopotamia (Stech 1999:66), on the Iranian Plateau (Pigott 1999b:73, 76), as far east as Pakistan, and as far west as Serbia. Many of the early agrarian sites with native copper have artifacts of copper ore as well (Stech 1999:61) showing simultaneous interest in and manipulation of both materials. Other possible locations of working native copper include the Sahel, Austrian Alps, and northern Eurasian forests (Craddock 1995:95). It is thought that fire setting was used to assist mining native copper, resulting in attaining the material in an annealed condition (Craddock 1995:99). Heating and hammering activities were probably used to acquire, shape, and harden copper from its earliest manipulation in the native state, whereas cold working is more commonly well attested in North America. Old World native copper working never reached the prevalence and elaboration that has been found in the New World.

Early Smelting Societies in the Old World

Evidence is accruing that the earliest smelting in the Old World may have occurred in polymetallic contexts. The earliest smelted metal currently documented is a lead bracelet from Yarim Tepe from the 7th millennium B.C. in eastern Turkey (Roberts 2011:139). The earliest smelted copper so far has been excavated at Belovode, Serbia, dating from the end of the 6th millennium B.C. (Roberts 2011:139; see also Bailey 2000) from a society undertaking mixed agrarian/hunting and gathering subsistence. Evidence for copper smelting dating from the early 5th millennium comes from villages in Iran (Pigott 1999b:77), and, it seems likely that comparably aged copper smelting will eventually be found in Iraq and eastern Turkey, where there is evidence of manipulation of native copper and copper ore by this time. Later in the 5th millennium, copper smelting is found in the Levant (Golden 2009) and central Europe (Roberts 2011:140) in "neolithic" contexts. This combination of neolithic societies smelting and working copper in some areas especially around the Mediterranean is traditionally termed the "Chalcolithic" or Copper Age. Current research indicates that polymetallic ores of copper with various other minerals including arsenic, antimony, silver, and others were often exploited and probably deliberately chosen to produce what some have called "natural alloys" (Thornton 2014). Lead appears in some areas as a contemporaneous early metal (e.g., southeast France [Roberts 2011:140]).

Copper smelting is also known to have been undertaken by some hunter-gatherers. Chernykh (1992:188) identified a clear example of 3rd millennium B.C. forest-dwelling hunter-fishers called the Karelian who exploited rich native copper deposits and smelted associated oxide ores. The Karelian culture northwest of Lake Onega and similar cultures east of the Urals appear to have developed copper smelting without outside stimulus (Chernykh et al. 2004:21).

Copper-producing societies were not all egalitarian, including some that used arsenical copper. "Chalcolithic" 4th millennium B.C. Palestine produced complex prestige/ritual objects (e.g., ornate, lost wax cast animals) of arsenical and antimonial copper and numerous small tools of smelted pure copper in a context that suggests that the arsenical copper goods were associated with the rise of an elite (Golden 2009). The 4th millennium B.C. copper-producing cultures of the eastern European steppe were mobile pastoralists, but their burials can be highly differentiated by grave goods (ceramics and ornaments, but not necessarily metal) and rare burials may have stone "scepters" and axe/hammers that suggest recognition of leadership (Chernykh 1992:44). Differential wealth and an extensive metal-driven trade network of the Carpatho-Balkan area during the 5th to 4th millennia B.C. suggest that copper production and long-distance metals trade with oreless areas could have facilitated considerable social differentiation (Chernykh 1992:50).

A focus on copper production is found in some regions when other regions already favored bronze. Most notably, even though the 3rd millennium B.C. Indus civilization is often discussed as a "Bronze Age culture" (e.g., Babu 2003), "Copper Age" may have been closer to the metallurgical reality. Some analyses indicate that the Harappan copper-base artifacts are predominantly composed of unalloyed copper (Agrawal 2000:72; Kenoyer and Miller 1999:114, 118). Of 129 copper-base objects tabulated by Kenoyer and Miller in their review (1999:114, 137–143), only 36 (28%) have 1% or more tin, and only 20 (16%) have 1% or more arsenic. Kenoyer and Miller's attribution of the Mature Harappan to the "Chalcolithic/Bronze Age" (1999:107) illustrates the problem of attributing societies to particular metallic "Ages" when the metals indicate one "stage" of development but the social context a different "stage."

Hoffman and Miller (2009) have since proposed that copper alloys may have been more common than previously thought in the Indus Valley, but the irregular compositions of the Harappan copper-base metals likely reflect their practice of importing copper ingots from neighboring copper-producing regions and fabricating metal finished products in decentralized networks of independent craftsmen or dispersed workshops (Kenoyer and Miller 1999; Wright 1996:128). Despite a highly differentiated urban social context, the craft of metalworking in Harappan contexts does not show clear evidence of incorporation into an administered or elite-dominated political economy.

One of the sources suggested for Harappan metal was the copper-smelting Ganeshwar-Jodhpura culture in Rajasthan, India. This society practiced a hunting-gathering subsistence strategy (Agrawal 2000:109; Rizvi 2007) perhaps including shifting cultivation (Kenoyer and Miller 1999:135). Their sites are extraordinarily rich in copper artifacts, principally arrowheads cut from sheet copper, and also fish hooks, bangles, chisels, and cast celts made from smelted local ores, which were abundant. Although this northwest Rajasthan culture has been proposed as a major source for the copper used in pre-Harappan and mature Harappan sites (Agrawal 2000:108), recent lead isotope studies indicate that the region was only a minor contributor of copper to Harappan cities (Hoffman and Miller 2009). Thus, it now appears that most of the copper product of the Ganeswar-Jodhpura cultural complex was for internal use, and the purpose of the extremely high number of arrowheads is considered to be for hunting (Rizvi 2007:91).

Summary of Early Copper Exploitation

In summary, societies falling within a logical definition of the Copper Age—"archaeological cultures…which produced tools and weapons from metallurgically 'pure' copper or alloys of accidental origin" (Chernykh 1992:15)—include most of the gamut of social "complexity": hunter-gatherer-fishers, early agriculturalists, mobile pastoralists, societies with incipient forms of social stratification, and those of considerable regional extent, with large urban settlements, extensive trade relations, and highly differentiated societies. For the simpler smaller-scale societies, proximity to accessible rich metal resources was a critical variable in exploitation of copper. The hunter-gatherer societies that did exploit copper had additional important attributes. The Archaic native copper-exploiting groups near Lake Superior, the Karelian hunter-fishers exploiting native and smelted copper, and the copper-smelting Rajasthan groups were hunter-gatherers who were relatively settled, could have ceramics, intimately exploited their region's natural resources (i.e., logistically rather than residentially mobile [Binford 1980]), and in some cases also practiced some plant cultivation. Participation in extensive exchange networks could characterize the simplest (Archaic and Ganeshwar-Jodhpura cultures) as well as the more complex copper-using societies.

At the other end of the social complexity spectrum, the Harappan copper-base artifact repertoire, while diverse in forms, technological treatments, and uses (Kenoyer and Miller 1999), is notable for the absence of huge and/or "regal" objects comparable in scale, casting difficulty, and corporate symbology (Hoffman and Miller 2009) to, for example, Shang bronze ritual vessels or Egyptian bronzes. This observation, coupled with the low proportion of deliberate alloys, especially tin bronze, known from Harappan contexts, and the absence of indications of state control over Harappan metal production, illustrates that copper metallurgy can be developed in complex societies outside of elaboration of managerial efforts and prestige good economies. Without managerial development, however, the product range may lack objects at the largest and most elaborate end of the spectrum. Even if future analyses continue to show that copper dominated Harappan metal assemblages, the wide range of metals and metal processing techniques present from Indus sites shows that metallurgy was not an incidental part of their technological repertoire. Its overall character reflects considerable and diverse regional economic interactions even if the emphasis was on unalloyed copper and even if metal production was not administered by a state or elites (Wright 1996). One wonders if the relative rarity of tin in the Indus tradition copper-base corpus, despite the availability of tin in Afghanistan roughly 500 kilometers northwest of the Indus River (Kenoyer and Miller 1999:108, 118), was due to low demand, and that the low consumer interest in tin bronze may have played a role in the Harappan copper-base artifact repertoire. One can imagine that managerial consolidation, which was apparently minimal and/or absent in Harappan contexts at least in regard to metallurgy, may have been needed both to acquire large, regular supplies of tin from this distance as well as to cast huge complex objects. Current knowledge shows that societies of various scales, complexity, and subsistence practices produced only or predominantly unalloyed copper. Moreover, the complexity range of copper-producing societies substantially overlaps the complexity range of bronze-producing societies.

Early Alloying Societies

As noted, the activities involved in smelting metal ores, producing alloys, and manufacturing metal artifacts require role differentiation and coordinated multi-step interactions among individuals and often among more than one community in the case of producing alloys whose ores likely come from more than one place. Hence, bronze production and artifact manufacture entail some degree of social differentiation, coordination, and integration, even if full-time specialization, overt social hierarchies, or centralized economies are not necessarily present.

The picture of the beginning of alloying is now known to be much messier than the conventional metallic sequence proposed by Wertime (1973:880; see also Wertime 1964; Tylecote 1992:xi) of unalloyed native copper, to smelted pure copper, to unintentionally alloyed arsenical copper, to deliberately alloyed tin bronze. Wertime (1973), Tylecote (1992), and others long regarded arsenical copper as not an intentional alloy, but an accidental one resulting from exploiting particular arsenic-rich zones of some copper ore deposits. The arsenical copper product had a variety of functionally desirable properties, particularly a capacity to harden when worked (hammered) that was supposed to have laid the groundwork for so-called true alloys (and hence, true metallurgy, according to some), namely the very desirable tin-plus-copper, true bronze.

However, recent studies (e.g., Thornton 2009a) have highlighted that a variety of alloys from the mixing of polymetallic ores produced copper-base metals with varied compositions in the 4th millennium B.C., not just arsenic, but also antimony, lead, tin, and others. It appears that a period of experimentation with many ores and alloys occurred before tin bronze emerged as the alloy of choice in some areas. Iran was one area of active metallurgical innovation during the 4th millennium B.C., according to Thornton (2009a, 2014), where diverse copper alloys were produced from what he argued were intentionally mixed ores. Similar experimentation in alloy production was also occurring in Anatolia and the Caucasus (Courcier 2007; Yener 2000). The societies engaging in these innovative activities were culturally diverse communities occupying highland zones, formerly thought to be peripheral to lowland centers like Mesopotamia. In short, it appears that the development of early alloys occurred in dynamically interacting but uncentralized sociopolitical contexts outside of the control of early urban centers and states.

Arsenical coppers became somewhat common in some but not all parts of the Old World (Tylecote 1992) before tin bronze became the alloy of choice after 3000 B.C. The focus on the alloy tin bronze only gradually consolidated during the 3rd millennium B.C., although it occurred sporadically in previous millennia. Tin appears in low percentages in copper-base assemblages in Anatolia and Mesopotamia in the 3rd millennium B.C. (Pigott 2011). Southwest Iran tin bronze appears in the early 3rd millennium B.C. and becomes common ca. 2500 B.C. (Thornton 2009a:309).

The degree and nature of social differentiation present in bronze-producing societies and the relationship of social stratification to bronze production are, as in copper-producing societies, quite variable. Although the bronze-producing Seima-Turbino group appears to be mobile hunter-fishers, most of the simplest bronze-producing societies relied on domesticated foods (including pastoralists). These societies did have social differentiation at least minimally evident in grave assemblages; examples include 3rd and 2nd millennium B.C. cultures in northwest China and the Eurasian steppes (Chernykh 1992; Mei 2003) and 2nd millennium B.C. cultures in Iberia (Rovira 2002). The most complex bronze-producing societies are overtly hierarchical states with state-controlled economies including control of metallurgy (e.g., Shang China).

The technological difference in smelting and alloying copper and tin from smelting copper alone would not seem to be significant enough to explain the somewhat differing (though substantially overlapping) ranges of social complexity of copper as opposed to bronze-producing societies. Possibly some more marked amount of social differentiation was necessary for the *social* coordination across communities needed to acquire and process two separate metal ores and combine the metals in regularized procedures, especially if the ores were distant from each other.

Bronze-producing archaic states like dynastic Egypt and Shang China illustrate what markedly hierarchical societies can accomplish in terms of the volume of bronze produced, the typological range and scale of bronze artifacts, and the potential intricacy of artifacts when elites are the primary consumers and control the production and distribution of the metal. But centralized control over metal production is not always present or exclusive for urbanized societies, as already noted for Harappan contexts (Kenoyer and Miller 1999; Wright 1998) and Mesopotamia (Zettler 1996). Moreover, complex societies without administrative bureaucracies and with fewer levels of hierarchy than states (often called chiefdoms) can also produce significant amounts and types of elaborate bronze artifacts (e.g., 2nd millennium B.C. transalpine Europe, and 4th century B.C. Dian civilization in Yunnan, China).

Some societies with modest and sometimes minimal levels of social hierarchy and differentiation are known to have maintained regular production of bronze and bronze artifacts. Transhumant pastoral societies could be significant producers of bronze, such as Okunievo, Afanasievo, and Seima-Turbino Eurasian steppe-forest societies in the Altai in southern Siberia, the latter dating from the late 3rd millennium B.C. (Chernykh et al. 2004). The Seima-Turbino societies produced a distinctive array of bronze as well as copper artifacts, including decorated socketed adze/axes, tanged points, fishhooks, bangles, and awls (Chernykh 1992:184).

When tin and copper ores are plentiful and not distant from each other or the society exploiting them, as was the case, for example, in 2nd millennium B.C. Spain (Rovira 2002), the need for social hierarchy to conduct the exploitation may be relaxed. If bronze production influenced social differentiation in such cases, it appears to have been muted. The Southeast Asian data in this volume also contribute to the discussion on the relationship of social differentiation and early tin bronze metallurgy in a context of abundant regional natural resources (Stech and Maddin 1988:173).

Early Iron-using Societies

Iron in both its meteoritic and smelted forms has appeared sporadically in archaeological deposits in Mesopotamia, Iran, Egypt, Anatolia, and other locations in the eastern Mediterranean, with the earliest example dated to 5000 B.C. (Waldbaum 1980). Eurasian steppe metalworkers experimented with iron in the Early and Middle Bronze Ages, according to Anthony (2007:336). He noted, as one example, a non-meteoritic iron knife with an arsenical bronze handle dating to 2500 B.C. from eastern

Ukraine. Early smelted iron was presumably done by the bloomery method. When iron is recovered from copper and bronze age contexts, it is apparent that it was treated as a rare but valued material; it was usually associated with hoards, temple grounds, graves, and, even when shaped as spear points or swords, is considered to have been ceremonial, not utilitarian, in purpose. The conventional idea that Hittites held the secret of iron production is outdated, as contemporaneous societies outside of Anatolia also yielded comparable iron artifacts. "Use of iron in Hittite Anatolia, though somewhat better documented than elsewhere, does not differ markedly in kind from that known in other regions during the Bronze Age" (Waldbaum 1980:81). It was not until the 10th century B.C., that, in some areas like Palestine, Cyprus, and Greece, iron implements began to outnumber those of bronze, and by the end of that century iron was in common use for both weapons and agricultural tools in the eastern Mediterranean. In short, iron was present long before it became the prevalent metal.

In central Europe, the occasional iron object has been recovered from 2nd millennium B.C. contexts, and most of these are considered imports from the Mediterranean region (Pleiner 1980:376). Not until the 8th century B.C. was there a significant occurrence of iron weapons in the Balkans, and from that time central and western Europe, though unevenly and gradually, saw iron increasingly present in some locations, even while bronze continued to be used for weapons and tools. Pleiner (1980:384) stated that the "true Iron Age" only began with Hallstatt C–D when bronze almost disappeared. Nonetheless iron was not a cheap or easily acquired material, as conventionally has been assumed; it was still recovered primarily from rich graves. In some cases, the technology and typology of the early iron artifacts suggest the sources of traders thought to have introduced iron metalworking to particular locations, such as Greek and Phoenician traders to the coasts of southern Spain and France.

In South Asia, outside of the greater Indus region, there seems to have been little significant copper alloy metalworking preceding the appearance of ironworking. Iron first appears in the early 2nd millennium B.C. in a few places of north central India just south of the Himalayan Range, including Uttar Pradesh (Tripathi 2015). Iron then was unevenly and slowly adopted, sometimes in a "Chalcolithic cultural milieu" (Tripathi 2015:98), across the region. As early as ca. 1000 B.C. iron was present in southern India (Gullapalli 2009:441). A source for the transmission of the technology is not clear, and arguments for indigenous development have been proffered (Gullapalli 2009:441). In some parts of India, early ironworking is associated with megalithic monuments. Products in various parts of India are primarily implements, and a florescence of the technology took place during the 1st millennium B.C., centuries after its initial appearance in the region. Iron technology is considered not to have had any rapid impact on South Asian society or economy.

In the territory of the PRC, iron appeared in small amounts before the "Iron Age proper" (Taylor and Shell 1988:206), and a changeover from bronze to iron implements took place over centuries. The earliest iron objects (some bimetallic) in central plains China have been excavated in southern Shanxi, dating to the end of the 9th century B.C. Three of these iron artifacts are meteoritic and others appear to be terrestrial iron (Han and Duan 2009:101). In Xinjiang and Gansu, iron was present probably by the 9th century B.C. and possibly in the 10th century (Gilmour 2009:137; Wagner 1999; Wu 2009). Another early iron object from the central plains of China is a sword buried with the king of Guo dated to the early 8th century B.C. (Wu 2009:113). Wu (2009:113) suggested that, "The people living in oases along what was to become the Silk Road or in the steppe region may have played an important role in this rapid transmission of iron technology."

The date for the appearance of cast iron is treated variably in the literature: Wagner (1999) said that casting of iron was an invention in southern PRC by the Chu or Wu, possibly in the lower reaches of the Yangtse Basin, by the 4th century B.C.; Gilmour (2009:137) stated it appeared in the 6th century B.C.; Han and Duan (2009:101) reported cast iron dated to the 8th century B.C. at Tianma in Shanxi. Cast iron came to be extensively used for agricultural implements, and the motivation behind the development of cast iron as a useful technology may have been interest in a metal cheaper than bronze in a region where copper and tin ores are sparse. The large-scale copper-smelting furnaces of the time could have produced castable iron metal byproducts, which could have provided the basis for further innovation to make it useable in several categories of object. It may be that while wrought iron could be produced and its technology transmitted via decentralized metalworkers, cast iron is more clearly associated with large installations, likely to be under the control of states with the social capacity to organize the large labor pool needed to effectively run the ironworks. On the other hand, both meteoritic and smelted early wrought iron in the central plain were found in contexts suggestive of a ceremonial role for the pieces.

Some Observations on Early Metal

From the above overview of the contexts of early development of metal technologies, three general points initially stand out. First, the adoption of a metal technology usually took centuries and even millennia, irrespective of location or metal, with the new metal often appearing only sporadically. Second, societies adopting metal technologies varied greatly in their social configuration. Third, societies adapted metal technologies in their own idiosyncratic ways to suit their own variable needs. One cannot generalize that native copper-using societies were characteristically simple, egalitarian or always stressed utilitarian or trinket uses (the Mississippian culture providing a counter example [Ehrhardt 2014]), or that complex societies always emphasized particular alloys or prestige uses or used metals to "control" anyone or anything (the Harappan civilization providing a counter example to the Conventional Paradigm [Hoffman and Miller 2014]).

Modeling Adoption

> *In order to model metallurgical adoption, the assumption of inevitability must first be discarded. (Roberts 2011:144)*

The terms "transmission" and "adoption" are replacing the now obsolete oversimplified terms "migration" and "diffusion" in discussions concerning how a technology such as metallurgy moves from one place to another. The newer terms are regarded as enabling more nuanced elucidations of the often ambiguous and multi-factored processes that were involved in movement of technologies among societies. Unless there is clear evidence that a substantial migrating metal-producing population rapidly settled uninhabited territory or abruptly displaced extant non-metal using inhabitants, archaeologists should not expect a uniform horizon of the earliest metal evidence in a region. Current metallurgical evidence discussed above shows that archaeologists should expect uneven occurrence of metal in a region even for hundreds of years before that metal becomes a common material. Therefore, lines of reasoning that if metal was present in a region it would be at particular sites (e.g., Higham et al. 2011a:244), and further concluding that if metal was absent at particular sites in a time frame, the technology must have been absent from the region (e.g., Higham and Higham 2009b), are inconsistent with the known uneven pattern seen globally of early adoption of metals (e.g., Roberts 2011). Assertions that "It has been possible to trace the expansion of copper-base casting across the steppes into China, and thence *inexorably* southward to Southeast Asia" (emphasis added, Higham 2014:136) reflect the outdated Conventional Paradigm.

In addition to the uneven presence of the earliest appearances of metal artifacts in a region, the likely ephemeral nature of the earliest direct evidence of metalworking in terms of identifiable hearth installations (Heeb and Ottaway 2014:170) or slag (Killick 2014:38) should be noted. "[T]he evidence for the earliest smelting processes...did not regularly produce large quantities of durable debris.... This situation had resulted in a long gap of several thousand years between the first metal artefacts and the first evidence for extractive metallurgy." (Craddock 2007:ix). For scholars interested in the initial *transmission* of metal technology into a region, it is clear that rare early appearances of metal are *essential* evidence to understanding the process. Occasional early occurrences are never irrelevant or peripheral. Moreover, it may take centuries before metal is incorporated into funerary grave goods. Kristiansen (1991:29) noted for northern Europe, "During the last part of the Neolithic [late Neolithic here dated about 2300–1700 B.C.], imported and locally produced bronze axes were numerous and gradually took over the role as prestige items from the flint dagger. The axes were not yet employed as grave goods." Bronze implements appear as grave goods apparently towards the end of the early Bronze Age in this region.

It is clear that to construct an argument defining the first appearance of bronze technology in a region based on a protocol of "valid Bronze Age contexts" (Higham et al. 2011a:230) consisting of "only copper-base artefacts that occur as primary and undisturbed mortuary offerings or inclusions (such as a bangle in place on a wrist), or the identification of an in situ mining, smelting or casting activity area" [Higham et al. 2011a:235]) will undoubtedly exclude the evidence most pertinent to the initial transmission of metallurgy into the region (see Radivojević and Rehren 2016). Such criteria might, however, be suitable for discussions of when metallurgy assumed a particular funerary role, or perhaps when the technology had reached the "take off" point in a local adoption process.

In most cases, initial evidence in a region for the presence of metallurgy is likely to be sporadic, small-scale, and deposited over a considerable period of time (consistent with slow adoption processes when transmission is by traders, moving metalworkers, and down-the-line exchange). Archaeologists need to differentiate the early adopter and take off points, realizing the evidence for different stages of the adoption process could vary substantially. The implications of these phases of metal adoption for "Age" terminology are discussed later in this chapter.

Independent Development

> *The utility of an emphasis on determining antiquity (and therefore "origins") has been exhausted. (Gullapalli 2009:455)*

Among current historians of technology, invention and discovery are less consequential than the development of the technological idea into an application that is sustained over time; i.e., the emergence of technological traditions is what counts (Schiffer 2011:13). Just as Leonardo da Vinci conceived of helicopters hundreds of years before a viable one was built, it seems likely that smelting ore to produce metal was discovered numerous times, in multiple places leading to no sustained implementation. What is of most interest to current scholars studying the emergence of early metallurgy is not the first invention of smelting, but rather the examples where smelting metal was practiced and applied in a sustained manner to produce products, i.e., examples from which technological traditions emerged.

As noted above, there are at least two examples of independent development of copper smelting technology. In the New World, copper-smelting technology emerged in the Andes during the latter half of the 2nd millennium B.C. (Lechtman 1980) and is proposed to have spread from this origin to some other parts of south and central America (Hosler 2009; Lechtman 1988:344). In the Old World, the case for the first sustained endeavor of copper smelting is more ambiguous as is the possibility that more than one location independently developed a copper smelting regimen.

Renfrew (1973:167–191) first made the case that even though the earliest evidence for copper exploitation (native copper and ores) occurs in the Near East (i.e., between the 11th and 9th millennia B.C. in Iraq and Israel [Roberts 2011:139]), the early metallurgical evidence in the Balkans provided evidence of a separate independent development of copper smelting. The region, Renfrew argued, shows a gradual technological development from working locally abundant native copper to melting copper and then smelting ore culminating in a flourishing tradition at a time (5th millennium B.C.) when, according to Renfrew, there is no evidence for interaction of that region with the Near East. Renfrew also showed that the Balkan cultures had other evidence of pyrotechnical skills, regional exchange, and social elaboration that could support production and demand for the material.

More data and new overviews of the Balkan evidence (Kienlin 2014; Radivojević et al. 2010; Radivojević and Rehren 2016) are available. It is now recognized that "there seems to have been occasional contact between the Balkans and northwestern Anatolia between 6200 and 5500 BC" (Anthony 2010:34), so Renfrew's original assessment of isolation of the two regions is not currently supported. Nevertheless, the earliest date for confirmed copper smelting in the Old World is in Serbia at Belovode at the end of the 6th millennium B.C. (Radivojević et al. 2010; see also Kienlin 2014:450; Roberts 2011:137). The earliest Near Eastern example is slightly later in the early 5th millennium at Tal-i Iblis in southeastern Iran (Roberts 2011:140; Roberts et al. 2009:1014).

The Balkans example has contributed to reviving the debate on polycentric origins for the development of smelting (see also Killick and Fenn 2012), and the region remains a strong candidate for an Old World location for indigenous development of copper smelting outside of the Near East proper (Radivojević et al. 2010; Radivojević and Rehren 2016). Proponents of a single Old World origin of copper smelting propose that earlier evidence may ultimately be identified in Anatolia (Roberts et al. 2009:1014).

At a minimum, the Balkan evidence provides a model for what a case of independent development might look like. An initial experimental/exploratory phase was comprised of a gradual sequence of interaction with and manipulation of native copper and copper ores in a region of rich copper resources. Small trinkets and tools such as beads, awls, fishhooks, and rings were fashioned by grinding, hammering, and annealing native metals prior to evidence for mining and later smelting. Other elements present in the culture include an interest in pyrotechnology, socioeconomic differentiation, and regional networking sufficient to provide the possibility of consumer demand, all before the appearance of smelting and casting of artifacts from the smelted copper. Once smelting and casting were adopted, larger tools like shaft-hole implements and more copper artifacts in general appeared (Kienlin 2014:451).

There are (so far) a few other candidate areas with possible but weaker claims for indigenous development of copper smelting that share some similarities to the Balkan example, such as southeast Spain (Ruiz-Taboada and Montero-Ruiz 1999), the Sahel (Craddock 1995:123), and Karelia (Chernykh 1992:188). These proposed areas appear to have been geographically and culturally isolated from contemporaneous copper smelters, were close to abundant local sources of copper, and development of copper smelting was preceded by a period of working and perhaps melting native copper. However, these proposed areas for independent development of metal smelting have been challenged on a variety of grounds including chronology (Roberts 2011:140; Roberts et al. 2009). Yet as archaeology continues to expand into the less explored areas of the world, particularly in the vast central parts of Asia, new potential examples for independent development of copper smelting may emerge in areas that have rich copper resources.

One can hypothesize that smelting might emerge in regions with evidence for an initial intrigue with unusual ores. If other conditions are present, such as advanced pyrotechnology, smelting technology might develop from ceramic technologies involving some manipulation of atmosphere and capability to produce temperatures in excess of simple cooking hearths (though see Craddock [2000] for a rejection of the connection between early ceramic technology and the origins of metallurgy). Nevertheless, the most readily appreciated cases for independent development of copper smelting would likely be supported by archaeological evidence for exploratory/experimental stages of metallurgy prior to regularized copper smelting—including working with native copper and other native metals, evidence of local mining for minerals, use of ores for items such as trinkets (malachite and azurite, are even today used for ornaments)—and contexts of sufficient age and cultural isolation that a source from another copper-smelting region is unlikely.

So far, despite its proximity to rich ore sources, Thailand (or even Southeast Asia) does not have any evidence for the kinds of experimental or exploratory behaviors that have been found prior to the emergence of copper-base metallurgy in places like the Balkans, Turkey, or Iran. Although the possibility of an independent invention of metallurgy in the region is still occasionally raised by a few scholars (e.g., Amzallag 2009; Higham 2002:116–117; Higham and Thosarat 2012:264), the region has not been regarded a candidate for independent invention for copper smelting by most regional archaeologists since the late 1980s (see White 1988:179). Until a technological sequence demonstrating pre-metallurgical experimental and exploratory behaviors comparable to regions like the Balkans is found, discussion that Thailand or Southeast Asia was a potential area for independent development of metallurgy should be dropped.

Alloys

Detailed discussion of the topic of the development of alloys, while very important in the history of metallurgy, is beyond the scope of this volume. Suffice it to note that once a society smelted copper, experimentation with different ores, with combining different ores in a single smelt, and with additives such as fluxes would result in products with differing properties. Some trace elements in ores, such as arsenic and antimony, if present in sufficient quantities, could produce alloys—even if somewhat unintended—with characteristics, such as increased hardness or improved castability, which may have been desirable depending on context and become sought after. Where to draw the line on what constitutes an alloy, an accidental alloy, and a deliberate alloy is rather variable in the metallurgical literature (e.g., Kenoyer and Miller 1999:114–115; Stech 1999:62).

An argument has recently been made (Thornton 2009a) that arsenical copper, long considered an accidental alloy, or a copper with a naturally-occurring impurity, might in some cases be better considered a deliberate selection and choice, more like a "real" alloy, especially when polymetallic or distinct copper ores were mixed. In the general history of Old World copper metallurgy, exploitation of arsenical copper often (e.g., in parts of Italy [Dolfini 2014:478]), but not always (e.g., possibly southern Siberia [Chernykh 1992:185]) preceded use of tin-copper alloys, especially west of the Urals (Chernykh 1992:14). East of the Urals, for example in the PRC and Thailand, arsenical copper-base objects appear occasionally as one of many alloys including tin bronze in the early assemblages except where sites were close to arsenical copper ores, in which case they occur frequently (Linduff 2000:16).

Tin bronze is usually considered a deliberate alloy, especially when tin is present in excess of 2% (Stech 1999:62), as tin and copper ores only rarely co-occur in the same geological deposit. Regular production of tin bronzes with tin in the 7–10% range in the Old World dates to about 3000 B.C. in Mesopotamia, but low-tin bronzes occur in a variety of early so-called Bronze Age assemblages (Tylecote 1992:18–19). In many Old World contexts, the source of the tin is problematic, and some have suggested that the florescence of tin bronzes in Mesopotamian centers reflected the capability of a stratified society to organize long-distance trade to get the tin, even from as far as Malaya (e.g., Muhly 1980:31; Selimkhanov 1984; Tylecote 1992:18). More recent work is finding tin closer to Mesopotamia, in locations such as the Taurus Mountains in Turkey and the Zagros Mountains (Nezafati et al. 2006; Pigott 2011).

Assimilating the implications of the ever-mounting exceptions to the "Standard View" of the development of metal technology into theories of archaeometallurgical change requires reassessment of fundamental ontologies of the discipline. The Standard View of metallurgical development coincides well with evolutionary theories of change of human societies prevalent in one form or another for more than a century. The resurgence of cultural "neo-evolutionary" theory in the second half of the 20th century, and elaboration of that theory beyond its 19th century predecessors, still influences much archaeological interpretation to the present day. Universalist stages, be they evolutionary, technological, or a combination, meshed well with archaeological goals in the 1970s to define laws of progressive change, be that a change from a simple to a complex society or from arsenical to tin bronze. While the exploration of the various permutations of cultural and technological evolutionary theories have stimulated much archaeological research for most of the past century, including much research into past metal technological systems, the limitations of these frameworks for understanding the full range of human socio-technical variability have become increasingly evident (e.g., Blanton et al. 1996; Schiffer 2011). The stage paradigm (Kuhn 1962) may have run its course in archaeology and archaeometallurgy.

Moving beyond "Origins of" Debates

> *The trends in global early metallurgy [are] away from concentrating on "origins" debates and towards the identification of mechanisms underlying technological transmission; the processes of adoption and adaptation; cross-craftsmanship and cross-material connections; and the role of metal in social dynamics... (Roberts 2011:137)*

Much of modern archaeometallurgy is following in the footsteps of archaeology generally by moving its focus away from grand narratives, notwithstanding that grand narratives still engender lively debate (cf. Amzallag 2009, 2011; Thornton et al. 2010; also Higham et al. 2011a; White and Hamilton 2009; cf. White 2017). The history of archaeology and especially of technology shows that grand "Origins of" statements (e.g., Higham et al. 2011a; Hua Jueming

2000; Li Xiandeng 2000; Pigott and Ciarla 2007; Rapp 1988) are doomed to short shelf lives as the ever-increasing archaeological record inevitably provides new data and reconfigured nuances. Superseding that 20th century grand narrative focus is an increasing interest in fine-grained studies of why individual societies chose to adopt metal or any other technology, how they adopted and adapted it, and how the technology was transmitted (see Roberts 2011; White 2017; see also Stark 2015).

The current generation of archaeometallurgists is calling for new models and new paradigms (e.g., Golden 2009; Gullapalli 2009; Hanks and Doonan 2009; Holl 2009; O'Brien 2015; Roberts 2011; Thornton 2009b, 2012). An outline for the emerging New Paradigm for archaeometallurgy and prehistoric technology is presented in Chapter 4.

However, before the new trends in archaeometallurgy thinking can be fully appreciated and applied to the prehistoric evidence in Thailand, it is helpful to briefly review the flaws and limitations in logic and lines of reasoning of the Conventional Paradigm.

Ontological Conundrums with Evolutionary Paradigms

ontology...a model for describing the world that consists of a set of types, properties, and relationship types... (Garshol 2004:384)

The typology band/tribe/chiefdom/state is too crude to be of much use in studying social evolution. (Rousseau 2006:28)

in parts of the world where the [Three Age System] model is still in common use, it needs to be accepted that...there never actually was such a thing as "the Bronze Age." (Connah 2010:63)

The Conventional Paradigm entwined social and technological stages together into a progressive sequence that has provided until recently an essentialized standard frame of reference for many Old World archaeologists. Yet the review of early metallurgy above demonstrates that extant data do not match the standard framework in much of the world. The Rousseau quotation above articulates a current archaeological consensus that social evolutionary typologies of progressive sequences of social types are no longer viewed as the fundamental framework for studying variation and change in past societies. The Three Age terminology of the Conventional Paradigm still hangs on in many regions, including Southeast Asia, despite its acknowledged flaws. As Connah (2010:63) has noted, "the three-age system is now so deeply entrenched in the public thinking that some archaeological writers have been trapped in an anachronistic mindset." An example in Southeast Asia is seen in the Higham (2014:15) statement: "In my view, the sequence since 2000 B.C. fits neatly into the so-called "Three Age System", which recognizes the important stimuli to cultural change resulting from the adoption of agriculture, of bronze casting and of iron technology."

Although the ontologies of 20th century archaeology are being questioned (e.g., Alberti et al. 2011), the vocabulary, assumptions, priorities, and lines of reasoning from progressive evolutionary paradigms, whether technological or social, permeate the archaeological literature of recent decades, including recent publications concerning Southeast Asian prehistory (e.g., Higham 2015b; Higham and Rispoli 2014; Higham et al. 2011a; O'Reilly 2007a).

In order to shift to new paradigms, problematic limitations in the anachronistic though still prevalent progressive evolutionary paradigm need to be summarized so that Southeast Asian prehistoric data including the data presented in this volume can be placed into a more contemporary perspective.

Pitfalls in Reasoning within Essentialized Progressive Paradigms

many approaches [to the topic of Spain's metal age social development] suffer from overly-ambitious readings of the archaeological record in order to validate predetermined models. (Rovira 2002:10)

A large literature has arisen over the past thirty years that has parsed and critiqued processualism and particularly Service's and other social evolutionary typologies and progressive interpretive frameworks (e.g., Chapman 2003; Earle 1991a; Feinman and Marcus 1998; Feinman and Neitzel 1984; Upham

1987). Nonetheless, archaeologists rely on typologies, be they social or material, as typologies remain practical steps to address the problem of at least initially organizing large data sets in many of our activities (Gnecco and Langebaek 2014). Problems arise when the types are treated as real and final categories having fixed essences, as opposed to being merely heuristic—a convenient device with no inherent reality. Typologies, whether they are of ceramics, societies, or "Ages," are ultimately an artifice to suit the needs of archaeologists. When archaeological needs change and a particular typology is recognized as a roadblock to understanding, that typology should be redefined or abandoned.

ESSENTIALISM: A CORE PROBLEM

A core critique of the processualist period in archaeology, and of other 20th-century social science disciplines including anthropology, is that their ontologies are based on essentialist thinking (e.g., O'Brien and Lyman 2002; Stark 2015:71; Sylvain 2014; White 2017). Essentialist reasoning argues that something is true because it is essential to the category, i.e., the type. Essentialist reasoning structures an argument overtly or covertly on the basis of types, which assume more reality and importance than the data themselves. The problem of typological approaches to social phenomena goes much deeper than merely recognizing the ontological issue that types are not real. Archaeological interpretations based on progressive typological paradigms, including those dealing with metallurgy, tend to be accompanied by structures of reasoning that can profoundly distort perceptions and interpretations of the past, even creating fallacious observations.

When a diverse group is subsumed into a type, one or a few variants tend to be treated as an "archetype" or idealized model against which other members of the group are compared. This *normative fallacy* can be recognized in statements like "the Indus did not have a fully developed Bronze Age" (Tylecote 1992:30), or that the Bronze Age in Southeast Asia was "anomalous" (Muhly 1988:16). Such statements assume that another Bronze Age is the universally recognized true or correct objective standard—the norm for what a "fully developed Bronze Age" is—against which other examples can be measured. Even though normative reasoning has many scholarly purposes, scholars need to keep in mind that their choosing of a norm is a subjective, not objective activity. There is no inherent reason why the type example for the Bronze Age could not be Ban Chiang, Qijia, or Ireland, and then the Shang Dynasty would be an "anomalous" variant.

Flawed reasoning develops further when the boundaries between types or categories such as stages are treated as necessarily meaningful, important, and definitive; other potential boundaries or thresholds not highlighted by the typological system are ignored or inappropriately minimized (see Feinman and Neitzel 1984). Arguments based on *boundary fallacies* may assert a conclusion based on an asserted boundary between adjacent reified types. This line of reasoning is illustrated in the following statement: "Neolithic settlements in Southern China and Southeast Asia…*by definition*…predate the establishment of bronze metallurgy in their respective areas" (Higham et al. 2011a:241; emphasis added).

Archaeologists now know there is a blurry line in most parts of the world between pre-metal purportedly "Neolithic" societies and the societies where initial appearance of copper-base metallurgy including even bronze occurs. In Europe, for example, copper-base artifacts commonly first appear in "Stone Age" contexts (e.g., Janzon 1988; Kienlin 2014; LoSchiavo 1988). European archaeologists have come up with terms like "late Neolithic," eneolithic, chalcolithic, and copper age to try to handle the first several centuries of metal use during the "Neolithic" (e.g., Keinlin 2014; LoSchiavo 1988). A different blurry line is revealed by Kienlin (2014:453–454) with regard to the "Early Bronze Age" in several parts of Europe, "Typically it is culturally, rather than metallurgically, defined as *its earliest stage refers to groups that did not yet use tin bronze*" (emphasis added).

East of the Urals where there is less evidence for a sustained phase of unalloyed copper use, copper-base alloys including tin bronze initially appear in cultures whose life-style was essentially "Neolithic," i.e., uncentralized kin-based societies of pastoralists and/or subsistence farmers (Chernykh 1992; Linduff 2000). Linduff and Mei (2014:790) have noted that copper and tin bronze first came into regular, if small-scale, production and use in "late Neolithic" village contexts in northwest PRC. The subsistence economy was mixed with plant cultivation, animal breeding, sometimes pastoralism, and hunting and gathering. Apparently local ore sources were exploited, demonstrating that the copper-base artifacts were not simply exotic trade items.

Other parts of Eurasia similarly have bronze technology first appearing in agrarian societies, with resultant terminological conundrums. Nelson (1993:110) has noted that archaeologists in Korea refer to the period when bronze initially appears sometimes as "late neolithic" and sometimes as "bronze age." This nomenclature struggle illustrates the problem of trying to adhere to the Three Age terminology when the reality was usually an uneven initial appearance of metal alloy technology in extant agrarian cultures.

Treating the boundary between the Neolithic and the Bronze Age as necessarily "definitive" can also result in flawed lines of reasoning regarding metal excavated from purportedly "Neolithic" contexts, essentially discrediting the context. Some scholars circularly conclude that the metal excavated from contexts they have defined as Neolithic must have been intrusive from later contexts since "by definition" Neolithic societies did not have access to metal. For example, Higham et al. (2011a:234) dismissed all of the earliest occupation evidence for metallurgy at Ban Chiang (and Ban Non Wat) because they asserted that due to bioturbation, "fragments of bronze, crucibles and moulds regularly find their way into [Neolithic] contexts *where they do not belong*" (emphasis added). From the perspective of Eurasian archaeology, the earliest appearance of metallurgy in Southeast Asia likely occurred in societal contexts that might otherwise be termed "neolithic." The methodological choice to exclude metals from so-called neolithic contexts from systematic examination ignores what is likely to be the evidence for the earliest appearance of metallurgy in the region.

Other logical fallacies and reasoning pitfalls associated with essentialized progressive frameworks are listed in Table 3.1. The *flaw of conflation* was a particular target in critiques of processualism and neo-evolutionary frameworks (e.g., Upham 1990a, 1990b). The extended critique of social evolutionary approaches has shown that various components of complexity that formerly had been conflated—assumed to co-occur and co-vary such as metal technology and social hierarchy—are now being "decoupled" and the components studied as independent variables (Upham 1990a). Related to conflation fallacies are *inferential fallacies* when the archaeologist can detect presence of one attribute and infers the presence of other attributes for which there is no empirical evidence—e.g., when wealthy graves are used to infer political control by elites.

Additional variants of faulty essentialized logic include the *dichotomous fallacy* which refers to a common practice of subsuming important variation into binary opposites, profoundly oversimplifying the topic under discussion (Rowlands 1988) and masking important and meaningful variability within societal elements. Chapman, among others, critiqued "the obsession with...dichotomous thinking" (2003:71), whereby archaeologists assert on the basis of a few pieces of data that a society was egalitarian or hierarchical, simple or complex, or that specialization was present or absent. Most archaeologists now recognize that each component of complexity that might formerly have been treated in binary oppositional terms encompasses vast diversity and variant expressions. Archaeologists are finding that almost every component can be identified in some form in societies at most every social scale (Cross 1993; Hayden 2001b); e.g., inequalities are found in supposedly egalitarian societies (Flanagan 1989; Kelly 1991); equalities are found in purportedly stratified societies (Blanton 1998); specialized craft production occurs in some hunter-gatherer societies (Cross 1993), household production can manufacture wealth goods in state societies (Rochette 2009), and so forth.

PROCESSUALISM: PLUSES AND MINUSES

Many methodological contributions from processual archaeology of the 1970s nevertheless remain fundamental to the discipline of archaeology today. Explicitly stating theories, assumptions, and hypotheses to be tested against data, rather than merely asserting facts about the past, continues to be a best practice in the field. Comparing examples of similar phenomena that are essentially independent of each other (e.g., Earle 1997) can still provide important insights if not covering laws. Without a scientific framework, archaeology moves towards sets of personal assertions and unconfirmed explanations about the past (Watson et al. 1971:27) and lacks the means to winnow the assertions that

Table 3.1 Common Logical Fallacies in Social and Technological Universalist Approaches in Archaeology

Logical Fallacy	Flaw in Reasoning
Normative	Assumption that ideal societal types or technological "Ages" and stages exist, and are the best way to define change in human societies over time. All societies can be assigned to a type, stage, or age, and societies not fitting into the typology are somehow aberrant or anomalous. Also termed reification or essentialism.
Boundary	Assumption that boundaries between normatively defined stages or ages were necessarily significant social thresholds; thresholds not coinciding with stage boundaries are ignored, minimized, or not perceived.
Conflation	Assumption that selected traits necessarily co-vary and do not occur independently of each other.
Inferential	Assumption that traits and types can be inferred from each other without supporting empirical evidence.
Dichotomous	Treatment of complex variables as if they only exist in two polarized states, such as presence/absence, thus masking internal variation in the trait, and not acknowledging variation that spans the dichotomy.
Progressive	Assumption that progress is normative and societies/technologies are best examined in their relationship to an idealized trajectory toward a preferred endpoint, such as states, civilization, or specific advances in technology.
Hindsight	Assumption that traits important late in a sequence were necessarily salient earlier in the sequence; traits outside of the progressive paradigm are ignored, dismissed, or not perceived. Fallacy often found in "Origins of" arguments.
Projection	Attributes a subjective quality to an entity with insufficient data support, often through the use of preferred vocabulary and unsupported assertions.

are more likely to be true from those that are just so stories. Ironically many of the logical flaws and problematic assumptions in processualism derive from the use of the evolutionary approach as a reality that is to be repeatedly rediscovered rather than as a heuristic structure or a model to be *tested*, revised, or rejected.

One core tenet of processualism, a belief in a "systems model" of human societies and social change, is now seen as fundamentally flawed (Kohler 2012; Upham 1990a). Systems models sought (Upham 1990a:2) "to link social and economic processes together through positive feedback loops. In such frameworks processes like population growth, competition over resources, agricultural intensification, and political centralization are seen as a suite of processes that consistently co-vary." Applying this approach, a large literature of formulaic narratives on past development of most of the world's regions was written in the last two decades of the 20th century. Variants of this "processualist formula" continue to echo through much of current Southeast Asian archaeology publications. As a recent example Higham and Rispoli (2014:2) clearly employ the processualist formula in their interpretation of directional cultural change in parts of Thailand in statements like: "It is in these powerful and wealthy Iron Age communities that we can identify the early transition into states with population growth, agricultural intensification, conflict and increased production and competition over salt and metal for exchange."

Due to the nature of archaeological data where important aspects of past societies may not leave any traces, archaeologists rely on inference to make statements about their subject of study perhaps to a greater degree than other social scientists. One could argue that inferential fallacies are a common, even a necessary, occupational hazard for archaeologists and even may be required for the field to advance. Incorrect inferences, often worded as assertions, can be weeded out over the long term when they are treated as hypotheses and not fact, and are repeatedly tested against the ever-changing archaeological record.

Processualism contributed particularly to logical fallacies that derived from the marriage of typological with progressive evolutionary approaches. This combination elevated one-way lineal, directional, stage-like interpretations that frequently turned out to be crude and/or inapt for the specific context under discussion. A value system was often imposed on data interpretation, elevating evidence that supported the progressive narrative and sometimes devaluing, ignoring, or treating non-fitting data as peripheral to the progressive narrative.

Progressive fallacies can be recognized when change is interpreted or evaluated relative to a normative or an idealized trajectory usually implying "advancement" of some desired sort. This fallacy emerges when a society is viewed primarily from a perspective of where it is assumed to have been headed, the next "stage." Thus, an analysis of a non-state society that prioritizes evidence that theoretically foreshadows the later stage of statehood in the region can miss salient attributes for understanding societies that lie outside the idealized trajectory. Such an interpretive approach lends a formulaic predestination or an "are we there yet" distortion to the interpretation of the archaeological data, potentially incorporating an array of biases and false statements that ultimately result in misinterpretation of the archaeological culture under study. In addition, societies that are viewed as leading to nowhere, or as being not the source for the focal end-goal are interpreted as peripheral, or a false start. An archaeometallurgical example is discussed by Thornton (2009a) who debunks a traditional view he terms the "Levantine Paradigm." That paradigm privileged lowland evidence into a progressive framework that relegated early metallurgy in the highlands of Iran as peripheral to the more advanced lowland metallurgy found in the Levant and Mesopotamia. Thornton showed the opposite to be the case. The copper-base metallurgy in the uplands was actually technically advanced, innovative, and dynamic. He showed that highland metalworkers were masterful innovators and made many technical breakthroughs before any evidence of them appeared in the more urbanized lowlands.

The related *hindsight fallacy*, which is a product of a rear-view mirror perspective on the past (White 2017), perceives societies primarily from the vantage of a social evolutionary endpoint. It can result in societies being assessed not in their own terms, but primarily in terms of selected traits that align with the perceived objective of progressive evolution—most commonly state societies. Such hindsight reasoning elevates, a priori, the importance of some data over other data that may in fact be more prevalent. The archaeologist is tempted to fill in the blanks and assert, infer, or imply the presence of missing data in order to fit the interpretation into the progressive narrative. Ultimately, these assertions may result in misinterpretation of what is significant in a given social context and sequence, when significant elements fall outside of the idealized normative trajectory. For example, if centralized economic specialization is identified in a region's early states, precursor evidence for economic centralization may receive undue elevated attention, and the evidence for decentralized household production may be ignored or glossed over, even if it was more characteristic and influential of the pre-state economies and remained prominent in state societies. Explication of the full role of household production in the earlier societies might help explain aspects of the later societies that might otherwise be missed or misinterpreted.

Another related pitfall can be termed the *projection fallacy*. The flaw of "presentism" mentioned earlier in this chapter would be one example, when an archaeologist projects a value important today, such as utilitarianism, in the use of metal in past societies, when another value, such as color, might have been the key value at that time and place. This fallacy is often evident in word choice that projects a current archaeological agenda. The 1980s and early 90s when Anglophone archaeologists were preoccupied with chiefdoms, power, and hierarchy, English language archaeological publications were flooded with a number of terms, reflective of assumed power differentials. A preferred vocabulary emerged with frequent use of words such as command, control, elite, exotic, prestige, strategic, surplus, restricted access, even when little evidence supported their use.

Yet as archaeologists move beyond oversimplified approaches to assessing their data, it quickly becomes evident why evolutionary approaches arose and have been so useful to archaeologists for so many decades. Evolutionary typologies and scenarios, even when only implicit in a progressive narrative, came to be used as a shortcut system of inference. If A (for example, burials of unequal wealth) was present, then B (a hierarchical society, a chiefdom) can be inferred (or more commonly asserted). Paradigmatic assertions (inferring from a model held as truth) are a convenient device for a discipline whose data have so many gaps and whose practice is so labor and resource intensive. To replace inferences with empirical evaluations of data will require more fine-grained and quantitative studies of archaeological data going forward (White 2017).

Paradigmatic Change

> *Objective approaches to craft specialization founded on systems theory are obsolete and should be replaced by more fine-grained approaches capable of dealing with human agents, social practice, and meaning. (Clark 2007:20)*

The path forward for archaeometallurgy can gain insights from the paradigm rebellion of sorts that emerged over the last few decades termed "post-processual" archaeology. While not encompassing a single alternative theory, post-processualism does advocate exploring a wide range of alternative points of view in studying past societies (Hodder 2012a, 2012c). Post-processual frameworks are compatible with the observed variability in social contexts for the development of early metallurgy and the poor fit and barrenness of progressive evolutionary frameworks for interpreting it, and these approaches recognize many of the logical pitfalls and fallacies of interpretations contained in the Conventional Paradigm.

Post-processualist archaeologists have come to be interested in areas formerly ignored or even taboo under processualist paradigms, topics such as the impact of personal choices and actions (often termed agency), symbolism, or random change on human societies.

One main shift is away from grand directional narratives toward more fine-grained examinations outside of frameworks that assume progressive directionality (White 2017). The focus is on smaller-scale contingent processes and bottom up views of data more akin to the study of genetic variation, mutation, and natural selection, rather than universalist progress through stages evolving inexorably toward states (Dunnell 1988; White 2017).

These more fine-grained studies require significant reallocation of archaeological effort. One trend employed by archaeologists moving beyond oversimplified approaches is systematic examination of variability within phenomena of interest such as craft specialization (e.g., Costin 1991, 2001) or warfare (Solometo 2006). These approaches are providing a rich array of concepts, terms, distinctions, and evidence to look for in the archaeological record, greatly enriching the tools with which to perceive and study the past, including ancient metallurgy. Ultimately quantitative assessments of multiple sets of data, not just assertions of presence/absence of a trait, are needed to make empirical assessments that can be compared with other examples.

Pertinent to metallurgy, assertions of warfare in a prehistoric society based on presence of metal arrows and spear points will be convincing only with quantitative support not just from the artifacts, but from other lines of evidence such as rates and kinds of traumatic injuries suffered by the population. (See Jantzen et al. 2011 discussing multi-proxy evidence for warfare during the European bronze age.) Rates and patterning of violent trauma on human skeletons should be presented and compared with quantitative evidence from other societies that have been established as having warfare. The absence or low rate of physical trauma consistent with perimortem and defensive injuries, or of other evidence indicative of violent behaviors (e.g., changes in a settlement system to emphasize defensive positions or defensive architecture), may support an argument that the arrows and spears were used primarily for hunting, gift exchange, and/or status and ceremonial display rather than for warfare.

Middle-range Societies

> *The variability of social forms in middle-range societies is of great theoretical significance. (Rousseau 2006:21)*

One concept that emerged during the process of critique and rethinking of evolutionary societal types since the 1980s (see overviews in Chapman 2003 and Trigger 2006) is deemed particularly useful for the study of metal age Thailand and the early development of metallurgy generally. That concept (not a type) is "middle-range societies" (Gregg 1991; Rousseau 2001, 2006; Saitta 1999; Scarry 1999; Spielmann 1998; Upham 1987, 1990a, 1990b). The term, introduced by Feinman and Neitzel (1984), classifies societies that are neither mobile bands nor bureaucratic states (societies formerly called tribes and chiefdoms) into one large heterogeneous group. The term subsumes intermediate level societies, pre-state sedentary societies, transegalitarian societies, complex hunter-gatherers, house societies, tribes, individualizing chiefdoms, complex chiefdoms, decentralized stratified societies and more. It is the variability in that group that is the object of study (Gregg 1991).

From the review earlier in this chapter of the social contexts for early metallurgy, it is clear that middle-range societies, not states, were the petri dishes of most metal technological development and transmission in prehistoric times. The contributions of states to metallurgical development were probably in scaling up production (e.g., dynastic China) and/or refining output for selected needs (e.g., Moche). Even so, they had no monopoly on large-scale products (e.g., Dongson drums). Moreover, middle-range societies were where the action was for many other topics of keen interest to archaeologists, including development of agricultural technologies, various manifestations of inequality, forms of craft specialization, and non-state political systems.

The societal grouping of middle-range societies builds upon the realization that social equality and inequality are not simple dichotomous states but vary with enormous complexity and permutations among non-state societies (Rousseau 2006). Varying expressions of inequality occur among hunter-gatherers, especially groups that are sedentary (Flanagan 1989; Hayden 2001b; Price and Feinman 1995b; Rousseau 2001). Uncentralized groups, traditionally called tribes and assumed to be egalitarian, turn out to have a range of expression of hierarchy and status inequality, including some cases of hereditary stratification (Ames 1995; Pearson 2007; Rousseau 2001). Centralized groups that traditionally have been called chiefdoms and even states turn out to have many egalitarian and heterarchical elements (flexible/contingent ranking, lateral differentiation, self-organization [e.g., Levy 1995; Potter and King 1995]). Archaeologists and cultural anthropologists have become intrigued with the interplay of hierarchy and egalitarianism at all societal scales (Blanton 1998; Flanagan 1989; Kelly 1991; Plog 1995; Rousseau 2001). Archaeologists have realized that the study of emergent social complexity cannot be confined to societies conventionally classed as chiefdoms (Feinman 1995). Middle-range societies contain the primary body of evidence in which to investigate organizational variation in social complexity as well as the diversity of dynamics and trajectories within complex societies through time, including the early development of metallurgy.

Middle-range societies provide a means for archaeologists to examine and discuss non-state and pre-state societies as a domain of study and avoid or minimize use of the problematic concepts of tribe and chiefdom (e.g., Hayden 2001b; Rousseau 2006:20–21). Since the entire prehistoric metal age in Thailand consists of non-state agricultural societies, the construct of "middle-range societies" is deemed a key concept for this volume. As the early development of copper-base metallurgy commonly has occurred in middle-range societies, one overarching contribution of this volume to scholarship is as a case study of metallurgical development in middle-range societies in one part of the world, northern northeast Thailand.

Beyond Parsimony

> *Although parsimony is often stated to be desirable when constructing scientific theories, theoretical economy is self-defeating if it ignores the diversity and complexity of what is being explained. (Trigger 2006:489)*

Elevation of parsimonious explanations by processualists has receded as contingent approaches and agency perspectives seek to elaborate more dimensions and vectors bearing on archaeological evidence. The inadequacies of oversimplified models reliant on social evolutionary, progressive assumptions of cultural change are increasingly recognized in archaeometallurgical circles (e.g., Hosler 1994:4–5; Hoffman and Miller 2014; Kienlin 2014), although perhaps at a slower rate than in other areas of archaeology. Topics and unparsimonious approaches not fitting comfortably in the processual and neo-evolutionary approaches of the 1970s and 1980s are beginning to attract students of ancient metals (e.g., Epstein 1993; Sofaer Derevenski 2000; Thornton 2009a).

There is beginning to be much exciting work pertinent to archaeometallurgy looking at the diversity of relationships among economic factors (agriculture, resource distribution, exchange mechanisms, specialization), political factors (such as leadership and warfare), social differentiation (along both horizontal and vertical dimensions), ideology, and regional histories in ethnographic and archaeological data from middle-range societies (e.g., Chapman 2003; Rousseau 2001, 2006). In archaeological contexts, organizational properties cannot often be directly observed, but with careful thought and analysis, structural interrelationships can be inferred from the material details and fine-grained evidence of the organization of the past economies. Because metallurgy can provide rich material detail on prehistoric economic relationships (Chapter 5), archaeometallurgy has great potential to make major contributions to archaeological scholarship.

What Is to Be Done with "Stages" and "Ages"?

> *search for correspondences between technical level and "stage" of economic organization does not seem likely to lead to a theory of the relations between technical systems and society, other than one so over-simplified and general that it quickly loses all interest. (Lemonnier 1986:151)*

Archaeological research over the past 75 years has shown that technological Stages cannot be equated with archaeological Ages in the universalist sense implied by Childe (1944). This realization is not new. Renfrew (1973:169) pointed out that "archaeologists, wrapped up in the old conventional classification of 'ages' of stone, bronze, and iron, have seriously overrated the importance of what were at first rather minor technological advances." Yet decades later the point continues to need reiteration. Kenoyer and Miller (1999:135) noted that "technological 'stages' like the working of metal cannot be equated with other levels of social complexity, such as urban/non-urban." Roberts (2011:137) observes that currently there is "widespread, and perhaps long overdue, recognition that the appearance of metal should not automatically be equated with the emergence of elites, and neither should it be assumed to be a cause of social change or even an important material within communities." Yet the assumption that there is a necessary connection between bronze technology and development of stratified societies has remained stubbornly in the literature (e.g., Earle and Kristiansen 2010:19–21; Higham 2011a; Kristiansen and Earle 2015).

What should be done with the "Age" terminology that is so entrenched in the archaeological literature, especially the "Bronze Age," now that it is clear that there are no necessary social correlates with bronze production and use? The intellectual approach underlying Ages and Stages, which long predates the advent of radiocarbon dating and the vast majority of the current knowledge base of prehistory, is flawed to the degree that the conceptual framework is hindering advancement of understanding the human past. Why? "Age" terminology reflects the desire of students of the past to identify periods of fundamental societal transformation. Attachment of Ages to technological and material change may have made sense to 19th century antiquarians or at least been extremely convenient before radiocarbon dating was available. But the data are so much richer in the 21st century and archaeological methodology is so much more refined, that the fit of material change and social transformation is demonstrably poor.

Ages and Stages encourage the archaeologist to assume correlations in their data rather than explicitly test for them, to create transitions where they may not have existed, to ignore data that do not fit the desired norm, and to measure their culture of study against illusory arbitrary standards, in other words, to essentialize. By emphasizing or even creating parallels and commonalities with the imagined archetype, building an empirical statement from a careful analysis of the full range of data from the past society is shortchanged or often even bypassed. Distortions endure and multiply. By providing a misleading oversimplified frame of reference, the Three Age System encourages archaeologists to employ essentialist lines of reasoning that stifle understanding the past.

Some prominent archaeologists seem to feel particularly free to play fast and loose with the Bronze Age, making global claims, which turn out to be parochial or idiosyncratic. One example that would exasperate archaeologists who work in China and many other areas in Eurasia is illustrated in Kristiansen and Larsson's 2005 volume *The Rise of Bronze Age Society*. The authors position this book as "a new book spanning the *Bronze Age world in its entirety, from Mesopotamia to Scandinavia*" (2005:xiii; emphasis added). Another illustration of a misrepresentative use of the phrase is in the title of Timothy Earle's (2002) book *Bronze Age Economics* in which two out of its three case studies are from the New World (Andes and Hawaii), which have *no* recognized Bronze Age. The book's preface reveals that the volume's actual subject is "chiefdom economics."

These examples demonstrate that the phrase "Bronze Age" has lost any coherent meaning in the broader discipline of archaeology. Since most archaeologists know that they can infer almost nothing about the technology and society of El Argar in Spain from knowledge of the nearly contemporaneous site of Anyang in central plains China, even though researchers in each region refer to the two sites as "Bronze Age," what can be done to move toward a clearer terminological system?

It is time for the Three Age terminology, particularly the Bronze Age, to be demoted from its position in Old World Archaeology from paradigm to region-specific rough chronological framework, if used at all. While the Three Age terminology is unlikely to be dropped in many areas such as Europe, it is essential to couch its use in regional terms, preferably detached from sociopolitical connotations. There is no all-inclusive Bronze Age, there are at best only bronze ages of Iberia, of Mesopotamia, of China, of the Balkans, etc. In several parts of the Old World, this demoting process has already begun. Thomsen's Three Age terminology is used as a pragmatic convention for delineating regionally specific temporal frameworks but not as a theoretical framework for interpreting the past in places like Korea (see Nelson 1993). Thus, "technological age" terminology has often been implicitly or explicitly redefined by regional archaeologists to suit regional needs for a general chronological periodization (Nelson 1993; Renfrew 1973:24). The terminology can be useful to delineate millennial-scale periods of time in macro-regions. One approach to keeping the terminology but divorcing it from sociopolitical presumptions has been to lower case the phrases neolithic, bronze age, and iron age (e.g., Nelson 1993; Renfrew 1973; White 2002).

Although at times the Three Age System has been used "only as a convenient shorthand" in acknowledgment of the absence of another agreed upon regional periodization at a comparable scale (e.g., Higham 1996a:7), in recent years several Southeast Asian scholars have reverted to fully essentializing the system (see for example Higham 2002, 2014; Higham and Rispoli 2014; Higham et al. 2011a; Rispoli et al. 2013). Reification influences perceptions and soon technologies become packaged with a linear progressive narrative that obscures or even rejects evidence that does not fit nicely with expectations derived from the narrative. Dates are discounted, not mentioned, or "adjusted" (e.g., Rispoli et al. 2013:136); associations are downplayed or ignored (e.g., the iron knife in Ban Non Wat's BA4 Burial 34 [Higham 2009:101]); the Childean urge to package metal technology with particular social visions and deterministic parlance becomes scholarly truth. For example, iron has been cast as the independent variable in the development of political centralization and social ranking in prehistoric Thailand, where the mere presence of iron tools "encouraged and enabled the construction of banks" and moats surrounding sites (Higham 2002:226–227). The narrative continues emphasizing purported evidence of political centralization, instances of iron "weaponry" implying increased conflict, presence of exotic raw materials that "necessitate[d] a new plane of expertise compatible with specialization," assertions of full-time craft specialization, and economic controls over production and exchange (Higham 2002:227).

Other voices, particularly from a younger generation of Southeast Asian archaeologists, are speaking up and viewing the evidence differently (e.g., Eyre 2006; Källén 2004). Their analyses of Southeast Asian iron age evidence reveal societies not so sharply different than preceding settlements and lacking a clear focus on warfare and political and economic centralization. With awareness developed from the post-processual literature, they also see other salient dimensions and currents in the data that have been missed by traditional frameworks. Research in central Thailand (Eyre 2006) found settlement patterns with variously sized sites but lacking in hierarchical spatial organization for both the bronze and iron periods. Evidence for social transformation and centralization occurred not at the beginning but at the end of the iron period. Working from an articulated awareness of the colonialist underpinnings of linear deterministic perceptions of the archaeological past, Källén (2004) explored iron production at the site of Lao Pako not as a function-first, prime mover type of phenomenon, but as a ritualized expressive activity integrated with subadult mortuary rites. Whether these kinds of evidence and interpretations are seen as "peripheral" (Higham 2002:193), or as illustrative of important regional currents of the time period, the next generation of Southeast Asian archaeologists will decide.

Currently some publications and archaeologists in certain regions reference regional sequences to the Three Age System in different ways. While many archaeometallurgists still use Three Age terminology, they have moved away from linking the terms to issues of social transformation. Many have also moved away from attempting to define normative and directional technological changes and instead emphasize variable metallurgical processes and regional patterns (e.g., Craddock 1995; Craddock and Lang 2003).

Chernykh (1992) is an example of an archaeometallurgist who refused to link technology and sociopolitical stages. His synthesis of the Eurasian metal age offers an important alternate organizational scheme. Chernykh (1992:11) rejected definition and subdivision of the metal ages based on criteria from other aspects of society, such as subsistence, social organization, etc. He argued that if a general Three Age periodization is to be offered, it should stick to criteria from the relevant technology, and let other components of the past, such as ceramics, subsistence, political systems, etc., use their own criteria and separate terminologies. In the realm of metallurgical technologies, he argued for the existence of supra-regional metallurgical provinces, "the largest historic production systems operating in antiquity, with their very extensive networks of...allied metallurgical and metalworking focuses" (Chernykh 1992:12). He defined temporally and technologically a set of discrete provinces, large territorial units spanning many individual cultures and periods of time but among which metal processing techniques and metal types were shared. Commonalities existed based on the sharing of technological knowledge, but simultaneously local cultural needs and resources affected local metal assemblages. Chernykh argued that technological stages may be definable not pan-globally or within the Old World, but within individual metallurgical provinces, the territorial unit within which technological knowledge is shared and transmitted. Although too little is currently known about a proposed "Southeast Asian Metallurgical Province" (White 1982, 1988) to advocate a similar province-specific periodization, in the future, when more data are available, such an approach may be usefully implemented as a way to relate the metal age development of Thailand to that of neighboring countries such as Burma (Myanmar). The theme of metallurgical provinces will be further addressed in TAM 2C (chapter 6).

For the purposes of this volume, the approach to the Three Age System is similar to that used by Renfrew (1973) for Europe, i.e., bronze and iron age may be selectively used, but in lower case to signal a disassociation with "Age of Mankind" presumptions. This usage and the thinking behind it are not new in Southeast Asian archaeology, a region where the contradictions of the Three Age System have long been a source of struggle (Dunn 1970:1050–1051; Higham 1989:28, 1996a:7; Stark 2015; White 2002) and where a variety of alternatives have been suggested (e.g., Bayard's [1984b] General Period A, General Period B, etc.; [see also Dunn 1970; Solheim 1972]). (It should be noted that calls to eschew the "19th-century technological model" have been long noted for the Near East and other areas [see e.g., Daniel 1965:245]).

This use of three age terminology is similar to that of Higham (1996a:7). In that publication, "Bronze Age" represented "the period from the first use of copper-based artefacts," and similarly for "Iron Age," which would be used "even if iron was later and less abundant in some areas than in others." In other words, in that publication, the stated intention of using metal age terminology was simply for a convenient, general, region-wide periodization based on the presence of a particular technology. Using the metal age terminology even in localities where the particular metal is scarce is similar to Chernykh (1992) and Coles and Harding (1979). Following

Chernykh's concept of a metallurgical province, locations within a defined province are assigned a period based on the overall nature of metal production, even if specific locations lack evidence for that particular metal or produce a variant atypical of the overall region. For example, a site or phase having only copper artifacts (e.g., Ban Non Wat) when elsewhere in the region bronze technology is present will not be termed a copper age site but rather a bronze age or, as discussed below, a bronze period site.

Although a lower case Three Age System has been among the variants used in Southeast Asian periodization (e.g., see Higham 1996a using neolithic), White (2002:xx) first commented on the rationale for lower casing neolithic, bronze age etc. for the first volume in the Ban Chiang monograph series. Other publications have also adopted the practice (e.g., Oxenham and Tayles 2006; Tayles et al. 2015).

Regional Periodization Used in This Volume

This chapter has reviewed the ambiguities and problematic usages of the Three Age System in general and the current practice among several Southeast Asian archaeologists to essentialize "Neolithic," "Bronze Age," and "Iron Age," even in the phasing terminology of sites such as Ban Non Wat. In the view of the authors, the current variable usages of the Three Age System in Southeast Asia so distort the interpretation of the archaeological record, that it promotes confusion in the region's archaeological literature and lends an anachronistic tenor to the region's recent publications. The "Bronze Age" in particular has become in Southeast Asian archaeology "an essentially contested concept" (Gallie 1955) for which scholars are unlikely to ever agree on a meaningful and consistent usage (White 2002, 2017).

The objective of this volume is not to define "ages of man" in Southeast Asia that coincide with particular technologies. The objective of this volume is to situate metal technologies in their prehistoric context from their earliest appearance in northern northeast Thailand. Therefore, in order to differentiate the intent and perspective of this volume from other recent publications on prehistoric metal-using periods in Thailand and Southeast Asia, and to help disassociate the perspective of this volume from the "ages of man" ontology, the following terminology is adopted.

Rather than use the "age" terminology that has become so loaded with irreconcilable connotations and conflicting usages, where suitable and possible, a more neutral terminology that focuses on chronology of technology is chosen. Agrarian societies prior to the earliest appearance of metal are termed "premetal" societies. The "bronze period" refers to the time from the earliest appearance of copper-base artifacts in the middle Mekong and Chao Phraya Basins until the earliest appearance of iron. The "iron period" begins with the earliest appearance of smelted iron in the region and extends to the protohistoric period. The "metal age" is a general term to refer to the combined bronze and iron periods.

The protohistoric period is defined here as beginning with the appearance of written documentation of the region, but with most of the data on the ancient societies still coming from material remains, both archaeological and iconographic. For much of Thailand, the material signal for the protohistoric is the appearance of "Dvaravati" remains from the 7th century A.D., although see Murphy (2010) for an argument that northeast Thailand be considered a development distinct from both the Dvaravati as well as Khmer.

Subdivision within regional periods is possible with the iron period, which can be roughly subdivided into an early, middle, and late iron period. The early iron period approximately coincides with the Ban Chiang Middle Period, although how early iron appears within the Middle Period is not certain. The Ban Chiang Middle Period spans the middle of the 1st millennium B.C. The middle iron period coincides with the Ban Chiang Late Period and extends from the last few centuries B.C. into the first couple of centuries A.D. The late iron period follows the Ban Chiang Late Period up until the protohistoric Dvaravati period.

Precise absolute dating for the broad regional periods is viewed as not yet determined and Chapter 2 presents the authors' current understanding and approach. The protocol used in some recent publications (e.g., Higham et al. 2011a) for defining and, therefore, dating the start of the "Bronze Age" in terms of in situ metalworking installations and/or graves with metal grave goods is viewed as a method that systematically excludes the initial period of metal evidence, including its earliest transmission and adoption. A transmission and adoption period prior to the placement of metal in graves or the archaeological recovery of unambiguous metalworking installations can be lengthy, normally lasting centuries in most parts of Eurasia. Transmission of metal technology is of great interest to the authors of this volume; therefore, all metal-related evidence is considered valuable and worthy of evaluation, not just that from grave assemblages and metalworking installations.

4

The New Archaeometallurgy Paradigm

Joyce C. White and Elizabeth G. Hamilton

The study of technology in society has become a much more complex and interesting pursuit than it was 30 years ago. A wide array of research tools to study materials is available, better knowledge of the range of human technologies over time and space is at hand, and conceptual tools for perceiving and explaining socio-technical relations and change have moved far beyond migration, diffusion, and independent invention. These advances in knowledge and methods have fostered changes in underlying paradigms for studying ancient technology. For archaeometallurgy, a few studies in the 1990s applied the new technological paradigms (e.g., Epstein 1993; Hamilton 1996; Hosler 1994), but a discipline-wide paradigm shift is still constellating (Killick and Fenn 2012; Roberts 2011; Thornton 2009a, 2009b; Thornton and Roberts 2014:4).

The process of a new paradigm emerging in scholarship can be contentious, accompanied by much academic *Sturm und Drang*. Proponents of the old paradigm may cling to the frameworks upon which they built their careers and ignore or try to eradicate competing ideas and views. Fortunately, the new archaeometallurgy paradigm for studying prehistoric metallurgy can borrow from ideas and terminology developed over decades by scholars pursuing studies of other materials, including both lithics (e.g., Bar-Yosef and Van Peer 2009; see overview in Soressi and Geneste 2011) and ceramics (e.g., Dobres and Hoffman 1994; Jones 2002; Roux 2003). These studies have produced a variety of methodologies, concepts, and theoretical frameworks that lay ample groundwork for the paradigm shift that is now occurring for the study of ancient metallurgy (Thornton 2009a, 2009b; Thornton and Roberts 2014). Of course, many components of older paradigms remain useful, as younger scholars inevitably stand on the shoulders of their predecessors, even as they bring new perspectives and data into the discussions.

Chapter 3 laid out many limitations of the Conventional Paradigm that was built upon a variety of assumptions centered on progress as the core value, the core occurrence in human history. At the very grand scale of technological development on the globe over thousands of years, i.e., from the bird's eye view (White 2017), the progressive narrative that forms the core of the Conventional Paradigm is an observable phenomenon. But on the scale of individual societies in particular places and times, explaining technological change primarily in terms of progressive development is rarely a good fit. As an analogy, biological evolution may be seen as a progressive development from single-celled entities to millions of species of complex organisms. The mechanisms of change, of evolution, however, are to be found in the processes of inheritance and recombination, natural selection, stochastic processes, gene flow, and other factors that at close view are not well conceived within a progressive framework (White 2017). Likewise, the newer approaches for studying and understanding human technological development focus less on the grand narrative and more on detailed processes occurring at the societal level.

These intellectual trends will increasingly affect studies of metallurgical development and are pertinent to research into pre-industrial technology generally and that of Southeast Asia specifically (Källén 2004:192–194). The localized perspective can be seen in at least two aspects of current pre-industrial metallurgical investigations that will be important for Southeast Asian archaeometallurgy: (1) the role of the meanings of and values for metals in particular societies; and (2) how particular local cultures and communities develop and use technologies of metal and metal artifact production.

Although most studies of metal technology in antiquity have been and still are narrowly focused on the technical qualities of the items (e.g., most chapters in Betancourt and Ferrence 2011), some archaeometallurgists are trying to understand what metals meant in societies that were less dominated by a utilitarian view of the material world than are modern western societies (Hosler 2009). The exploration of the meanings of metal in particular societies *beyond* creation of more effective and elaborate tools and weaponry or somewhat vaguely termed "prestige items" entails sophisticated assessments of both contexts of metal finds and their material properties.

A Not-so-new Paradigm

Broadly speaking, the new archaeometallurgy paradigm (hereafter the New Paradigm) can be said to fall under an umbrella of scholarship termed the "anthropology of technology" (Dobres 2000; Lemonnier 1992; Pfaffenberger 1988, 1992, 2001; Schiffer 2001a, 2001b). Since the 1970s, Michael Schiffer, among others, has developed a large body of work with accompanying terminology, concepts, and synthetic observations from examining historical and modern cases of technological behaviors ranging from 19th century electric light houses to 20th century garbage dumps (Schiffer 2011). Ethnographic studies of ceramic production and use in traditional contexts have provided another large body of literature about technical behaviors in societal contexts (e.g., Gosselain 2000, 2008; Wallaert-Pêtre 2001).

Ethnographic studies, particularly in Africa, are shedding light on the ways in which the act of producing metals and metal artifacts can hold profound social and religious meanings that are inextricable from the act of creation and are often more important than utility (Childs and Killick 1993; Collet 1993; Herbert 1993; Rowlands and Warnier 1993; Schmidt 1996). The power of technical knowledge may be more important than the material itself. As an example, Reid and MacLean (1995) showed that in Karagwe, an East African kingdom, the power of metals lay less in their possession, than in the control and manipulation of the knowledge and ritual of their production. Budd and Taylor suggested a similar schema for much of early metalworking. "In the earlier Bronze Age copper-alloy artefacts may have had little 'utilitarian,' i.e., mechanical function, but the ability to put on a show of colourful, transmogrifying pyrotechnics may have commanded considerable respect" (Budd and Taylor 1995:139–140). In support of this idea, many of the richest graves in the central European Early Bronze Age also contained smith's equipment, suggesting an elite dominion of metallurgical skills in that cultural context (Sherratt 1994b:260) and an association of magic, metal, and authority. Whether the control of ritual craft knowledge was vital to the maintenance of elite status in Southeast Asia (Spielmann 1998) will be an important concept to assess.

This set of approaches is particularly pertinent to Southeast Asian archaeometallurgy and this volume because they provide conceptual frameworks to examine specific technological choices within specific sociocultural contexts. This more fine-grained approach to analyzing technological systems seeks to provide a bridge between technical analyses often undertaken by materials specialists and contextual social evidence, usually the domain of the archaeologist.

Anthropology of technology grew out of several intellectual traditions: (1) Anglophone anthropology, in particular the neo-evolutionists (White 1949), and the post-processualist emphasis on material objects as text (Appadurai 1986; Knappett 2005; Tilley 1990); (2) the long-standing French tradition of the study of the interrelationships of society, technology, and the body (Lemonnier 1986, 1989, 1992, 1993; Leroi-Gourhan 1943; Schlanger 1990); (3) the modern historians and sociologists of the technology of industrial societies (Bijker 1995a, 1995b; Hughes 1987; Pinch and Bijker 1987; Smith 1981; the journal *Technology and Culture*); and (4) work of archaeologists using scientific analysis of material remains in order to cast light on questions of culture contact and social organization (e.g., Ehrenreich 1991a, 1991b, 1995; Golden 2010; Hamilton 1996; Hosler 1994; Jones 2002; Kingery 1996a; Rehren et al. 2007; Weeks 2007). Approaches to technology and technological change include the following: evolutionary (Kuhn 2004), behavioral (Schiffer 2001b, 2004, 2008, 2011; Schiffer and Skibo 1997; Skibo and Schiffer 2001), and what David Killick (2004) called "social constructionist," a wide-ranging group including proponents of agency (e.g., Dobres 2000, 2001), practice theory (Dobres and Hoffman 1999), ethnoarchaeology (Gosselain 2000; Schmidt 1996), technological style and choice (e.g., Bleed 2001; Epstein 1993; Hosler 1994; Lechtman 1977; Lemonnier 1993; Sillar and Tite 2000; White 1992), and even symbolic anthropology (Pfaffenberger 2001). This disparate group, according to Killick (2004), is united by the assumptions that: (1) for any given purpose, there is usually more than one technological practice that will produce a satisfactory result; and (2) the choice of technological practice is usually governed less by considerations of efficiency, market forces, evolutionary fitness, or the unstoppable force of progress than by the beliefs, social organization, and past technological choices of the group being studied.

The object of study in this set of approaches can be summarized as technologies in social contexts, or what Pfaffenberger (1992) terms "socio-technical systems." The schematic in Figure 4.1 illustrates the focus of socio-technical studies, namely the zone where the material world and the social world overlap. Proponents of anthropology of technology approaches would say that technologies do not exist outside of the societies and individuals that are choosing, prioritizing, organizing, and producing everything from the "look" of an artifact, be it a flaked stone or a cell phone, to the activities in which the technology is used. The social uses for any artifact cannot be assumed but must be studied. A flaked stone axe can be used for butchering and/or tree ringing. Modern cellphones have a number of uses including photography, communication with groups via social media, acquisition of information via the internet, and conversation with a single person. Anthropology prides itself as being a holistic discipline where nothing in the human experience is off the table for consideration. Thus, anthropology of technology approaches strive to study technology holistically.

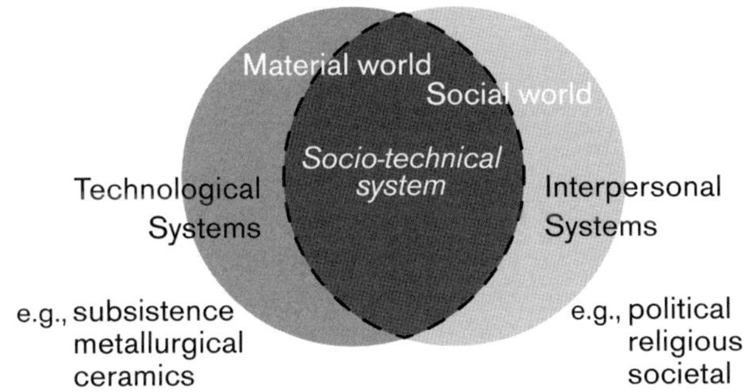

Figure 4.1 Socio-technical systems are domains where technical systems and social systems overlap.

Fundamental Concepts

There are many concepts, phrases, and terms used by proponents of anthropology of technology, and often a particular phrase, such as "technological style," is used somewhat differently by scholars. A complete review of all the terminology and conceptual distinctions used by the schools of thought that have emerged among scholars interested in anthropology of technology approaches is beyond the scope of this volume. However, in the following sections, several general foci common in anthropology of technology scholarship are reviewed, providing a background for how they are used in this volume. Five general foci are: (1) materials properties; (2) technological systems; (3) life history frameworks; (4) technological choices, styles, and traditions; and (5) technological change. Additional useful, related terminology is introduced as these foci are more fully discussed below.

Materials Properties

To understand a technology, one must understand the material properties of the technology's components, and this knowledge base is needed whether the technology of interest is rice cultivation, ceramic vessel production, or iron forging. The materials used to make an adze head, an arrow shaft, or a pot are not arbitrary. The material properties of stone, wood, clay, and metal—properties that are derived from each material's genesis in the physical world—are what enable manipulation of the material to create an object with a desired purpose. The material's native properties provide both possibilities and constraints for how it can be manipulated in a technological application.

Basic information on the properties of materials used by early human societies provides a foundation to understand craftsmanship and technology. As an example, generically, rock may have material properties of hardness and granularity that enable the creation of sharp cutting edges, in contrast to, for example, wood. The material properties of glassy rock such as obsidian, which fractures concoidally due to its lack of crystalline structure, enable the creation of cutting edges by the activity of flaking (knapping). However, the brittleness of obsidian makes it a less desirable material for groundstone technology, for which a more granular (macrocrystalline) rock such as basalt or rhyolite is suitable. The material properties of different kinds of rock affect the selection among rocks for making flaked or groundstone tools such as axes and adzes. Some rocks such as pumice and vesicular basalt have properties that discourage and, in some cases, eliminate their potential use for edged tools of any sort.

Of course, the prehistoric craftsperson is not a 21st century materials scientist. Knowledge for the prehistoric craftsperson was primarily 'know-how'—practical, experiential knowledge of material behaviors and the skills to use that knowledge to accomplish a task. Prehistoric practitioners would have known, for example, to remove the metal from the hearth when it is white-hot or harvest the rice when the seed heads bend down and bear edible kernels, rather than remove the metal from the hearth when it is 960° Celsius, or harvest the rice when the ovule is mature and capable of reproducing another rice plant. Keller (2001; see also Crown 2001; Minar and Crown 2001) indicates that not only observation and instruction but also long practice, often beginning in childhood, were required for practitioners to acquire expertise and motor skills to successfully pursue a craft. While the modern-day scholar cannot know the exact cues used by the prehistoric practitioner to manipulate a material, modern knowledge and studies can reveal the capabilities of the prehistoric practitioner, thereby contributing to the inferential base upon which understanding of the ancient society is built. Thus, the mere presence of evidence of bronze manufacture in a society demonstrates that practitioners were capable of creating certain temperatures and manipulating atmospheric conditions, if not exactly how they accomplished it.

Appreciation of technical know-how is heightened when dealing with compound technologies that involve multiple materials, for instance wood, string, and stone in a bow and arrow. The properties of each contributing material would need to be understood not only on its own, but also in relation to the other materials. The wood for the bow stave needs to be flexible so it can be drawn, but it must spring back to its original shape with sufficient power to throw the projectile the required distance. The string for the bow must have long and strong fibers twisted or twined in a manner that they do not break, stretch, or shred while taut or under increased tension given the size and weight of the particular bow while the arrow is cocked and then abruptly released. The wood for the arrow must be light, strong, and straight, sized in scale with the particular bow and arrow point. The stone for the arrow tip must be shapeable to a point that can pierce hides at the velocities achieved with the bow thrust; its weight must balance with the weight of the bow stave so that the arrow flies true and reaches the prey; and it must be attached to the shaft with cord and/or an adhesive so that the arrow remains intact until effectively piercing the target. Technologies like the bow and arrow that require several materials and interlinked technologies are termed a complex technology in this volume.

PERFORMANCE CHARACTERISTICS

A construct common in anthropology of technology literature used in many discussions of materials properties is that of the performance characteristic. The term derives from engineering, where it refers to behavioral and mechanical characteristics of a piece of equipment when it is used, e.g., a performance characteristic of a BMW M235i Coupe is that it can go from 0 to 60 mph in 4.6 seconds (http://www.bmwusa.com/bmw/2series). Hally (1986), Klemptner and Johnson (1986), and Schiffer and Skibo (1987) brought the term into archaeology. As an example, a performance characteristic of bow staves is flexibility. While material properties may determine many performance characteristics, they are not the same. The material properties of a wooden bow versus a composite bow differ, but the performance characteristic of flexibility is the same. Defining performance characteristics of artifacts in a technological system provides a baseline for examining the functioning of a technology on its own terms first, beginning as a set of material relationships.

Although related to an object's material properties, the performance characteristic construct helps to focus attention on those materials characteristics that "matter" in behavioral contexts. A cooking pot has a variety of measurable material properties, such as color, that may be unimportant to its function of cooking, although color may be important to other activities such as ethnic signaling or exchange. To perform its cooking function, however, the pot needs the performance characteristic of some degree of thermal resistance, at least under the conditions for cooking in the society using the pot. It may have many material properties contributing to its thermal resistance, from fabric, to shape, to surface texture (Schiffer and Skibo 1987, 1990; Schiffer et al. 1994). The craftsperson must understand and be able to exploit and modify the performance characteristics of natural materials in the environment, as well as man-made materials. For example, not only do the performance characteristics of clays used in forming a pot need to be understood and manipulated with respect to temper, hydration, wall thickness, etc., so that the unfired vessel holds its shape when drying, but also in consideration of how that pot, once fired, will perform its intended function such as cooking or storage.

For copper-base metallurgy, different alloys have different performance characteristics such as fluidity, color, and hardness that may (or may not) be significant in individual technological systems. In a given cultural context the redness of a copper implement might satisfy an aesthetic or symbolic need that a more golden tin bronze implement would not. For example, in some sub-Saharan African societies like the Burundi, conventional implements like hammers, axes, and weapons could be made entirely of copper when they served as regalia (Herbert 1984:248–250). On the other hand, a tin bronze axe would perform better as a chopping implement in comparison with an unalloyed (softer) copper axe in cultural contexts where effectiveness in chopping superseded color.

The terminology derived (especially by Schiffer) from the common use of performance characteristic also can be helpful in discussing technology. Some performance characteristics may be critical not only to an artifact but to the functioning of the overall system. Determining what characteristics are critical both technologically (e.g., a furnace that reliably achieves certain temperatures to smelt an ore) and for the cultural system under study (e.g., an alloying protocol that reliably achieves certain colors or sounds considered essential in metallic ritual paraphernalia [Hosler 1994]) requires thoughtful analysis. Some performance characteristics may be critical in one culture but could be present and not critical in another culture. Nevertheless, it is useful to consider that a successful bow must meet the performance requirements of flexibility (capability to be bent) and spring (tendency to rapidly return to the shape existing prior to shape modification of drawing). That any particular bow meets these performance requirements is a product of both the raw material (e.g., the particular wood selected) and modifications undertaken by the craftsperson (e.g., shaping the thickness, length, and longitudinal configuration of the wooden bow stave) that will enable it to perform effectively in the activity of shooting an arrow. Two separate bows may vary in how they meet the requirements, such as one might have less flexibility and another more flexibility. Hence the bows meet the performance requirement but have different performance characteristics; both may be usable but the less flexible bow may be preferred by a stronger archer.

Schiffer (2011:28) pointed out that the critical performance characteristics for an individual artifact can vary based on its use over the course of its life history. A critical performance characteristic of clay paste being formed into vessels is its capability to hold its shape through drying; once fired, a critical performance characteristic of the same artifact in its new state and new activity, if it is a cooking pot, is to resist thermal shock. It is also useful to keep in mind that an artifact's functions can change throughout its life as it passes through many activities and can also vary by reference group. A bronze adze to the maker may be a good to exchange for food, but to the consumer it may be a tree-cutting implement and/or a status symbol.

The performance characteristics are established not in the abstract, but in the interactions of people and artifacts in specific activities, including product manufacture (Schiffer 2001b). Any artifact or modification is a result of compromise. For example, coarse ceramics are functionally better for cooking, but the harsh tempered clay can be difficult and unpleasant to shape, particularly if using a fast wheel technology. The performance characteristics of a cooking pot could include not only its thermal resistance, but its ease of manufacture, resistance to mechanical stress, behavior of its clays, the available and preferred temper, the finishing treatments such as burnishing and resin-coating, the marketability, the color, the style, and the ease with which it could be turned into grog for tempering the next pot, among many other factors. The performance characteristics of an artifact are not abstract qualities but are embedded in the social context.

There may be characteristics affecting performance in neutral or negative ways, possibly creating *performance problems* (Schiffer 2005:486). A performance characteristic of a satellite television receiver is that its signal reception is interrupted during storms. Such a performance problem may be critical for some viewers such as sports fans, but only of nuisance value to others, not preventing them from acquiring the satellite receiver system. Efforts to solve performance problems can be major stimuli to technological innovation. Due to the interlinked nature of components in complex technologies, often change in one component necessitates design changes and innovations in other components for the technological application to endure (Schiffer 2005). Adoption of new technologies may in part be a response to performance problems of older technologies. Therefore, availability of fiber optic technology in a region may result in former users of satellite television receivers switching to the newer technology, if the cost/benefit balance is perceived as desirable.

Supplementary actions employed by producers to influence performance characteristics of artifacts can be a fruitful area for technical analyses. Technical analyses of finished products and equipment used to manufacture finished products may reveal that after the essential artifact was formed, additional steps were taken to modify properties of that form—such as working tin bronze artifacts to harden them or coating a basket with resin so that it can hold water. These steps are here termed post-manufacturing *treatments*. In this study, post-manufacturing treatments have been found in some metal artifacts and in crucibles. Evidence for this type of treatment should differ from modifications occurring from use of the artifact, here termed *use-modifications* (e.g., unintentional changes in materials resulting from use and exposure to technological processes, for example vitrification of clay after having been exposed to very high temperatures).

THE ROLE OF MATERIAL CULTURE ANALYSIS IN THE NEW PARADIGM

Traditionally, the discipline of materials science has been used in archaeological settings to address questions of composition, functionality, source of material components, level of technological capability, and material constraints. The anthropology of technology framework expands and reorients the role of the materials sciences in archaeological research programs and increases the demands upon such studies. The emphasis moves away from assessment of "what technological level had society X reached" (Did they have the fast wheel? Did they make alloys?) towards broader questions (What was the role of ceramic or metallurgical technologies in society X? Why did society Y choose production system A in contrast to production system B for making bronze spear points? Where did precursor, antecedent technological systems come from?).

To answer the kinds of questions posed by scholars of anthropology of technology approaches requires comprehensive analyses of material remains, for only by such analyses can one ascertain how the artifact was made, or the performance characteristics of any artifact, tool, or process, or material evidence for changes in artifacts through use. Materials analyses can reveal not only the composition of an artifact, but also the sequence of its manufacture and the various choices possible in the manufacture or technique (Kingery 1996a, 1996b). They can reveal the mechanical properties and their relationship to non-mechanical properties (Jones 2002), for example, studying how important thermal shock resistance was in the manufacture of pots (Schiffer and Skibo 1987). The researcher should strive to learn what choices were possible and available in a technology within its societal context. The lines of inquiry explore potential variability and potential effects of variation in choices. How was this axe hafted, how many ways could the axe have been hafted, and what effect do the various ways of hafting have on the look, the usefulness, and the meaning of the axe? Copper melts at 1083°C, but what combinations of furnace, fuel, crucible, and air supply enabled that temperature to be reached in the society of study? How did the labor organization in a society influence the choice of refractory system used to smelt metals?

A handful of technical analyses will not provide the answers to these questions. For the prehistoric period, only analyses of archaeologically provenienced remains will be able to address questions involving societal context. What is needed first and foremost is quantitative assessment of assemblages, with clear sampling strategies in cases of large populations of artifacts. To move the conversations about the past beyond presence and absence of phenomena requires quantitative assessments of the variability present (Van der Leeuw 1993:243). A tiny sample of elemental analyses of green metal artifacts may show some use of tin bronze, but only a thorough analytical program will be able to evaluate if tin bronze was the alloy of choice or a minor, perhaps even incidental player in the metal technological system of a past society. More thorough analyses of assemblages may reveal potentially significant variations in alloys by time, function, style, or other dimensions, and the degree and patterning of variability in alloys may reveal much about production and exchange in alloys, base metals, and finished products. Moreover, analyses must be conducted not only on finished products, but also on fragments, residues, byproducts, manufacturing artifacts, and deposits associated with production activities. In the absence of these analyses, archaeologists must temper their conclusions to reflect this lack in their data.

Many past materials analyses in archaeology have focused on the questions of functionality of an artifact or origin of the raw material(s), because these are much easier to establish in the laboratory than to determine how important cultural valuation of a material or technique was. These laboratory data, however, do provide a baseline against which cultural factors can begin to be assessed. Evaluating the technology means considering the total environmental, economic, social and ideological context, as well as that of other technologies used in the culture. As Killick (2004:574) noted, "[s]ocial interpretations of technological acts will always be much more credible if it can be shown that the same pattern recurs in different contexts."

Implementing this research agenda will not be easy (Muhly 2003). Determining the technical analyses needed is the first challenge. These analyses can be expensive and will require hard-to-find specialists, preferably ones who function well in a collaborative environment. Sillar and Tite (2000) explored what would be required to go forward with materials analyses within an anthropology of technology framework for archaeology, a prospect they viewed as an exciting phase for materials sciences. Cross-training in archaeology and the sciences is producing archaeological scientists who can address technology in cultural contexts. They saw long-term, large-scale research excavations as the venues that could provide the level of funding and constellation of specialists for the ongoing collaboration required to meet these agenda. Experimental archaeology where processes indicated by materials analyses are replicated to test the viability of the reconstructions were also seen as central, leading to new insights into practical, social, and technical requirements for carrying out the technological activity (Fasnacht 2009; Heeb 2009; Jackson 2009; Ottaway and Wang 2004; Pryce 2009; Rovira et al. 2009).

Carrying out such ambitious research programs using prehistoric data is not yet routine. Most studies implementing anthropology of technology approaches have been performed on modern, historic, or ethnographically accessible topics (Killick and Fenn 2012). For archaeological examples, as Killick (2004) noted, the richer the material culture, the better the excavation, and the finer the temporal control, the more believable a social reconstruction of technology will be. Few studies have yet met this standard. Some that successfully use material analysis to reconstruct sociocultural systems include Hosler (1994), who used metallurgical analysis to reconstruct a world in West Mexico where the sounds and colors of metals were their most meaningful properties. Pétrequin (1993) used the extraordinary preservation of organic materials in the Swiss Lake dwellings to reconstruct technological choices and social boundaries. Jones (2002) used comprehensive analyses of pottery assemblages including petrography to reconstruct patterns of communal, household, and individual food consumption in Neolithic Orkney in order to shed light on social relations at local and regional scales. Epstein (1993) analyzed a notably "inefficient" pre-Columbian copper-smelting technology in the Andes, which illuminated social meanings of technological acts at Cerro Huaringa, Peru. These studies provide some guidance for scholars seeking to make substantive contributions to understanding human-technology relations in prehistoric societies.

Technological Systems

When discussing technologies in social contexts, it is important and useful to differentiate a technology (the application of knowledge of the materials properties of physical things to achieve practical purposes) from a technological system (the way technological knowledge is implemented in a specific context) (White and Hamilton 2009:360). For example, during the 2nd millennium B.C., societies that had the technology to produce bronze artifacts existed in many regions of Eurasia, from Iberia to the Pacific Ocean. However, the technological systems by which they produced the bronze artifacts—the ore selections, smelting equipment and procedures, designs of refractory equipment to hold the molten metal, molds, labor organization, and so forth—could show dramatic differences. Contrast, for example, the central plain of China's early states, with their piece mold cast vessels weighing as much as 875 kilograms (Chang 1980:37; see also Barnard 1980:221) to southwestern Europe's array of simply cast, small implements and the occasional ornament (Roberts 2009a, 2009b). Each region had knowledge of bronze technology in the general sense, but the technological systems, the performance characteristics of all of the equipment and products, the specific processes and activities for producing and using bronze artifacts in the two regions, and the know-how to carry out the activities to implement the technology in each system, were markedly different. Practitioners in the two technological systems were unlikely to be interchangeable.

An attraction of the general concept of a technological system is that, for the purposes of deeper analysis, it facilitates isolating subsets of materials and behaviors pertinent to the functioning of a technology from the totality of a sociocultural system. Anthropologists and archaeologists rely on similar constructs for exploration and analysis, be it a settlement system, a subsistence system, or a symbolic system. Such constructs are open subsystems artificially but usefully isolated from a larger whole for the purposes of analysis and understanding. A technological system is a logical, and, for the archaeologist in particular, extremely useful subsystem of analysis.

When studying a metal-producing technological system, and particularly when comparing technological systems from two or more societies, it is helpful to appreciate that production of metal artifacts is a complex technology requiring many interlinked technological subsystems. For example, pre-industrial bronze-producing systems normally required competence in ceramic technology as well as metal technology. Crucibles, containers used to hold molten metal often made of ceramic, must meet the minimum performance requirement of neither melting nor breaking when filled with contents heated to hundreds of degrees centigrade. Clay fabrics vary in their response to thermal stresses, and ceramic bodies resistant to thermal stress commonly have constituents with low coefficients of thermal expansion (Rice 1987:105). This specification can be achieved in various ways with clay selection and additives (temper). Thus, it is clear that the design and successful production of ceramic crucibles represent a technology in and of itself that is critical to the success of individual metal technological systems (Martinón-Torres and Rehren 2014). Beyond the basic requirement of holding molten metal, many other performance requirements of crucibles will depend on the specifics of other components of the technological system. For example, smelting different ores or melting different alloys occur at different temperatures and with different atmospheric requirements. One performance specification for a crucible used in smelting lead, possibly the earliest smelted metal in the Near East (Roberts et al. 2009:1013), is that it stays intact with contents at only 328°C. A smelt of copper ore such as malachite requires a container with specifications that can perform (remain intact) at temperatures ranging from 700°C–1200°C, the temperature range in which copper can be obtained from a pure carbonate ore (Glumac 1991:217). The minimum refractory specifications for the two smelting systems will differ, which will affect how the containers for the two systems (with respect to clays, additives, even shape) are made.

Another major factor in a molten metal technological system is the scale (volume) of product produced per smelting or melting episode. Differences in episode scale will have important ramifications for the design and technology of the interacting components of the overall technological system. Barnard (1980:221) discussed some Shang period crucibles containing more than 1000 kg, versus a kilogram or less for the Iberian example. The refractory specifications for furnaces and crucibles designed for processing large quantities of metal must consider that the episode could take many hours to complete (Craddock 1995:147). Rehren (2003:213) noted that "a furnace with tapping holes...may run for days." One refractory performance requirement of the large container is that it remains intact at high temperatures for a much longer period than would be required of the crucible processing small quantities over a few hours. These different performance requirements will likely engender variation in the design of the crucibles for the different scale systems, from the clay fabric recipe to the morphology of the crucible (Barnard 1980). Moreover, other equipment and materials must work with the containers. How is heat applied? What are the atmospheric requirements and how are they achieved? The practitioners undertaking the production of copper-base artifacts would need the practical knowledge and skills (know-how) to produce the necessary manufacturing equipment of various materials (e.g., ceramic crucibles, stone molds) to the practical specifications required to accomplish all the steps from mining a raw material to creation of finished product, in a *specific* technological system. (Further discussion of refractory technology is undertaken in TAM 2B, chapter 5.)

COMPLEX TECHNOLOGICAL SYSTEMS

The example of how crucible design must relate to other components of a technological process introduces another construct with heuristic value for this volume: the complex technological system. Schiffer (2005:486) defined a complex technological system as "any technology that consists of a set of interacting artifacts; interactions among these artifacts—and people and sometimes environmental phenomena—enable that system to function." Thus, in a complex technological system, the critical performance characteristics among several artifacts involved in manufacturing finished products are tightly interrelated (Arnold 2007). To produce functional artifact A requires producing artifact B that effectively functions with artifact A, which cannot be made or does not perform adequately unless artifact C is effectively produced, and so on.

However, the conception in this volume of a complex technological system lays greater stress on the interactions of *people* than that of Schiffer. This volume strives to explicitly integrate the roles of materials and society. A complex technological system is here defined as a system involving suites of interacting *technologies* comprising more than one material and requiring multiple production steps and social cooperation for implementation (modified from Schiffer 2005; see also Costin 2005:1054). Technological know-how is applied to the linked manipulation of two or more materials to attain a finished product, and group effort and usually a division of labor are required to conduct the full sequence of steps to implement the technology. A complex technological system engages not just at an individual level, but at the societal level.

A given technology is not inherently performed by a complex or non-complex system. The bow and arrow would qualify as a complex technology based on its integration of multiple materials, but if individuals in a society make bows and arrows from start to finish, their production would not qualify as a complex technological system, as conceived here (but cf. O'Brien and Bentley 2011). On the other hand, a bow and arrow factory employing several practitioners among whom there was a division of labor engaged in the production of these artifacts would qualify as a complex technological system.

However, it is important not to essentialize the construct of a complex technological system, as there is, more realistically, a continuum of complexity across technological systems. The construct is useful in this volume as a reminder to archaeologists to look for evidence of labor organization and its changes over time in reconstructing any technological system. Some technological systems can, in theory, move along a continuum from simple to complex. Thus, at one village at one point in time, production of ceramic vessels may be undertaken entirely by an individual potter, who conducts each activity from mining the clay to firing the finished products. At another point in time or at another village, potters working in the same technological tradition—same methods of preparing clay, building vessels, firing, etc.—may incorporate a division of labor (Joyce White, field records from Udon Thani, Thailand, 1979–1981). The labor distinction is akin to Schiffer's (2001b:5) contrasting of multi-actor versus single actor technologies. This point is important for archaeologists seeking to gauge trends towards social differentiation and complexity, even though in many archaeological cases structures of division of labor may be hard to prove.

Among prehistoric technologies, metallurgy may be among the most complex systems of interacting artifacts and technologies developed by prehistoric societies. The smelting of copper and tin, and the production of bronze artifacts, are parts of a complex technological system ipso facto involving multiple materials and multiple steps in: (1) the acquisition and manipulation of ores and fuels for smelting; (2) the creation of refractory ceramics; (3) smelting and refining of molten metal; (4) alloying; (5) fabrication and manipulation of molds often made of stone and/or ceramic; (6) casting; and (7) undertaking post-casting treatments such as working. When one appreciates the amount of knowledge, skills, and steps involved in even the simplest metal production system, it is difficult to imagine how and why one individual would or could perform solo every step from ore prospection to working the final cast object; and indeed, there are no ethnographic examples of single-actor metal production. The most plausible inference for the archaeologist is that more than one individual was involved in producing any single bronze artifact.

Life History Framework

The life history framework is fundamental to the study of technologies in social contexts (e.g., Skibo 1999). "[T]he life history framework reminds us that artifacts begin as raw materials, and are altered and assembled, transported and exchanged, used and reused, maintained, and eventually discarded, abandoned, or ritually deposited" (Schiffer 2011:23). A related concept is the "cultural biography of things" (Kopytoff 1986) where, after manufacture and before discard, an object may have an extended and changing life of meaning, use, and value.

This framework is important for archaeologists in that it helps to specify what parts of a life history particular evidence illustrates, and what parts remain to be identified or studied. For example, excavation of a pot from a grave illustrates, at the very least, that the vessel was ritually interred with a deceased individual. Whether that pot had a prior use for cooking requires identification of evidence for soot, residues of cooked foods, use-wear consistent with cooking activities, etc. Further questions on where the vessel was made require petrographic and chemical analysis. Taphonomic analysis may be needed to determine the artifact's history at and after disposal. Thus, the excavation data (pot recovered from grave) is just the first step of potentially a long investigation into what that pot could tell about the past. It reveals only one small part of the vessel's life history or its full behavioral chain, and to answer questions about other parts of its life history requires additional studies targeting evidence for other stages. Keeping a life history framework at the forefront helps archaeologists avoid becoming like blind men describing an elephant, insistently generalizing about the whole from their tiny piece of the evidence.

CHAÎNE OPÉRATOIRE

A related concept, often used to refer to a subset of the life history framework, is the concept of the *chaîne opératoire*. In essence, the *chaîne opératoire* focuses on that portion of the life history relating to manufacture of the artifact. This concept again is very useful for archaeologists because often artifacts have evidence of manufacturing processes embedded in them that illuminate the technologies used in creating them. For example, a pot may have evidence of horizontal joins indicating it was formed with a coiling technique. A copper-base artifact may have microscopic strain lines indicating it was hammered at one stage.

The *chaîne opératoire* concept, introduced by Leroi-Gourhan (1943), has been widely used by archaeologists interested in technology, especially lithic technology (reviewed in Bar-Yosef and Van Peer 2009), and it is also common in ceramics studies (Balfet 1991; Jones 2002; Roux 2003; Kolb 2011 has a useful overview). The term has more recently appeared in the archaeometallurgy literature, beginning with Pfaffenberger (1998; see also Bauvais and Fluzin 2009; Ehrhardt 2009; Hoffman and Miller 2009; Jackson 2009; and more recently Dolfini 2014; Kienlin 2014; and Stöllner 2014).

As Pfaffenberger (1998) pointed out, comprehensive reconstruction of the *chaîne opératoire* of a production system has great value for understanding past technologies, economies, and social systems, including ancient metallurgy. An emphasis on the explicit reconstruction of individual steps of the production sequence makes clear what evidence is lacking (e.g., do we know how the crucibles or working tools were made, how many people would be needed to smelt ore using that furnace structure, how the product was formed, etc.), what alternative steps would have been possible for the craftsperson, and how closely this sequence of steps matches those of areas from which this technological system perhaps was derived. To truly understand the spread of technologies, we must look not at the spread of finished products, but at the spread of *chaînes opératoires*.

TECHNOLOGICAL ACTIVITIES

Another term useful in discussions using a life history framework is activities (Pfaffenberger 2001; Schiffer 2011:23). Technology is not a one-time event, or a cognitive pursuit, but to exist requires recurring sequential activities in the physical world. A product goes from activity to activity during both its manufacturing and use-life. For example, a pot engenders specific activities to create the vessel such as shaping and firing and additional activities during its use-life such as cooking and storage. The activities undertaken to implement a technology bring into articulation the social, geographic, and artifactual components of a particular technological system.

Technological activities may be considered that subset of activities related to manufacturing and maintaining all artifacts interacting within a technological system. Archaeologists may find concrete evidence for many technological activities, such as a kiln for the activity of firing pottery, or a cluster of flakes that can be refitted, demonstrating the activity of shaping and sharpening stone tools. Technological activities commonly exist within behavioral chains where activities occur in characteristic sequences in *chaînes opératoires*.

That technological activities usually occur in sequences has implications for their relationships in space—sequential activities often occur in proximity—and time that are very useful for archaeologists inferring past technological systems. Suppose at a site evidence for activity A, for instance mining copper, is identified, and also contemporaneous evidence for activity D further down the *chaîne opératoire*, perhaps casting a copper-base alloy, but no direct evidence is excavated demonstrating the intervening activity of smelting copper. Since it is unlikely that the mined ore would be removed far from the mining site for the activity of smelting and subsequently brought back to the mining site for casting, the best inference considering the empirical evidence for activities and the overall *chaîne opératoire* for copper production is that smelting activities also occurred at or near the mining site, but empirical evidence for smelting has not yet been identified in the archaeological record.

In this volume, production aspects of metallurgy as a fully functioning complex technological system are conceived of in terms of two discrete stages: the *production of metal* and the *production of products made from metal*, or *metal product manufacturing*, each with its own activities and finished products. Metal production encompasses all processes, activities, materials, manufacturing equipment, etc., used to extract metals from ores. Various shaped and sized masses of metals are the usual final products of this step, and, if the masses are of a regular and standardized shape and size, the masses are termed *ingots*. The manufacture of finished products from metal comprises all processes, activities, materials, manufacturing equipment, etc., used to create finished products from extracted metals. Sometimes metal production is termed "primary production" and metal product manufacturing is termed "secondary production" (e.g., Roberts 2011:144).

The distinction between the two kinds of production is useful, particularly because the activities for the two steps are often undertaken in different locations. The distinction can be artificial, however, in that ingots may have their own social/symbolic role as items of trade. Overall though, when viewed in the full social context within which metals are circulated, there are no known societies that smelted metals and simply used unformed miscellaneous-sized metal masses as objects in their own right. All known societies that smelted metals have *added value* to those metals with subsequent shaping into deliberate forms (in other words, fabricated metal products), even if different subsets of the societies undertook the two production steps.

Over the course of product life histories, the technological activities involved in producing final metal products in either stage, metal production or metal product manufacturing, can result in a range of evidence left in the archaeological record (see Tables 7.1 and 7.2). Technological activities can be identified in the archaeological record by remains such as:

(1) *manufacturing equipment* including molds and crucibles used in the creation of finished products;

(2) *manufacturing byproducts* or artifacts that exist because of human activities but are not themselves intentionally created, such as casting drippings, and manufacturing residues including mineral products left on artifacts that form from chemical reactions during manufacturing, such as dross;

(3) *technological features* or discrete installations created for technological activities, such as casting hearths;

(4) *technological deposits* or incidental remains (i.e., not in situ installations) resulting from technological activities, such as middens of slag or spreads of crushed host rock;

(5) some *technological events* can be inferred, for example, the volume of an intact crucible suggests the size of a discrete occurrence of a casting activity.

In addition to these types of archaeological remains, archaeologists may find evidence for:

(6) *technologically modified landscapes*, i.e., modified during the functioning of a complex technological system, such as evidence of mining, deforestation (e.g., to produce fuel for pyrotechnologies), or reconfiguration of hydrological systems (e.g., creating sluices), that provide further indications of technological activities.

The spatial relationships among the various kinds of evidence for technological activities can provide abundant information on the functioning of the technological system, as well as how that system was integrated with other parts of past societies, and, in fact, how technological activities themselves integrated past societies (Pfaffenberger 2001). For example, whether evidence for technological activities is segregated from or integrated with residential evidence, or if some activities are documented at some sites, and other activities at other sites, these observations are important for inferring the role of social organization in the implementation of the technological system. Also, temporal changes in the spatial organization of production should help develop inferences about societal changes.

Finally, Schiffer (2011:34) pointed out that technologies themselves have life histories, alternatively called technology life cycles. The stages and processes of technological invention, application and product development, adoption, commercialization, adaptation, and senescence are extremely important for archaeologists to consider when studying past technological systems. A society's initial adoption of a technological system (considered in greater detail below), versus adaptations the society makes to a previously acquired technological system are likely to reflect distinct social processes as the stakeholders' needs (*performance requirements*) change. For example, bronze often appears as a minor constituent in a society's material culture when it is first adopted, such as in Longshan and contemporaneous cultures found in the People's Republic of China. In the central plains, a few copper-base tools (e.g., awls) and personal ornaments (a small bell) have been recovered from the pre-dynastic period (Linduff 2000:13–14). Metal had no evident prestige value during its initial adoption stage there. After a few hundred years in this minor role and after centralized political systems developed, stakeholders transformed the application of metal, adapting it to prestige uses and developing many innovations in manufacturing metal products to create finished products, such as wine containers for use by the elite (Underhill 1996). Therefore, the performance requirements of bronze products in the Erlitou period changed from the minor symbolic and functional performance requirements of bronze products during the preceding Longshan period. This example illustrates the importance of not conflating life cycle stages when interpreting a region's technological trajectory.

Technological Choices, Styles, and Traditions

Another suite of concepts common in the anthropology of technology literature derive from appreciation that cultural values, traditions of production, and social and practitioner (i.e., agent) contexts strongly influence the configuration of any technological system.

TECHNOLOGICAL CHOICE

The notion of technological choice refers to the reality that any technology may be implemented in a number of ways. Numerous raw material and design variants exist in the broad category of bow and arrow, for example, and societies, communities of practice, and individual practitioners choose among sets of options. Many proponents of anthropology of technology approaches have embraced and highlighted the importance of this concept (e.g., Bleed 2001; Lemonnier 1993; Pétrequin 1993; Pryce et al. 2007; Sillar and Tite 2000). As Sillar and Tite (2000:3–4) noted, "The physical properties of matter provide a vast range of technical possibilities, only a few of which are ever realized....It is only through a consideration of the overall context that affects the availability of resources as well as the valuation of alternative techniques that we will be able to explain why particular technological choices were made and what material and social effects they had. This overall context spans the environmental and technological constraint, the economic and subsistence base, the social and political organization, and the ideology or belief systems of the people making the choices."

Technological choices—whether by societies or individuals—can take place in five areas, in the view of Sillar and Tite (2000):

(1) Raw materials (e.g., choice of clay, choice of ore, treatment of slag—will it be resmelted to extract residual metal or discarded?).

(2) Tools or equipment that work the raw materials (e.g., mauls, picks, crucibles, hands).

(3) Energy sources to alter the raw materials (e.g., fuels, human and animal muscle, the sun).

(4) Techniques to bring the raw materials, energy, and tools together to produce the desired result. Most of these techniques are not in the realm of formal knowledge, but rather in the form of body knowledge, the know-how (e.g., Crown 2001). Anyone who attempts to tie his or her shoes using only written instructions can appreciate the difference between knowledge and know-how.

(5) The operational sequence, the order of production steps or the *chaîne opératoire*, in which the preceding four areas are linked together to produce usable results. While the technology may structure many parts of a sequence (copper must be acquired before the alloy bronze can be produced), some aspects of any technological sequence may have alternative means (e.g., smelted copper can but need not be refined before alloying). The *chaîne opératoire* includes not only the specific steps to making the product—a bracelet, a plowed field—but also the frequency of repetition and the locations where the activities are performed. For example, does casting take place seasonally near consumers or year-round near smelting locales?

Technological choices are constrained by many factors, from the materials themselves, preceding choices in the technological system, physical limitations of the human body, social needs, labor organization, etc. Evidence for both choices and constraints may be found in assemblage variability; trial attempts may be differentiated from the ultimately successful product that endures and becomes prevalent. Some constraints on choices will be relatively easy to determine, like natural resource availability, others will be difficult, such as individual preferences. Nevertheless, Bleed (2001:159) argued, "Still, identifying both internal and external constraint patterns should be a part of the description and assessment of any technology." For example, if an ore was not present in the environment, it would not have been among the ores that could have been chosen by the ancient metal smelter. Identifying the range of constraints in any context will facilitate identification of choices.

The phrase "technical choice" is also in the literature (e.g., Hosler 2009). Although some scholars use technical choice and technological choice interchangeably (e.g., Lemonnier 1993), it may be useful to conceive of the two as overlapping concepts, with technical choice focusing on specific alternative techniques possible in a production sequence, whereas technological choice (Hoffman and Miller 2009) encompasses a broader range of technological behaviors, especially those characteristic of a technological system or tradition. This distinction may inform, for instance, when noting that, in an assemblage of copper-base adzes, a small percentage were worked. Therefore, occasionally a technical choice to work adzes was made by the practitioner. However, the overall assemblage suggests that unworked adzes were the prevalent technological choice in the system.

TECHNOLOGICAL STYLE

Heather Lechtman, a materials scientist and archaeometallurgist, was one of the key early proponents of the concept of "technological style" (Lechtman 1977). She observed that most new technological choices are constrained by past technological choices, not just due to physical materials properties, but to cultural rules on how to manipulate materials. These cultural preferences and patterns in how a society implements a technology constitute the style of the technology in that society. Thus, many specifics of technological systems can be products of a given society's technological style. Past technological choices often produce an unconscious "this is how things are done" when an individual is considering alternative actions. A technological style is a way of handling materials that is characteristic of a particular group or culture.

Technological styles can extend to many different technological systems in one society. For example, the manufacturing technology of Shang and Zhou bronze ritual objects was strongly influenced by how ceramics were made prior to the appearance of bronze (Barnard 1980; Franklin 1983, Wang and Mei 2009:387). Metals were cast into elaborately carved ceramic piecemolds, i.e., the technology was adapted from previous habits of carving and incising ceramics. In another example, the depletion-gilding techniques of the Inca were a result of a cultural preference that decorations be integral to the manufactured object, so that a golden appearance was produced by depletion gilding rather than plating, and similarly colorful designs on textiles were produced by weaving rather than appliqué (Lechtman 1977). In addition, the use of technological choices in systems and techniques to mark social boundaries, either within a group or between societies, has been observed in ethnographic and other contexts (e.g., Mahias 1993; Pétrequin 1993; Stark 1998).

Awareness of the impact of technological style on the configuration of any technological system is essential for archaeologists and archaeometallurgists attempting to explain choices that emerged in trajectories of socio-technical change (Ehrhardt 2009; Hegmon 1998; Hoffman and Miller 2009). Patterns of change are not random. The configuration of concurrent and predecessor technological systems is among the key elements influencing the course of technological change over time and space and is one of the most central topics to archaeological scholarship.

TECHNOLOGICAL TRADITIONS

In turn, technological choices and styles in technological systems help define *technological traditions* (Chilton 1998, see also discussions of technological phylogenetics, e.g., O'Brien et al. 2008). The technological tradition is a system of technological knowledge and practice that is passed among practitioners from one generation to the next and endures over time. The concept of a technological tradition is again extremely helpful to archaeologists seeking to explain spatial and temporal relationships among technological systems. These traditions may be transmitted to neighboring regions (horizontal transmission) under various conditions. As the knowledge and practice are culturally transmitted over time and space and adopted into new cultures, innovations may emerge that introduce variation into the practices and knowledge base.

Comparison of technological traditions across regions can provide evidence of great interest to archaeologists regarding the relations among communities. Shared decorative styles provide relatively weak evidence for substantive social interaction because of the ease with which superficial traits can be copied (Gosselain 1992). Shared technological traditions, in contrast, can produce more robust evidence for ancient social networks because transmission of expertise likely required some amount of face-to-face training and/or artisan-to-artisan sharing of know-how. As an outstanding example, Hosler (2014) provides a comprehensive argument, based partly on compelling technological similarities, that the early copper-base metallurgical system in west Mexico must have been introduced by metalworkers traveling with maritime Ecuadorian seafarers, who traversed distances of 3000 or more kilometers between the two regions.

However, ethnographic research, particularly of pottery production, has shown that technological traditions are not discrete isomorphic lineages that endure over time as unchanging entities. Different components of a *chaîne opératoire* may (or may not) be transmitted independently. This can create the situation where, for example, a tempering tradition may have a particular distribution in time and space that overlaps but does not precisely correspond to a forming or a firing tradition (see Gosselain 2011 for examples). The probability of the differential transfer of different steps in a *chaîne opératoire* shows the importance of large and comprehensive data sets for demonstrating not only the spatial and temporal distributions of technological traditions, but also the specific processes of divergence, convergence, and "co-relation" (i.e., when components of a *chaîne opératoire* commonly co-occur).

Technological Change

How and why technologies appear and change is another important topic for proponents of anthropology of technology approaches, as it is for archaeologists of many persuasions (e.g., Geselowitz 1993; Killick 2004: Kuhn 2004; Roux 2003; Schiffer 2004, 2011; A. White 2008). The frameworks for examining the issue of technological change have moved far beyond the conventional diffusion, migration, and independent invention. A set of terms and approaches has been developed to discuss the mechanisms of change, including cultural transmission, learning frameworks, communities of practice, and adoption processes. Before considering these terms the role of discovery in technological change needs to be reviewed.

DISCOVERY, INVENTION, AND INNOVATION

In approaches to studying technological change that come from the Conventional Paradigm, much attention is focused on the origins of the technology, its invention, who discovered the idea first, and similar concerns. Adoption of the technology into other societies is conceived in terms of diffusion and/or migration. In situ change once a society acquires a technology is of less interest, but more or less adaptationist and progressive models are inferred (e.g., more efficient procedures and higher productivity developed over time are expected).

In contrast, anthropology of technology proponents are less interested in singular discoveries—the discovery that a key material property can be manipulated. The first discovery of a piece of knowledge is considered archaeologically unknowable. As a result, anthropology of technology proponents would downplay the importance and singularity of the discovery that metals can be smelted from ores, perhaps even maintaining that these types of discoveries could have occurred numerous times.

While there is recognition that trial and error activities by individuals can produce new knowledge and new ways of doing things, the key piece of the process of technological change is what happened *after* the discovery of, for example, a material property. In many and probably most cases, nothing happened. If no technological system emerged that exploited the observation and applied the new knowledge systematically, the discovery resulted in no technological change. In some cases, additional inventions and innovations were made so that the key discovery acquired an *application*. If, after a discovery is made of a material property or a material behavior, a prototype of a product is created, the next step has been taken on a path toward a technological change, although the process may again, stop at the prototype stage without any significant impact or technological change (as in the case of Leonardo Da Vinci's flying machine).

The development of a technological system, including the innovation of manufacturing equipment and procedures so that a desirable product is created and can be replicated on a regular basis, is the key to materializing a discovery. This later stage in the discovery-to-full-technology continuum is the focus of interest for proponents of the anthropology of technology. They are interested in why and particularly how the enabling technological components were developed. The societal contexts that created viable production systems that could and did repeatedly and reliably manufacture a desired product (such as a refractory system for smelting copper as well as other production equipment for casting the molten metal), serve as the key to grasping technological development and change.

In short, the creation of viable technological systems is of greater interest and importance than the original idea or discovery. It is also the technological system and its accompanying know-how that are usually transmitted, adopted, and adapted to new social contexts, not just the idea or prototype. Development and transmission of a technological system, however, may be stillborn if there is no consumer demand for the product. Therefore, examinations of consumer desires and behaviors are integral to understanding trajectories of technological change.

Problems and needs within societies often provide contexts where inventions and innovations result in technological systems that can replicate products that address the society's needs. Schiffer (2011:36–37) noted that problems addressed by inventions can include satisfying curiosity and alleviating boredom, as well as responding to drought or warfare or similar practical needs. In short, the study of technological innovations is inextricable from the study of their social context.

According to Schiffer (2011), many inventions and innovations occur during the development of a technological system that commercializes a product after an application has been envisioned. A technological system is developed often in the context of a project that seeks to take an idea or prototype and find a way to produce and replicate a product, making it reliably available to a society. During this phase, one or more innovators take a new idea, either invented locally or from an outside source, and, building on prior existing knowledge, integrate the new idea "into the spectrum of existing skills, activities, symbolic schemes and ideological structures of that society" (Ottaway 2001:88–89). Often an innovation leads to additional innovations, as solving one problem (perhaps designing a furnace that can attain a particular temperature), necessitates other innovations to address new performance problems (perhaps a new way to remove the product from the furnace that minimizes the chance of injury for the practitioners). One innovation can therefore result in a cascade of inventions and innovations (Schiffer 2005, 2011:57) before a sufficiently stable and reliable technological system is designed.

CULTURAL TRANSMISSION

> *Scholars analyzing the "spread" of metallurgy have tended to rely either on the migration of a people...or...the all-encompassing and vague concept of diffusion....The transmission of metallurgy can be addressed more systematically by analyzing the metallurgical knowledge, skills, and equipment that would be required to perform each identifiable transformation from ore to metal... (Roberts 2014:429)*

Many instances of technological change in societies of interest to archaeologists occurred because the societies adopted technologies, or more accurately technological systems, from other societies. The study of cultural transmission, i.e., how knowledge, cultural behaviors, etc., are passed from one individual to another and one social group to another, is critical to the study of technological transmission—the transmission of technological knowledge and behaviors among individuals and societies (see Stark et al. 2008).

Thus, the topic of technological transmission is a subset of the larger study of cultural transmission, the movement of any cultural phenomenon among societies, such as values, skills, knowledge, and practices. Cultural transmission refers both to *vertical transmission* (i.e., the passing on of practices through generations of families or social groups) and *horizontal transmission* (i.e., the passing of practices among contemporaneous social units). Even though interest in cultural transmission dates from the 19th century (as reviewed in Lyman 2008), seminal publications from the 1980s stimulated a renewal of interest in this topic (e.g., Boyd and Richerson 1985; Cavalli-Sforza and Feldman 1981). Recent literature on cultural transmission in archaeological and anthropological scholarship is growing (e.g., O'Brien 2008; Roberts and Linden 2011) as archaeologists increasingly focus on assessment of cultural variability (Eerkens and Lipo 2005; O'Brien and Bentley 2011; Shennan and Wilkinson 2001). There are many varying viewpoints and emphases among scholars pursuing this topic (e.g., Arnold 2011; Eerkens and Lipo 2005, 2007; Henrich 2001; Jordan and Shennan 2003).

Technology is a special case of cultural transmission, in part, because a key question is: How are entire systems of mental knowledge and physical practice, not just ideas, beliefs or symbols, transferred? Transmission of metallurgy or any complex technological system is ponderous because interlinked technological systems in aggregate create the finished product. The multifaceted and precise systems of metallurgical knowledge and behaviors, the mastery of lengthy production sequences, and the required labor coordination are not likely to spread to another culture by passive observation, exposure to trade goods, or to the idea of smelting. Therefore, transmission of sophisticated metal technology to a society with no prior experience of working with metals could not occur without direct instruction (Roberts et al. 2009; White and Hamilton 2009).

It is now recognized that simple assertions of diffusion and the "spread of ideas" are woefully inadequate answers to how and why metallurgy spread, whether in Thailand or elsewhere (Gullapalli 2009:455). "Diffusion theory obviously rests on a simplistic physical model, such as the ripples that emanate outward when a pebble is dropped into a pond, that is unrelated to the actual processes that lead from a technology's inventors to its adopters" (Schiffer 2011:17). So how does an anthropology of technology approach address technological transmission?

LEARNING FRAMEWORKS

The first fundamental concept in current transmission theory is the "learning framework" (along with its related term "teaching framework" [see Schiffer 2008; Schiffer and Skibo 1987; Wallaert-Pêtre 2001]). Technological transmission must first occur among individuals in a society, such as parent to child, or master to apprentice. There is a growing body of literature on how learning frameworks affect technological transmission (e.g., Crown 2001; Minar and Crown 2001; Wallaert-Pêtre 2001). Among the considerations particularly pertinent to transmission of technologies is that transmission of technological know-how is highly related to acquiring motor skills. The skills are acquired usually through a combination of instruction and practice, sometimes lasting years before competency is attained. In pre-industrial societies, the supervised practice in craft-making begins in childhood, often between the ages of six and twelve. Depending on the learning framework and the complexity of the technological system, training can extend into adulthood, with skills potentially ever improving and/or changing throughout young adulthood before an individual becomes a master of the craft. It is the cross-generational transmission that maintains (some would say reproduces) the technological system in an ongoing society.

Much ethnographic work has been undertaken regarding technology transfer in traditional societies, especially regarding pottery production (e.g., Gosselain 1992, 2008; Wallaert-Pêtre 2001). This research is revealing many considerations important for archaeologists. For example, Wallaert-Pêtre (2001), working in Cameroon, found that different apprenticeship programs resulted in some potting communities producing highly conformist products, and other apprenticeship programs creating potting communities open to new ideas and techniques. The latter type resulted in high rates of change in decorative style, vessel morphology, and clay recipes. Moreover, during a craftsperson's lifetime, he or she may reconfigure their technical repertoire, for example, if they join a new potting community (perhaps by relocation for marriage) that practices techniques different from those learned in childhood. The learning framework incorporates not only the apprenticeship phase, but also learning that takes place in adulthood. This later learning, also termed "post learning" (e.g., Gosselain 2000:200), is highly influenced by adult socialization patterns such as marriage, mobility, and engagement with local and regional economies (Gosselain 2011).

Some archaeologists may consider that prehistoric learning frameworks are beyond the ken of the discipline. However, this limitation cannot be assumed with the current level of available methodologies (e.g., Crown 2001). Examinations of mortuary sites that include petrographic analyses of grave pottery along with isotopic studies of skeletons have the potential to situate pottery production and pottery producers explicitly in landscapes. With enough data, pottery-producing communities may be located and their technological styles evaluated. Isotopic data of human skeletons may produce evidence of marriage practices (endogamy/exogamy [e.g., Bentley et al. 2005]). Judicious comparison of potting communities and the distribution of potting practices across landscapes, including careful comparisons with ethnographic examples, should at the least lead to the formulation of testable hypotheses. For example, evidence of high rates of female post-marital relocation (as may be determined by strontium isotope analyses) might correlate with high rates of pottery diversity and/or pottery change (cf. Herbich and Dietler 2008). Since adult females tend to be the pottery producers in many hand-building pottery traditions, movement into a new community may expose them to other operational sequences than the one they learned in apprenticeship and expand their repertoire, as discussed above. Conversely, little female movement during life (including female potters) could result in low exposure to alternative operational sequences. Correspondingly low rates of female post-marital relocation might correlate with conservative pottery traditions. These are hypotheses testable with sufficient, well-analyzed data.

COMMUNITIES OF PRACTICE

Building upon the concept of learning frameworks is the concept of "communities of practice" (Lave and Wenger 1991; Wenger 1998). Individuals engaged in crafts such as metallurgy or pottery production usually do not practice their vocation in isolation but rather in group settings, such as villages, guilds, families, workshops, or neighborhoods. These productive social groups overlap and are embedded with other social contexts and groupings, ranging from households to regional alliance systems. Technological practice commonly fosters communities of practitioners who share knowledge and resources specific to the technology, and the shared practice engenders a shared identity. The practitioner identity may or may not correlate with other identity groupings, such as villages or subregions, depending on the practices of the pertinent societies. If practitioners are not constrained by a rigid social system, such as a caste system, communities of practice over time can multiply, subdivide, join in various ways in accordance to the fluidity of the society and, in the case of a technological community of practice, in accordance with how the practitioners relate to each other in the activities of the technological system.

There is now a large body of literature that provides theoretical underpinnings to the role of practice in social theory (e.g., Bourdieu 1977; Dietler and Herbich 1998; Wenger 1998) and that explores practice theory in archaeological or ethnoarchaeological contexts (Dobres and Hoffman 1994; Peelo 2011). There are, of course, many variations within this general perspective, but delineation of these variant points of view is beyond the scope of this work. The important point for this volume is to appreciate the usefulness of the community of practice concept for studying prehistoric technological systems (e.g., Sassaman and Rudolphi 2001).

TECHNOLOGICAL COMMUNITIES AND REGIONS

Investigating technological communities can assist in specifying regional social relationships. "Because it is through social relations of production that technologies are enacted, empirical aspects of prehistoric raw material modification can be studied as a 'window' into those social relations" (Dobres and Hoffman 1994:216). In other words, archaeologists can use their studies of prehistoric technological systems to learn about *social networks* of the past.

Technological ways of doing things requires ongoing transmittal of knowledge, materials, behavioral sequences, and training experiences from person to person and generation to generation. The areal extent of a technological system, therefore, has significant social meaning. *Networks of skilled practitioners* may be distinct from other social networks and groupings operating in the larger region, such as political units, settlement units, alliance configurations, or kinship groups (Hegmon 1998). Studies of sub-Saharan potters have shown that communities of technology practitioners often span boundaries maintained by other societal systems (Gosselain 2011; although cf. Graves [1994] for Kalinga evidence for co-occurrence of social and technical boundaries). The key points are that congruity between technological regions and other kinds of spatial social units such as polities, confederations, ethno-linguistic groups, or village alliances must be demonstrated and not assumed and that technological communities may cross other social boundaries.

Chernykh (1992) was perhaps one of the first archaeometallurgists to recognize that coherent metallurgical traditions imply the past existence of networks of practitioners. The theme was revisited by Roberts (2011:143) who stated:

> The transmission of copper production expertise throughout Eurasia would therefore have to involve the movement of a sufficiently skilled individual or a group to a new ore source. This introduces two interrelated mechanisms for metallurgical transmission—the movement of metal workers and the learning of metallurgy from experienced metalworkers. This would have created an extensive yet fragile network of metallurgical expertise, potentially over substantial distances.

Chernykh's (1992) comprehensive overview of Eurasian metallurgy also demonstrated that technological provinces with shared technological traditions can span vast terrains at continental scales and endure for millennia. He stated (1992:9) that "the morphological and technological unity of the different centres of the [metallurgical] province depended mainly on close and continuous interactions between different population groups, as well as between the associations of artisans..." Therefore, the spatial extent of any technological system is an important object of study for anthropologists and archaeologists, and the extent of complex technological systems that require multiple actors and division of labor is particularly informative about a technological community of practice. Distinguishing the extent of practitioner knowledge and maintenance of a complex technological system distinct from the extent of the distribution of its products to consumers is an important index of knowledge interaction as opposed to economic exchange.

Introducing division of labor coordinated to steps in a production process may enhance the potential for appearance of technological innovation, economies of scale, and other changes in the system. Addition of different brains and sets of hands to the process may initiate intended or unintended variability to the endeavor, variability that evolutionists might argue could be subjected to processes of selection (Basalla 1988; Kuhn 2004).

ADOPTION

For a new technology to be transmitted to a new cultural context, individuals or groups in the society must chose to adopt it. They must decide to manufacture the products by acquiring the technological systems and skill sets and they must decide to use the products. Adoption processes, the how and why individuals or a society choose to take on a new technological system, and the degree of success in doing so, is a topic of great interest to anthropology of technology proponents and should be of central concern to archaeologists investigating the prehistoric dispersal of metallurgical technologies. Pfaffenberger (1998), among others, has shown that archaeologists cannot assume that mere exposure to a new technology ipso facto results in its adoption.

A large body of theory on cultural transmission and adoption has arisen in anthropology in the last two decades, as archaeologists and cultural anthropologists have sought to not only document but also explain variation of artifacts within and among cultures in more nuanced ways beyond mere assertions of diffusion or migration (Bentley et al. 2004; Bettinger and Eerkens 1999; Eerkens and Lipo 2005; Jordan and Shennan 2003; Mesoudi and O'Brien 2008a, 2008b; Roux 2008; Schiffer 2008; Shennan 2008; see especially Eerkens and Lipo 2007 for an overall review). Some of these studies investigate the transmission of technology-related traits, though many focus on stylistic variations (Eerkens and Lipo 2007). Rather than assuming that humans conduct a coolly rational cost-benefit analysis when faced with a choice among behavioral—including technological—alternatives, it now seems apparent that contextual biases can influence who or what is copied and how a copy is made. For the purposes of this volume, some awareness of several factors influencing adoption processes can help put archaeometallurgical data into clearer and broader perspective.

Adoption and Choice: Who Is the Decider?

The rate and patterning by which a technology is adopted is affected by a number of societal influences. Studies have shown that, at the individual level, not all individuals immediately embrace a new technology. Layton (1989) examined the adoption of new agricultural technologies in a modern, relatively egalitarian French village and isolated four roles individuals played: (1) The *initial adopters* (about 2.5% of the group, according to Rogers [1962], as cited by Layton [1989]) are those with sufficient resources to allow experiments that might be unsuccessful, and whose individual personalities inclined them both to be interested in technological innovations and interested in cultivating contacts outside the village. These initial adopters are admired by some but derided by others, so they pay some social cost for their innovations. (2) The *opinion leaders*, a group that sometimes includes the initial adopters, but more often encompasses people who are generally respected and better integrated into the group than the initial adopters. The majority of the group feels comfortable going to them for information and advice. (3) The *majority* of the group, having seen the innovation being tested and judged by the opinion leaders. (4) The *laggards*, those who refuse to adopt and are, in consequence, somewhat marginalized. It is important to note that opinion leaders can be but do not necessarily mean the elite of the whole group.

Defining the "chooser" of a new technology can be difficult, but potentially very important for understanding the dynamics of cultural transmission. Some models assume that the unit of choosing is the society or group and is dependent on group needs or group consensus (Kim 2001). Others assume that the unit of choice is the individual, and technological changes are a result of more and more individuals autonomously adopting the new technology, in a manner similar to a beneficial mutation spreading through a breeding population (e.g., Layton 1989). Determining the unit of choice requires analyzing the power relations among the elements of society—who will benefit from the new technology and who has the power not only to decide, but in some cases enforce the decision (Kim 2001). The choosers may well not be the practitioners; what benefits elites may not benefit the actual workers. What may be the consensus of one group in a society may not reflect the feelings and needs of other groups. Likewise, autonomous individuals have their decisions constrained not only by the preexisting technology and cultural rules for appropriate ways of doing things, but also by the decisions of opinion leaders or those who control the flow of information and resources (Kim 2001).

Adoption and Context

Another important consideration in examining adoption processes concerns preexisting contexts at various scales. An important example for archaeometallurgists is provided by Wade (1989), who looked at the adoption of lost wax brass casting among the Fali of northeastern Nigeria. He proposed a schema of four essential components of the context of adoption and innovation. The first is the *macrocontext*. This component refers to the availability of the requisite knowledge and raw materials. In the Fali case, brass was available and there was widespread knowledge of casting.

Wade's second component consists of the *immediate parameters*, or the enabling environment. In the Fali case, the society was relatively safe—warfare was mostly ritualized—settlements were nucleated, the land and farming technology was productive enough to produce surpluses, households were largely independent and were the focus of much ritual and economic activity, and the society, notwithstanding the presence of kings with little direct power, was "essentially nonhierarchical" (Wade 1989:228).

Wade's third component is the *immediate context*, e.g., the presence of already existing systems of craft organization, exchange, forms and artifacts in other valued materials, and roles for ornamental and ritual goods that must assimilate the new material. Before the adoption of lost wax brass casting, the Fali already possessed a system of craft organization based on endogenous clans, and a system of wealth generation and media of exchange based on sorghum and the production of cloth. Cloth was exchangeable for brass and vice versa, so that the production of brass goods reinforced and elaborated an already existing system of exchange and valuation. A system of distribution, based on casting to order, intercommunity trade, and travel by especially admired crafters, was already in place from other crafts. The Fali already had an elaborate material culture devoted to personal ornament and most of the artifacts cast of brass were skeuomorphs.

Wade's fourth and final component is the *active social component*, the actual reasons why the Fali purchased brass goods and the "patterns of use and signification relative to groups and individuals in Fali society" (Wade 1989:235). The first three components set the stage but do not provide explanations for the adoption of the new technology. According to Wade, brass casting technology was adopted because this new material, perceived as being beautiful by the Fali, was a powerful tool, both in itself and in the form of artifacts, for the assertion of wealth and personal and lineage power. "The brasses constituted the most elaborated and precisely articulated set of Fali material symbols: status, prestige, wealth, and power symbols, and tools of self-praise" (Wade 1989:236).

Interestingly, the way that the Fali adopted brass casting was not mirrored in nearby related societies. Other societies of the Biu-Mandara peoples, of which the Fali are one, responded very differently to exposure to the technology. The degree of the adoption of metallurgy was variable, and even some societies proficient at iron smithing never adopted brass casting. In these cases, the last three of Wade's (1989) components—the immediate parameters, the immediate context, and the active social component—were radically different from the Fali example. Historical contingency is clearly a major player in any adoption story and any account of the adoption of a new technology needs to explicitly consider all four components of the model. The Nigerian study also illustrates why a ripple diffusion model almost certainly will be inapt and oversimplified for discussing the transmission of metal technologies.

Adoption and Risk

The adoption of any new technology involves costs and risks. These costs and risks can be formulated in economic terms (Kim 2001), which can be applied to pre-industrial societies if time, effort, social disruption, and ritual and personal status are considered to be economic elements. Adopters of any new technology usually possess imperfect information about its costs and risks. Frequently the benefits of a new technology appear only after a long time spent achieving proficiency, while the costs are immediate. One might predict that transmission of a metal technology into a region with no preexisting metallurgy would result in a slow and uneven adoption process, and the review of transmission evidence in Chapter 3 demonstrates this to be the case (see also Roberts 2011).

The initial cost, as perceived by the adopters, is often the crucial factor in choosing to adopt a technology (Kim 2001) and is one reason for the disjunction between knowledge of a new technology and actual adoption of it. Returning to the case of the Fali, the presence of an already existing system of craft production and specialization, exchange, household production, and status systems that could be asserted using material objects made the initial cost of the adoption of lost wax brass casting relatively low. As a rule, the more divergent the new technology is from the technology already in place in a culture, the higher the costs of adoption (Kim 2001). Potential adopters can seek to reduce the risk by sharing the cost and risk with others and by conducting only limited trials. Adoption also depends upon the degree of urgency of the group's needs; if the need is critical, adoption will be swifter. The speed of adoption is related to the benefits of the already existing technology, the expected benefits, costs, and risks of the new technology, the need for the new technology, and the availability of capital (time, resources, status) to invest in it.

An application of these models can be seen in Kim (2001), who compared the adoption of ironworking in 1st millennium B.C. Denmark and southern Korea. The macrocontext of both areas was similar. Both had access to iron ore, both were familiar with bronze working, and the knowledge of ironworking was available from adjacent regions. The immediate parameters differed somewhat. The economy of southern Korea was based on rice agriculture, while Denmark relied primarily on stockbreeding. In both cultures, the main subsistence tools were of stone. Both societies were hierarchical, but not state level. The late bronze age in Korea was a time of increasing nucleation and increased competition for land. The immediate context in Denmark was one where deposits of elaborately worked bronze goods in hoards were associated with ritualized leadership based on manipulation of prestige goods. Iron artifacts appear as early as 900 B.C., but, for several centuries, were used only in small ornaments and personal tools, some bimetallic, such as rings, pins, and razors found only in elite graves. In Korea, iron was first used in weapons, agricultural tools, and woodcarving tools, and appears not only in elite graves but in many non-elite contexts as well, unlike goods of bronze. In Denmark, iron was not used for tools until the apparent collapse of the prestige-goods system based on the control of supplies of imported bronze and the emergence of a subsistence system with a greater reliance on cultivation. The active social component in both societies was based on the different needs of the elites. In Denmark, the widespread adoption of iron technology, though iron ore was abundant, was delayed for centuries because elite legitimation relied in part upon controlling and monopolizing a scarce and prestigious resource, bronze. Iron did not contain the symbolic weight of bronze and was of little economic use in a stockbreeding economy. Costly investment in an ironworking technology, therefore, would have given the elite no more power than before. In Korea, on the other hand, elite power rested upon the accumulation of agricultural surpluses and military control, in both of which iron tools and weapons would have been advantageous. Hence there was considerable elite advantage in quickly adopting iron technology and spreading iron artifacts widely to the non-elite farmers and soldiers.

Wade's model of the contexts of adoption, Kim's analysis of power relations and the differing motivations of elites and non-elites and his consideration of innovation as an activity with costs, and Layton's analysis of how innovations actually spread within a particular society or segment within society can together form an encompassing model that incorporates the broader social and economic contexts, competing interest groups within societies, and the differing initial responses of individuals within any particular group to new technologies. These factors and perspectives have great potential for helping archaeometallurgists more accurately model metal technology transmission in prehistoric societies.

Adoption and Technological Change

How technological systems change during the adoption process is a topic of keen interest, particularly among proponents of Darwinian models of cultural evolution. When a technology is adopted by another social group, there is potential for change in the donor technological system due to a variety of selection and what are termed "bias" mechanisms (Boyd and Richerson 1985; Cavalli-Sforza and Feldman 1981; Shennan 2008:76). These mechanisms can help account for many changes in a technological system over space and time (Eerkens and Lipo 2005, 2007) and potentially could shed considerable light on the patterning of transmission of metal technologies across space. References in this discussion can be reviewed for a detailed understanding of this aspect of transmission theory (Eerkens 2000; Henrich 2001) and only a few brief highlights are discussed here.

As noted above, the learning framework in which a technological system is transmitted can influence the resultant production system and finished products in a variety of ways. Numerous variants can produce distinctive outcomes. Learning contexts where each individual replicates parental models and then evaluates or learns through individual trial and error (figuring out things on one's own) have been termed "guided variation" (following Boyd and Richerson 1985), or "environmental learning" (Henrich 2001). A learning framework based on guided variation has been called "unbiased" in the sense that the individual's learning is considered not biased by his larger social context; individual evaluation of cost/benefits is seen as the driver of change (Henrich 2001:995).

When learning and adoption occur within a sociocultural context during group-to-group interaction, more potential for technological change may be introduced. If practitioners from a donor community move to a recipient community and carry on the production program of the same products, passing the know-how on to their children, the highest fidelity (low variation, lowest copying error) relative to the source technological and manufacturing system will be maintained. However, under different economic and learning scenarios, i.e., different modes of transmission (Bettinger and Eerkens 1999:238), technological changes may arise. Practitioners from the donor community directly training practitioners in a recipient community can result in higher rates of copying error than if they kept production within their families, as effective communication could be lessened due to greater social distance between master and apprentice. Demand from new consumers for different products will also introduce change, especially if the new products require modifications in the production system. When a recipient group tries to copy the system without direct instruction from the source practitioners, greater error is likely to be introduced, if in fact the donor system can be reproduced at all. The recipient society may not understand the nuances in the manufacturing know-how or be able to figure out the fine motor skills needed to perfect the production system. A high amount of variation is expected when practitioners merely attempt to copy a product from another context, as the copiers use trial and error (guided variation) to approximate the donor products. Of course, new consumer demands in recipient societies can also stimulate innovations in product range and correspondingly result in innovations in *chaînes opératoires*.

Adoption and Social Biases

Observations (Eerkens and Lipo 2005; Henrich 2001) suggest that various forms of cultural bias can be more influential in the diffusion of cultural traits than cost-benefit analysis, trial and error, or random variation. Socially biased transmission occurs when the learning framework is socially based with less emphasis on individual payoff evaluation. In other words, individuals change behaviors based on who else is using the behavior. A variety of different biases have been proposed.

Direct bias results from the interaction of beliefs and opportunities with the putative demands, past or present, of natural selection; for example, the general human preference for sweet, fatty, or salty foods, it has been suggested, is a consequence of the need for calories, fat, and salt in ancient environments (Henrich 2001).

Prestige bias results from humans copying traits or habits of successful, prestigious individuals, even when the traits or habits have nothing to do with success (Eerkens and Lipo 2005; Henrich 2001).

Conformist bias is the human predilection for imitating ideas and behaviors held by the majority of the group, rather than those of a minority (Eerkens and Lipo 2005; Henrich 2001).

Results bias refers to a situation where people look at what other people do, for example, the crops they plant, compare the results with what they are doing themselves, and then change what they do because the other way of doing things seems to be more effective (Shennan 2008:77). In this case the adopters do conduct a cost-benefit analysis but draw from observing the experiences of others in the society.

Cases of a *pro-novelty bias* have also been proposed (Eerkens and Lipo 2007; Shennan and Wilkinson 2001).

In addition, simple chance, or what Shennan (2008:77) called "the cultural equivalent of genetic drift," can produce technological variation and change. Henrich (2004) pointed out the importance of demography and culture contact in the maintenance, loss, and increase in technological skills in a society. The Tasmanians, isolated as they were since the end of the Pleistocene, would be an example of how the loss of numerous important complex technologies can be mathematically predicted by their limited population size and lack of contact with a wider world.

Adoption Trajectory

The processes by which a cultural trait is adopted has been argued to affect the rate and trajectory of adoption. Henrich (2001) argued that most ethnographic studies have shown that "s-curve adoption" is the predominant adoption trajectory. With this trajectory, the innovation initially is adopted only gradually (forming a long tail), followed by a phase of rapid adoption of the innovation, and then by a flat phase of equilibrium.

The "r-curve adoption" is an alternative trajectory. Unbiased transmission (i.e., when each generation simply replicates the behaviors of the preceding generations) plus the individual's evaluation of environmental information, as well as copying errors (Eerkens and Lipo 2005, 2007) will result in guided variation. Over generations, guided variation will, according to mathematical models, produce what Henrich (2001:996) described as the r-curve of trait adoption. This pattern of adoption, which begins with its maximum rate of growth and tapers off to equilibrium, is rare (Henrich 2001:993). To account for rare and extremely rapid r-curve adoptions, the social data need to be scrutinized to account for why and how this unusual adoption trajectory occurred.

Summary

In summary, appreciating the embedded social nature of technology, anthropology of technology approaches do not see technological adoption or change as a straightforward or necessarily simple affair. The determinist view and the ripple diffusion model of the Conventional Paradigm (and the "Standard View" of Pfaffenberger 1992) is that technological change is inherently unilinear and desirable, progressively functional, and increasingly complex. Proponents of an anthropology of technology viewpoint, based on ethnographic and historical examples, disavow these simple sequences and favor less linear, more complex and messy—and not necessarily functionally logical—narratives. However, products, their variability, their spatial and temporal distribution, and empirically observed rates of adoption should assist archaeologists in inferring these different kinds of bias and more complex adoption processes in the archaeological record (e.g., Eerkens and Lipo 2005; Shennan and Wilkinson 2001).

The transmission of bronze technology to Southeast Asia and its subsequent transmission over time and through space within the region is a case that should benefit from anthropology of technology perspectives (White and Hamilton 2009, 2014). Previous interpretations (e.g., Higham et al. 2011a; Pigott and Ciarla 2007) have followed many assumptions of the Conventional Paradigm and have essentially advocated a gradual spread of bronze technology from central plains China, reminiscent of a ripple diffusion model, without a close examination of technical or social details. Given that it is almost certain that the bronze-making technological system was transmitted from outside of Southeast Asia as an intact complex (i.e., not locally copied from trade goods), the details of how it was transmitted, from where, and why it was adopted become paramount archaeological questions. Even the simplest bronze-producing system requires skilled know-how to find and smelt two or more ores, combine metals into viable alloys, make crucibles and molds that will not disintegrate with contents at temperatures in excess of 1000°C, and cast artifacts that are desired with sufficient regularity so that the needs and demand, and hence the system, are maintained over generations. The complexity of the initial adoption of that technology by a previously non-metalworking society is a major statement of commitment. Reasons that a society would choose to adopt the technology cannot be assumed.

Once a basic bronze and bronze artifact production technological system is established in a region, in situ change within the system becomes a further topic of study. How does innovation arise and become incorporated, or not, into the technological *system*? Why do some innovations succeed and others die? Changes in needs and performance problems among interacting artifacts (Schiffer 2005) may stimulate innovation, even cascades of innovation. But the socially embedded nature of technology, as well as environmental contexts, will shape and/or constrain trajectories of change. Technological style, derived either from the donor society or from already existing technological habits, shapes the formation of artifacts. The economic dimensions are then important to address: How was production organized? How were products distributed? Who were the consumers? The organization of labor is profoundly dependent not only on overall social organization, but also on the purpose of metal and metal artifacts in the wider society and the technological system employed (see Chapter 5 for more on economic dimensions).

Archaeology, Metallurgy, and the Anthropology of Technology

The development of methods and theories for examining technological change at the societal level is beneficial for archaeology. Archaeological data provide stronger evidence relevant to technology than for other domains of human culture, such as belief systems or kinship structures. Because archaeological data cover a variety of spatial and temporal dimensions, important contributions can be made to understanding technological change at scales quantitatively and qualitatively different from, for example, an ethnographer studying one or a few villages over a period of years, or an industrial sociologist studying one industry or one factory over a decade or two. Archaeologists will establish those contributions by examining their evidence in more intimate and thorough ways. Thankfully, there now exist many concepts and approaches suitable for undertaking a study of prehistoric metallurgy under the umbrella of the New Paradigm. Applying anthropology of technology approaches to prehistoric societies is becoming more common, even though preservation biases in archaeological data mean that many aspects of a technological system directly observable by an ethnographer may remain imperceptible in the archaeological example.

The important points in this brief introduction to the New Paradigm for investigating prehistoric metallurgy include that the paradigm has no inbuilt theory of causality and no integral directionality. It holds no technological system as a norm against which other systems will be evaluated. It shifts away from grand narratives of change and towards investigation of processes and mechanisms of innovation and adoption along with the details of change society by society. For archaeologists seeking to understand the place of particular technologies—including metal—in past societies and addressing questions concerning social interaction across regions as well as trajectories of regional development, it is extremely helpful to investigate and articulate these processes. The archaeometallurgy of Southeast Asia will surely benefit from this approach, once systematically applied.

Applying the New Paradigm

> *we cannot expect the results of scientific analytical techniques to be interpreted in the absence of any form of theoretical framework. (Jones 2002:175)*

How can the New Paradigm be applied to specific archaeological examples of early metallurgy? At a general level, there are two domains of phenomena contributing to these studies: the physical and the social/behavioral. The domain of overlap (Fig. 4.1) is the main focus, in that only some properties of any material will be relevant to a society's use, and only some activities of a society will engage a particular material. A variety of methods and theories can be used to study this domain of overlap, ranging from technical studies aiming to elucidate materials properties of artifact assemblages to an array of conceptual frameworks that seek to evaluate socioeconomic dimensions such as production and consumption. Combining the material and social evidence can produce a rich understanding of a technology and its place in a society, i.e., Pfaffenberger's socio-technical system. The core aim of this volume embodies that objective—to examine the metal-related socio-technical systems at Ban Chiang and related sites and articulate their significance in regional context. Ultimately, comparisons among socio-technical systems across time and space can help address broad questions, such as: What contextual factors influenced the rate of adoption of bronze technology in Eurasian societies, including those in northeast Thailand? How do societies choose among alternative production systems for producing copper from ores? How was a particular casting or alloying technology transmitted from society A to society B in contrast to society C? It is the stance of anthropology of technology approaches that these larger scale questions cannot be addressed without a thorough foundation in the material and contextual details of specific socio-technical systems.

Outline of an Archaeological Study

A study of archaeological materials from an anthropology of technology framework would begin by identifying the technologies of interest from a particular place and time. In the case of this volume, metal technologies from four sites in northern northeast Thailand during the prehistoric period are the technologies of interest. The study would entail three general components: (1) identify all the relevant artifacts recovered from the excavations under study, including manufacturing equipment and byproducts, and assess the materials properties of these relevant artifacts, such as copper, tin, alloys, iron, and technical ceramics; (2) reconstruct the life history evidence of how these artifacts were produced, used, and deposited to gain a view of how they fit in the local economies; and (3) compare evidence with other assemblages to address variability and larger questions such as technological tradition, regional patterning, transmission processes and sources, role in political economy, etc.

With regard to component 2, the reconstruction of life histories would include: (a) basic archaeological assessment of all materials (full assemblages) pertinent to the technologies recovered from the sites being studied (typology; counts of finished products both intact and fragmentary; manufacturing equipment and byproducts; distribution of pertinent artifacts spatially, stratigraphically, depositionally, and chronologically); (b) technical studies of representative samples of artifacts from all artifact classes to retrieve data on composition, manufacturing, and use; and (c) contextual assessment to provide data on the activities involving the technologies, such as manufacturing activities, consumption activities, use activities, and deposition activities.

These steps would allow the archaeologist to begin to reconstruct the pertinent technological system(s) present at the sites of study, along with the technological choices and technological styles active in that system, and the technology's role in its social context. Note that study of only selected intact final products from selected contexts such as graves and decontextualized (i.e., looted) artifacts will not be meaningful in the framework advocated here. With the starting point in the contextualized technological system, not the "Age" or the "Stage," the archaeologist can assess the sociopolitical and regional place of the technology from a non-directional standpoint. These aspects of the technology at the sites of study, i.e., the system, choices, and styles, will facilitate understanding the technology in its social context, allow meaningful comparison with other archaeological data, and enable the larger questions of interest, such as the place of origin of the technology and how it compares with other global examples of the technology to be addressed.

The aim of this volume is to demonstrate the archaeological study outlined above and position the significance of the prehistoric metallurgy from Ban Chiang, Ban Tong, Ban Phak Top, and Don Klang not only regionally, but also within the general context of understanding prehistoric metallurgy globally. The examination of the metals at these four sites within an anthropology of technology framework is essential to achieving this greater perspective.

5

Metals in Economic Context

Joyce C. White and Elizabeth G. Hamilton

Technologies exist in specific socioeconomic contexts that frame what products are created with the technological know-how present among practitioners. In Chapter 3, the usefulness of the term *middle-range societies* was noted for examining variability in the social, economic, and political manifestations in non-state societies. Adoption and early development of metallurgy took place in the context of middle-range societies in Thailand and most other parts of the world. Examining this adoption and early development from an economic point of view is one of the objectives of this study. This chapter provides a background in how archaeometallurgical studies conducted from the viewpoint of the New Paradigm can shed significant light on prehistoric economies and their associated societies.

Economic Background for the Study of Early Metallurgy

As reviewed in Chapter 3, archaeology over the last century has shown that the roles of metals in ancient societies were highly diverse, with no universal, cross-culturally uniform place in or effect on socioeconomic systems. Moreover, the roles of metals and of the material world generally in societies are increasingly recognized to have many dimensions beyond economic ones. Irrespective of this variability of purpose and place, the means by which resources were mobilized, including metals, is a basic task in the reconstruction of any past society. The place of metal and metal products in prehistoric economies in Thailand is just beginning to be studied from current frameworks that examine the economics of craft production in pre-industrial contexts (Pigott 1998; White and Pigott 1996). This chapter reviews current thinking concerning the three fundamental components of economies—production, consumption, and exchange—as background for understanding early metals in economic contexts.

Craft Production, Specialization, and Early Copper-base Metallurgy

The ability to create a vast array of products is a key attribute of being human. Humans can create products used just by themselves or their families and also for use by non-family members. When a society regularly obtains through exchange a product primarily produced by a smaller subset of individuals, a threshold of economic specialization has been crossed. Although discussions of specialized production have often been associated with examinations of social complexity, egalitarian and hunter-gatherer societies also are known to produce crafts for exchange, which can be important in reciprocal economic relations, social identity definition, and maintenance of egalitarian relations (e.g., Cross 1993; Sassaman 1998).

Because objects that humans have deliberately produced are among the most important remains of human activity left in the archaeological record, issues of craft production have been a particular concern for archaeologists. But craft production—the manufacturing of objects—is only one genre of productive activity, albeit one that can leave much evidence behind for the archaeologist in contrast to say, production of plant foods. Production of crafts involving inorganic materials like metals and ceramics is also of central interest to archaeologists because their remains preserve well. Metal and ceramic production can leave material remains not only of the final products but also of manufacturing processes, and the manufacturing activities of their *chaînes opératoires* can often be traced archaeologically. It is important, however, to keep in mind the biases of the archaeological record when discussing prehistoric economies. Not all productive activities leave traces in the archaeological record (e.g., evidence for basket-making is rare), and some productive activities may leave a quantity of remains that are disproportionate to their actual economic role. It would be unwise to infer simply from the volume of slag relative to the volume of plant food remains that metal production far outweighed food production in economic importance at a prehistoric site.

Society and Craft Specialization

Craft specialization as defined by Costin (1991:3) is "the regular, repeated provision of some commodity or service in exchange for some other," excluding exchange and production within single households. Thus, the creation of craft objects for exchange is inextricably interwoven with social interaction, interdependency, political relations of power, access to and control over resources and labor, and the transfer of materials. "Production is a social activity because participation in craft production shapes participation in society and access to goods and services, creates interpersonal ties and obligations, and represents differences in status and power" (Costin 2001:274). Production is also the way that concepts are transformed into concrete existence and material objects become imbued with ideas about status, membership, and individual relations. "The function and meaning of objects cannot be understood fully without understanding who made them" (Costin 2001:275).

The processes involved in producing metals and metal artifacts, by their very complexity, are unlikely to have been undertaken by everyone who actually used the artifacts. From the mere presence of metal production and metal product manufacturing, one can infer the presence of some form of specialized production in Costin's sense. Anyone interested in the social history of metallurgy must also be interested in the study of economic specialization and organization of production.

Childe conceived of metallurgy as inherently a full-time occupation that required the metalworker to produce objects in exchange for subsistence (Childe 1951:97). However, archaeologists and other scholars now recognize that metal production can be a part-time activity, often seasonal and taking place during non-agricultural periods (Childs 1998; Golden 2009; Roberts 2009b; White and Pigott 1996). Archaeologists are increasingly aware of the great variability found associated with economic specialization, even if many of Childe's views still have influence overtly or covertly.

Simple assertions that craft specialization was present or absent, full-time or part-time, might have sufficed in Childe's day. Now a wide range of theories, parameters, concepts, terms, and criteria for examining specialization in its various manifestations is available (see articles in Costin and Wright 1998; Hirth 2009a; Hruby and Flad 2007; Wailes 1996b). While a comprehensive review of specialization literature and schemes is not undertaken here, awareness of some of the terminology, contrasts, and issues is useful for understanding the place of metallurgy in prehistoric economies of northeast Thailand.

Control and Craft Specialization

Many discussions of craft production over the last few decades have been couched within larger discussions of power and control, or political economy, i.e., economy at the service of power relations (e.g., Reents-Budet 1998; Wattenmaker 1998; Wright 1998). Responding to Childe's interest in relating craft specialization to the roots of socioeconomic differentiation, much attention has been paid to the topic of who had control over products, producers, and manufacturing activities, especially in relation to elites (Brumfiel and Earle 1987). One simple distinction often cited in the literature is that between attached and independent specialists (Brumfiel and Earle 1987; Earle 1981). According to the traditional theory, independent producers manufacture utilitarian goods intended for a broad consumer base; attached producers produce high-status, special items for the elite or a state and are under the control of the elite/state, as "control over production translates into straightforward control over distribution" (Costin 1991:11). Note that both Costin and Earle made distinctions between the producer—attached versus independent—the intended consumer—elite or general populace—and the type of goods made—utilitarian versus high-value prestige goods.

The attached versus independent producer dichotomy is now recognized as over-simplified (e.g., Costin 2007; Flad 2007; Flad and Hruby 2007). It does not account for a variety of cases that have emerged, such as: independent artisans producing utilitarian goods for a broad consumer base as well as high value goods for elites (Rochette 2009); elites themselves producing prestige goods for which they also control the distribution (Inomata 2007); elites controlling production and distribution of basic commodities like salt (Flad 2007); or situations such as the Tarascan Empire (Maldonado 2009), where multiple production/distribution mechanisms co-occurred, and mining and production of utilitarian objects were intermittent independent domestic activities, with the state/elites procuring sumptuary objects as tribute from a corvée labor force.

Nonetheless, Costin (2005:1073, 2007:152–153, 155) argued that as very general heuristic concepts, attached and independent production remain useful to distinguish *social* contexts of specialization important for understanding political economy and societal controls as broadly conceived. Clark and Parry (1990:298) proposed: "When craftspersons retain rights of alienation, specialized production is *independent* craft production" (emphasis in original). Independent craft specialization is defined by the ability of the producer to control the distribution of the goods, but not necessarily the type of goods made because consumers create demands for selected items. In short, control is not a simple attribute of elite-dominated specialization but has turned out to be a multifaceted phenomenon with diverse socioeconomic configurations that can differentially affect the specialized craft at several points in its life cycle, from design through production and distribution to consumption (Costin 2007:152). In general, attached production overtly manifests sociopolitical differentiation and inequality in that the producers are not free. Independent production, though specialized, does not function primarily to promote institutionalized inequality, but instead tends to promote social solidarity through exchange relationships, including producer-elite relations. Discussions such as these will surely continue and new distinctions will be proposed such as "ritual production," indicating manufacture of ritual items was undertaken by discrete, perhaps religious, practitioners who passed on this knowledge and associated performances (Hruby 2007) outside of an economic system; or "embedded production," where members of a corporate kin group manufactured crafts for that corporate group (Janusek 1999).

Costin's Classification of the Organizational Variants of Craft Production

The more closely specialized craft production has been studied over the years, the more permutations have been found in its organization. To provide a sense of order among all the variations of specialized production and to facilitate comparison among examples, typologies of specialization and additional concepts beyond dichotomous approaches have been proposed. An early influential attempt to expand the conceptual system applied to discussions of production organization was proposed by Cathy Costin (1991).

Costin's approach to specialization is a matrix of variability rather than a sequence of development. Her matrix framework considers four aspects: (1) context, or the degree of elite control; (2) concentration, or whether the producers are dispersed or nucleated; (3) scale, whether the production takes place in a factory or in small, family-based units; and (4) intensity, whether part-time or full-time. It also considers the identity of the consumers.

In Costin's seminal treatment (1991:8–9), her proposed classes of specialization are:

Individual specialization: autonomous individuals or households producing for unrestricted local consumption;

Dispersed workshop: larger workshops producing for unrestricted local consumption;

Community specialization: autonomous individual or household-based production units, aggregated within a single community, producing for unrestricted regional consumption;

Nucleated workshops: larger workshops aggregated within a single community, producing for unrestricted regional consumption;

Dispersed corvée: part-time labor producing for elite or government institutions within a household or local community setting;

Individual retainers: individual artisans, usually working full-time, producing for elite patrons or government institutions within an elite or administered setting;

Nucleated corvée: part-time labor recruited by a government institution, working in a special-purpose, elite, or administered setting or facility;

Retainer workshop: large-scale operation with full-time artisans working for an elite patron or government institution within a segregated, highly specialized setting or facility.

This 1991 classification is a good starting point in any discussion of the organization of craft production, However, as happens with virtually all typological approaches (i.e., essentialized) to the study of cultural phenomena, discovery of variability that does not fit defined types spawns new types, new typologies, new concepts, and discussions of the limitations of any type, typology, concept, model, or approach. As Clark (2007) and Costin (2007) pointed out, this process of assessment and revision is expected and desirable, and Costin herself has revised and expanded her own thinking (e.g., 1998, 2001, 2005). Nonetheless, as a classic presentation, the Costin 1991 matrix remains a common point of departure for discussions of the organization of specialized craft production (e.g., Maldonado 2009:234). One advantage of Costin's categorization is that it is not inherently evolutionary. While there is a progression of intensification, there is no suggestion that these eight categories are stages that every society will pass through as it increases in social complexity and centralization.

Her recognition of community specialization, where independent households within a community specialize in the manufacture of similar product lines, such as ceramics or cloth, for regional consumers independent of a governing elite, has been particularly useful for characterizing craft specialization and metal production in prehistoric Thailand (Pigott 1998; White and Pigott 1996). Community specialization has been well attested ethnographically in Southeast Asia, for example, among Kalinga potters (e.g., Stark 1991).

Many earlier schema of craft specialization were evolutionary, especially discussions that have used economic arguments of increased efficiency and output, cost-effectiveness in time and raw materials, and market-like supply and demand, such as are common in the Conventional Paradigm (Evans 1978; Torrence 1986; Tosi 1984). These discussions presupposed that processes found in modern market economies drove production systems in prehistoric societies, even among relatively unstratified societies (cf. Cross 1993; Dalton 1977).

In contrast, there is another body of literature that discusses a range of economic mechanisms prominent in pre-industrial economies where optimization strategies are not necessarily prioritized. Cross (1993) suggested, for example, that in small-scale societies, especially in situations where a particular resource such as fine flint may be available to all, specialized production emerged as a way to create and maintain social ties. It is in this social function that the origin and maintenance of craft specialization in less-complex, middle-range societies may best be situated, rather than in a putative greater economic efficiency. Moreover, Shennan (1999) argued that specialization in regional economic systems even when resources are widely distributed can promote economic growth for all, though benefits may not be distributed equally (similar to "a rising tide lifts all boats"). Cross (1993) is an early example of the more recent scholarly appreciation of the complexity of the interactions among technology, systems of production, and social relations (Costin 2001; see also contributions in Hirth 2009a). Post-processual interests in identity formation, agency, ideology, and ritual also are being brought into the discussions of craft production and specialization, expanding the topic beyond its usual focus on economic relations in complex societies (e.g., Costin and Wright 1998;

Hruby 2007). These less evolutionary approaches are relevant to the examination of craft specialization in the community-based production systems of prehistoric Thailand.

Part-time Specialization

> *In the archaeological literature, full-time specialization has been viewed as a threshold: once crossed economic relations are transformed as part of a self-amplifying spiral that leads to greater institutional control over specialists and their products....By contrast, part-time specialization has received comparatively little attention from archaeologists; the social relationships that support part-time specialization are often interpersonal rather than institutional in nature. (Cross 1993:64)*

The degree to which specialists rely on their craft as a means to exchange for subsistence resources has been shown to be highly variable. Part-time specialization turns out to be remarkably common among societies all along the complexity range. Using the Human Relations Area Files, Clark and Parry (1990) surveyed 53 societies, ranging from the hunting and gathering Copper Eskimo to the state-level Inca and Romans, and discovered that all societies have part-time independent specialists. A wide range of variation in specialized craft production in non-elite contexts has been recognized (Cross 1993). Such specialization can be intracommunity (e.g., some people produce pots for their neighbors in exchange for cloth) or intercommunity (e.g., a village near salt sources mines the salt and exchanges it for iron produced at another village). Different degrees of specialization can be present based on the rarity of the skill set. If there are many producers of a commodity relative to consumers, the degree of specialization is low. If there are few producers relative to consumers, the degree is high. Truly full-time craft specialists who satisfied all their subsistence needs by exchanging their products appear to be rare in early complex societies.

Part-time specialization is not always independent and can have at least partially attached variants. Societies with high levels of social stratification often have demands that are met with a variety of production frameworks. One variant is *patronized* craft specialization, i.e., a patron commissions an item and sponsors the craftsperson during production (Clark and Parry 1990:299). Patronized (commissioned) craft specialization tends to be semi-attached and semi-independent with the craftsperson producing for both elite and non-elite consumers. A variant form includes cases where the elite themselves participated in the production of high-status goods, such as the production of jade items in neolithic China (Liu 2003), or weaving and garment making, as illustrated by Penelope's and Helen's labors in the *Odyssey* (Barber 1994:226–227). Clark and Parry hypothesized that patronized craft specialization may be instrumental to the emergence of some ranked societies: "patronized craft specialization provided one avenue for ambitious social climbers to convert surpluses of perishable foodstuffs into primitive valuables and status goods, produce goods for extra-regional exchange…and accumulate wealth with which to attract a following or build a faction" (Clark and Parry 1990:322). How frequently producers in pre-industrial societies used their products to attract followings has yet to be demonstrated. The presence of full-time specialization and attached specialization may be correlated with highly stratified societies, but part-time, independent, and patronage craft production can co-occur with attached production in diversified economies (Stein 1996; Zettler 1996).

Other data have emerged that undermine several tenets of conventional expectations for how craft specialization developed in pre-industrial economies. That attached specialists in early Mesopotamia often had lands they would farm (e.g., Zettler 1996:19) shows that they did not necessarily exchange their products and labor for full subsistence, but perhaps derived other valued benefits, such as protection. Seasonal (intermittent) craft production, including manufacture of valuable items, commonly took place in domestic contexts in periods of low agricultural activity in pre-industrial societies, such as the Maya; thus scheduling of the two kinds of production could be complementary (Hirth 2009b). More rigorous assessments of evidence for production contexts and a re-orientation of analytical frameworks beyond time commitment are needed to position the role of craft production within its socioeconomic framework. "The analytical focus should not be on the amount of *time* spent on craft production, but rather on the economic *strategies* that domestic units employed and how craft production was integrated into the domestic economy" (Hirth 2009c:13; emphasis in original).

In order to move beyond the inadequacies of the full-time/part-time categories, a distinction between *product specialization* and *producer specialization* has been drawn (Flad and Hruby 2007) to differentiate economies where individuals rely on exchanging their products for some or all of their subsistence needs (producer specialization) from those where individuals do not rely on exchanging their products for subsistence needs, but nevertheless do produce products (have the expertise, skills, and role) for non-kin use (product specialization). Costin (2007) noted that distinguishing these two genres of specialization in the archaeological record is not a simple task. An evidentiary argument may require quite rigorous analysis of a variety of data sets, including identifying discrete artisan households and determining their consumption sources as well as production activities. Nevertheless, a distinction between product specialization and producer specialization in the past could have important ramifications for disentangling dependency relationships among craft and food producers and consumers. These in-depth analyses could, in theory, help delineate power and network relationships among different social agents, including both consumers and producers, depending on the role of the product in society (Costin 2007:148).

CRAFT SPECIALIZATION IN THE ARCHAEOLOGICAL RECORD

> *It is one thing to assert that craft specialization was present in a given archaeological context and another to demonstrate that this was the case. (Cross 1993:62)*

A variety of lines of evidence can be examined in order to infer the presence of specialization in past societies. Archaeologists have often sought clues to specialization in the objects themselves. Proposed correlates of craft specialization can include evidence for special expertise. Skill in making particularly elaborate or beautiful objects is often used as a criterion for specialization; as such, skill is unlikely to be widely distributed in a population that is also predominantly involved in subsistence activities (e.g., Clark and Parry 1990; Wells 1996). Objects requiring lengthy *chaînes opératoires* and the mastery of complex production sequences using resources such as metal ores that are not widely distributed in the landscape have also been argued to have required specialist producers (Roberts et al. 2009:1018; White and Pigott 1996).

Uniformity, "repeatedly achieving sets of proportions or combinations of traits within an artifact population," or standardization, "repeatedly achieving the same or similar value for a single variable of form or technology," (both from Cross 1993:71) of products can be indications of craft specialization, both because specialist skill can reproduce forms consistently, and because it is assumed that the greater the homogeneity of a class of artifacts, the fewer the number of people who produced them (Anthony 1996; Costin 1991, 2001). Costin (2001) noted, however, that producers can have a number of reasons—stylistic, functional, and economic—for deliberately producing relatively uniform products. Wellbourn (1985) cited a number of ethnographic studies that demonstrate that domestic production can result in a high degree of standardization; much uniformity can result from the importance of the artifact type as a marker of group identity. Budd and Taylor (1995) pointed out that for metallurgy, reuse of molds and the use of older items as patterns for new molds could lead to a remarkable degree of uniformity; therefore, this criterion needs to be used cautiously. It should be noted that discussions of the context of production or specialization based on the study of only the final products are considered inadequate (Costin 2007:154, 155; Flad 2007).

Efficiency, the use of the least amount of time, energy, and materials in the production of artifacts, has also been used to judge the degree of specialization (Costin 1991), though often this is based on the application of Western, market-oriented assumptions associated with the Conventional Paradigm (Cross 1993). The level of efficiency is deduced by assessing as far as possible: (1) the degree of routinization of production; (2) the expense of, or capital investment into, production equipment, because it is more cost-efficient to keep expensive technology in use full-time; and (3) the amount of training needed to produce the goods, assuming that if training is long and expensive, it is more cost-efficient to train fewer, full-time workers than many part-time workers (Costin 1991). Costin (2001) warned that the criteria that archaeologists use to find craft specialization in the archaeological record are often subjective. (See her article for an in-depth discussion of the complexities of determining craft specialization in the archaeological record and drawing social inferences from that evidence.)

It is now recognized that arguments for specialized production are best built upon multiple lines of evidence. Careful analyses of the avenues that furnish information on manufacturing processes and producer-consumer relations are generally considered to provide a stronger basis for arguing for specialized craft production, in its various forms, than examination of only final products in isolation from the full production-consumption context (e.g., Costin 1991; Flad 2007). The most persuasive evidence for some degree of craft specialization in a society is the differential distribution of manufacturing debris, tools, and unfinished artifacts within a site and within a region, along with site locations near rare, nucleated, and dispersed resources, such salt, obsidian, and ore (Costin 1991).

The relative scale and investment in manufacturing equipment can also reveal how production was organized. Modest portable equipment can support periodic efforts by small workgroups to create small quantities of a product, while large, permanently installed equipment and facilities like the furnaces at Tonglushan (Zhou Baoquan et al. 1988) can enable continuous efforts and large workgroups to produce larger quantities of a product. Production of quantities demonstrably beyond what local communities might need or show evidence of using should also be considered. Another clue to the existence of craft specialization may be the placement of specialized manufacturing tools within particular graves, implying that the person interred had been especially associated with that craft (Wells 1996). Labor recruitment and organization when production begins to require coordination of non-kin workgroups is a topic needing further investigation (Costin 2005).

Another possible dimension of specialized production in middle-range societies is segmentation, the breaking-down of complex processes into segments performed by different sets of people. This process has been cited as a feature of economic specialization that is potentially observable in the archaeological record (Cross 1993). Segmented production can be revealed by spatially segregated caches of items at similar stages of forming, as well as clear evidence for activities from different stages of a *chaîne opératoire* undertaken in different locations by different communities. This bears some theoretical resemblance to modern-day industrial division of labor, or what Franklin (1983) called "prescriptive processes" that are more characteristic of state-level societies; but in middle-range societies, segmented production may be most common where economic activities are instrumental in the creation and maintenance of societal ties (Cross 1993).

SPECIALIZATION AND METAL PRODUCTION

Notwithstanding the difficulties in interpreting the archaeological evidence, some assumptions can be made about metal production in any society. One is that, by its nature, some form of specialization will be involved; metal production will always entail "production for use by others" (Costin 2007:150). Even in areas where native metal or metal ore is available, winning the metal will involve travel, expertise in identifying workable deposits, and a considerable degree of skilled physical labor both to extract the metal and then shape it into objects (Ehrhardt 2009). Smelting ore requires even greater specialized knowledge about more processes and materials, including prospection and handling of different ores and teamwork, as does metal product manufacture. To cast a bronze artifact, for example, the metal must be alloyed, charcoal made, thermo-resistant crucibles fashioned to contain the molten product, molds carved, wax shaped (if lost wax casting is used), air blown into the smelting and melting fire to achieve and maintain sufficient temperatures, and molten metal carried and poured into the molds.

However, there is no need to assume metal production was necessarily industrial in scale or undertaken by a discrete artisan class, even when considerable quantities of crucibles, waste rock from mining, or slag from smelting have been found (Pigott 1996). As Budd and Taylor (1995) noted, any metal extraction carried out at a locality over a period of centuries will result in a large volume of debris, even when the exploitation consists primarily of seasonal visits from several different villages for the purpose of obtaining enough metal for a few bracelets or awls. As an example, the copper mines at Mount Gabriel in Ireland (ca. 1800–1600 B.C.) show large and widespread spoil heaps of waste rock, despite metal production estimated at no more than 15 kg a year (O'Brien 1994:197). Moreover, ethnographic evidence supports the commonality of part-time specialization of metalsmiths and suggests that in non-state societies demand for metal products tends to be seasonal (e.g., de Barros 2000; Rowlands 1971, 1976). In short, there are many dimensions to craft specialization that can and should be explored beyond the mere presence/absence of specialization that can be assumed with metal production. Detailed examination of the archaeological evidence for metal and metal-making activities in their sociospatial context needs to be the basis of statements about the role of specialization in ancient societies.

The important conclusion in the preceding discussion of craft production and metals is that there is ample literature outlining methodologies and lines of reasoning for *how* to examine empirical archaeological evidence to make economic inferences about producers and the organization of craft production activities. Such inferences are important not only in themselves and for the reconstruction of past economies and metal production specifically, but also in concretely demonstrating particular portions of past social networks.

Demand for and Consumption of Early Metals

characteristics of demand must be empirically demonstrated, not assumed or asserted without substantiation... (Costin 2005:1047)

Consumer demand shapes many aspects of the products manufactured by any technological system, including product lines and volume of output (Costin 2005:1047). The producers, whether conceived of as craftspersons, artisans, practitioners, or even slaves, are creating and manufacturing products and product lines for consumers, i.e., all who use the products. The consumers may be patrons who need an item to display status, or neighbors who need a tool for everyday use or desire simple ornaments. No production system can be understood without a thorough understanding of the consumer side of the economy. As Roberts (2011) pointed out, with regard to metallurgy, investigating the impact and role of the consumer side of the system has been underdeveloped relative to the conventional focus on the production aspects.

The Conventional Paradigm for archaeometallurgy has implicitly assumed that the consumers of early metals, especially those of bronze, were elites, even patrons. This assumption remains prevalent among some scholars (e.g., Higham 2015b; Keswani 2005). Although some consumers of bronze in bronze-using societies may be suitably conceived of as elite patrons acquiring or commissioning particular product lines not intended for ordinary folk (e.g., the Shang state), it is now clear that this was not universal among societies in the early stages of adopting bronze technology, even in civilizations like the Indus (Hoffman and Miller 2009). The consumer side of early metal economics has been oversimplified, due in part to an overemphasis on the producers as well as the unexamined assumptions of elite consumption (Fowler 2013:88–89; Roberts 2011:145).

For products made of any material, Costin (2005:1047–1048) saw assessment of demand as including examinations of: (1) uses and functions of products; (2) distribution of products in the archaeological record spatially, contextually, and quantitatively; and (3) characterization of consumers as to their social roles, identities, and relationships to producers. The first line of evidence to evaluate consumer demand is examination of the product lines and their uses. Personal ornaments, functional implements, and regalia each connote distinctive consumer demands. The proportions among the different product lines in the archaeological record can also reveal much about the consumer base driving production. If among items made of copper-base metal in an archaeological culture, tools are rare, personal ornaments plentiful, and regalia absent, an inference might be made that these proportions reflect primary consumer demand for individualized ornamentation and a secondary demand for implements, but no demand for metal items for the sole purpose of collective religio-political display.

As for the second line of evidence for interpreting demand, the distribution of products in the archaeological record can help determine the restriction, if any, in the consumer base, referring to Costin's (1991) distinction between restricted and unrestricted consumption. Are certain objects only found in elite contexts (elite graves or households)? This could imply elite patronage of certain product lines. If the same products are found in both elite and non-elite contexts, even if there is a concentration of that product in elite contexts, consumption cannot be argued to be restricted to elites. Differences in wealth may allow greater accumulation of the product by wealthy individuals and families, but presence of the same products in non-elite consumer contexts in the same society or region demonstrates that consumption of that product was unrestricted in Costin's sense. Restricted use or access can reference not only high-status groups, but any subgroup in a society, such as age cohorts like the elderly, individual clans or lineages, or ritual specialists like shamans or priests. These subgroups defined by criteria other than inherited ranked status may still have exclusive access to product lines through their ownership of certain culturally defined characteristics, including age, skill set, or group affiliation. Therefore, assessment of restricted access must consider societal subdivisions beyond hierarchical ones.

Conversely, unrestricted consumption does not imply widespread use or utilitarian product lines. A rare product in circulation can have high value, i.e., be a "valuable" (Dalton 1977), but not imply exclusionary circulation or access. Distribution of products in the archaeological record can shed light on the society's consumer valuation of the objects. Are objects exclusively found in ritual contexts such as graves or temples? Are broken objects found in occupation debris? This is a particularly important question in the case of copper-base objects that could have been remelted and recycled if the metal was deemed precious.

Costin's third criterion for examining the consumer base—identification of the consumer identities and their relationships to producers—is essential to determine if the economic framework for product distribution is to be assessed. The acquisition of goods by elite consumers from independent craftspersons connotes a different economic relationship from elite consumers acquiring the products from attached specialists. Assessment of consumer identities employs the methods of evaluating identities generally, with mortuary data providing a primary source of evidence.

Of course, all the variations and permutations among product lines, consumers and producers, and the circumstances for use of the products can be vast, requiring a thorough and circumspect evaluation on the part of the archaeologist. The implication of the various dimensions of consumer access for archaeologists trying to assess a product's social value is that consumption patterns of products such as metal objects must be carefully ascertained from the full range of available evidence (see Hoffman and Miller [2009] and Maldonado [2009] for good models), not just asserted from a few examples.

Prestige Goods versus Primitive Valuables

A useful distinction to bring into the discussion at this stage is that between a "prestige good" (Earle 1991b) and a "primitive valuable" (hereafter a "valuable"; Dalton 1977), as bronze and other metal objects have often been assumed by archaeologists to be prestigious (see Chapter 3). Before elaborating on this distinction, it is necessary to note that these and related terms such as luxury goods, exotic goods, and wealth objects have become employed in a variety of overlapping and imprecise ways by archaeologists. Are "prestige good," "prestige item," "prestige object," and "prestige valuable" interchangeable (e.g., Hayden 2001b; Higham et al. 2011a)? Or might it be not only useful but important to terminologically differentiate, for example, Shang bronze ritual wine vessels from copper adzes excavated from metal age Thai sites? Over-liberal use of the word "prestige" with no or only vaguely defined criteria has rendered the word so ambiguous in archaeological literature that it has become close to meaningless.

This volume returns to a more rigorous distinction between a valuable in contrast to a prestige good to help to differentiate the social, political, and economic implications of exchange contexts, including exchange contexts for early metals in Thailand. According to Earle's classic definition (1991b:10), prestige goods are either: (1) special craft products produced for elite patrons who controlled the craftspersons and products; or (2) goods acquired from long-distance exchange only accessible by elites. In other words, prestige goods are products whose consumption was restricted to social elites (see also Flad and Hruby 2007:9). The function of prestige goods in prestige good economies is to promote exclusionary elite status in societies with evidence of social and settlement stratification and centralization, also termed complex chiefdoms and states. Prestige goods may be detected in the archaeological record by their appearing not in general habitation contexts, but only in restricted locations such as elite graves that contain other artifacts whose quantity or exclusiveness of distribution signifies elite status of the individual or group. Prestige goods may also be found in contexts plausibly defined as elite—temples, elite residences or caches in the foundations of major buildings. The key task for the archaeologist is to show from the totality of the evidence that the purported prestige good was in various ways (contextual, artifact type) used in an exclusive manner by an elite portion of the populace. This volume demonstrates that, during the metal age of Thailand, metal artifacts overall do not satisfy the exclusivity criterion of prestige goods by Earle's (1991b:10) definition because metal is found in a wide array of contexts including occupation contexts and ordinary as well as wealthy graves.

"Valuables," on the other hand, serve integrative social functions, particularly in non-state societies, according to Dalton's classic definition (1977). Even though more highly ranked individuals may have greater access to valuables, the valuables' primary function is to assist in the conduct of social and political transactions via their strategic dispersal and circulation. Distribution (exchange) of valuables may occur during ceremonial or ritually sanctioned occasions, such as rites of passage, death compensations, establishment of alliances, or periodic festivals. Dalton also mentions their use in "emergency conversions" such as in exchange for food in times of famine. In essence, valuables have a role well-articulated in Mauss's *The Gift* (1966, but first published in French in 1925), and further discussed by Levi-Strauss (e.g., 1943), Sahlins (1972), and Earle (2002:19–42), among others, as integral to the creation and maintenance of networks of obligations that cement and integrate non-state societies, including many hunter-gatherer societies (Cross 1993).

The valuables themselves are culturally and contextually defined and can range from strings of beads to bolts of cloth. Livestock such as cattle and pigs may function as valuables, and unmarried women could serve the function of valuables in intergroup marital exchange. Valuables have some degree of limited access (rarity or scarcity or requiring large, long, or specialized labor inputs) plus desirability within their cultural contexts to function in these transactions.

Chapter 2 in Earle (2002) provides an extensive and useful discussion of variation in the roles and uses of valuables among complex stateless societies that can assist archaeologists in interpreting potential valuables and how they may have functioned in specific prehistoric societies. He points out (2002:31) that some valuables can be used both on occasions requiring exchange payments for social and political purposes, and on other occasions as utilitarian goods such as food and tools.

Valuables are not crafts produced by specialists exclusively attached to or controlled by elite patrons. Items used in gift exchanges can be crafted items, perhaps finished in a distinctive or unusually accomplished manner, as Cross (1993:68) describes for the Susquehanna bifaces, or of materials with limited availability such as a special stone. One way for individuals and/or groups to attain a viable currency with which to engage in social networks and the transactions that commonly underpin survival and conflict resolution in non-state societies (Cross 1993; Dalton 1977; Sahlins 1972:182) is to specialize in producing items that can be used as valuables in reciprocity/gift exchange socioeconomic networks. This study finds that the social and economic contexts and the distribution of metal artifacts in prehistoric Thai sites indicate that overall the metal artifacts have more in common with Dalton's (1977) "valuables" than Earle's (1991b) "prestige goods."

It is important to examine the evidence for the roles of metals in prehistoric economies, including those in Southeast Asia, to determine if metal goods served fundamentally as valuables in reciprocity/gift economies, or as prestige goods in prestige good economies, or even potentially as commodities in market-based economies. It is also important to understand that any overall individual economic system may include parts that behave with characteristics of markets, parts that act like reciprocity/gift exchanges, and parts like prestige good economies. How did metals fit into the economies of the middle-range metal-age societies of Thailand? There are several avenues to pursue to discriminate the economic roles of products such as metals in these different kinds of economic frameworks, including examination of consumption as well as production evidence (Cross 1993).

CONSUMER DEMAND FOR EARLY COPPER-BASE METALS

The impact of variable consumer demand on the product lines and distribution of early copper-base metals is increasingly evident as the global evidence for the adoption of copper-base metallurgy is being revealed and understood. Roberts (2008, 2009a) reported that the first copper-base metals in Western Europe were not of a standardized type, such as awls or beads, but instead appeared in regionally specific forms. Metal was a rare and valued material—in some areas it could be used in adornment of the dead—but its distribution and other qualities are consistent with roles as valuables rather than as prestige goods. Rarity implied not preciousness but limited production that was likely undertaken sporadically with the product lines and the proportions manufactured specific to each region. Copper flat axes were the most common early metal objects in Ireland; they imitated the forms of stone axes and were found in similar contexts. In southeast France, metal was just one of many materials used to make beads, and copper beads represented the vast majority of early copper finds, though daggers, awls, and flat axes were present. In southeast Spain awls were the most common early metal artifact. Roberts (2008) related these regional variations to differing choices among regional consumers who were acquiring metal from small-scale, part-time producers.

For the Indus civilization, one current assessment of copper and copper alloy consumption patterns (Hoffman and Miller 2009) indicated that there was widespread access to copper-base metals, which were made predominantly into tools, ornaments, and vessels by independent craftspersons. There is no evidence for elite control over production or exclusivity in access or consumption. Hoffman and Miller described such uniformity of distribution of domestic copper-base artifacts as well as prominence of tools like chisels in productive contexts that one is tempted to conclude that copper-base artifacts in the Indus were in most cases commodities circulating in an unrestricted, market-like, supply and demand exchange system.

In contrast, primary consumers of objects made of copper alloys in west Mexican complex societies were elites and perhaps ecclesiastics, judging from ritual paraphernalia and elite status ornaments including vast numbers of bells, rings, discs, shields, and crowns. According to Hosler (2009), there was no centralized state control of production. Maldonado (2009) argued for a segmented and geographically multifaceted production organization based primarily on independent producers and possible limited state involvement for some stages and in some locations of the *chaînes opératoires*. Although copper was alloyed with tin, arsenic, silver, and gold to produce ritual and elite artifacts, Maldonado (2009) also provided evidence for concurrent small-scale production of utilitarian copper and bronze products like awls, axes, needles, punches, and fishhooks for unrestricted consumption.

The Shang period in the central plains of China is a good example of metal demand and consumption developing within a markedly hierarchical political patronage system (Linduff and Mei 2009). Exclusive access by an elite fostered product lines of copper alloy ritual vessels and weapons by the Erlitou period. State involvement in production and restriction of access to copper-base metals to specific high status sociopolitical groups is unambiguous, and the evidence is consistent that metals were prestige goods in a prestige goods economy. Interestingly, early copper-base metal evidence from surrounding contemporaneous cultures in Gansu, Inner Mongolia, and Shandong evinces a focus on copper alloy ornaments and simple tools and weapons, whose archaeological distribution indicates an unrestricted consumer base.

In summary, the examples given above attest to the early adoption of copper-base technology within a wide range of socioeconomic contexts with no single producer-consumer framework. The evidence for and trajectory of early metallurgical technologies in any region reflected consumer choices and the socioeconomic and political systems of the recipient cultures. There is no simple correlation of complex societies incorporating copper alloy technology into a centralized production mode or a prestige goods economy, with the Indus civilization evidence providing an unambiguous counter-example to that dated assumption. A patronage model may be relevant in only a few cases of copper alloy technology adoption and development, such as central plains China. In the future, the task for archaeologists interested in the adoption of early metallurgical technologies is to collect and analyze detailed evidence for each pertinent area and culture to ascertain the economic context (production, distribution, and consumption) specific to each adoptive context.

Trade and Exchange in Metals and Metal Artifacts

Trade and exchange in pre-modern economies is a complex topic of central importance to discussions of the early development of societies and dispersal of technologies. Simply demonstrating that a commodity has moved from its place of manufacture to where it was found in the archaeological record can be challenging when its constituent raw materials are widely distributed. Other topics of great interest to archaeologists include the distances covered, the intricacy, scale, and nature of exchange networks, and the regularity, intensity, and regulating mechanisms of exchange activities. Such an inquiry requires a multifaceted approach, including both conventional morphological and archaeological evidence, technical studies of the physiochemical properties of artifacts, and a range of theoretical tools and ethnographic knowledge.

Metals have the potential to provide an extraordinary window into the spatial dimensions and dynamics of past economic systems. Because metal raw materials are not ubiquitous and instead occur in dispersed but nucleated contexts, their very presence in a past society usually indicates that exchange of some sort occurred. The primary exploiters are likely to be groups in proximity to the resources, but for metals to move beyond the immediate areas of the source raw materials, as they commonly did in prehistory, there must have been societal mechanisms for distributing metal objects. Because metals are tangible objects that tend to preserve well in archaeological contexts, usually have physiochemical evidence of manufacturing processes and, hence, technological systems, and can be studied from a variety of methodological approaches, they have great potential for providing a material framework through which to examine the existence, scale, and workings of past commodity distribution networks, and in turn, intra- and intersocietal interactions in the past.

Although the full range of methods, theories, and ramifications of trade and exchange in metals are too many to be addressed here, a minimal appreciation is needed for two aspects of this study. First, how can trade/exchange in metals be demonstrated and spatially tracked in the archaeological record? Second, what societal mechanisms were involved in the distribution of metals in the past, how can the mechanisms be inferred from archaeological data, and what do these inferences reveal about the organization of past economies? Both stylistic and technical analyses are usually involved in demonstrating movement of metals and metal artifacts. Pre-modern and non-state economies usually incorporated non-market exchange systems; therefore, a variety of other kinds of exchange and valuation systems must be explored.

Style and Commodity Movement

> *The study of style often forces archaeologists to wrestle with complex social concepts. (Hegmon 1998:266)*

Most studies that examine trade and exchange of any craft begin with morphological and stylistic discussions, because such formal variation is macroscopically observable and can often be traced to cultural norms at specific times and places. If an object with a characteristic morphology and/or decoration is common in one area but appears as a rare item in a neighboring area, that rare occurrence may be cited as evidence of exchange, trade, or at least contact between the two areas. Using a study of ceramics as an example, Alden (1982) marshalled a variety of conventional archaeological evidence to argue that pinch-rimmed bowls of the late 4th millennium highland Iran were made at a single site, and that selected variants were traded throughout a valley of at least 1500 km^2 by a small group of distributors via market supply and demand mechanisms.

Metal artifacts often have culturally distinctive forms that provide useful starting points to attempt to trace exchange activities and cultural contacts (e.g., Champion [1985] on fibulae workshops and spread of fibulae in early iron age transalpine Europe). However, the mechanisms that create formal variation are many, and because styles and shapes are often easily copied, it is well recognized that stylistic similarity by itself provides only relatively weak evidence of commodity movement. Although at a general level, style can provide a useful starting point to scope out broad cultural relationships, the complexities and limitations of stylistic analysis become more evident as the research questions become more specific and more probing (Hegmon 1992). Attempting to relate formal variation (e.g., archaeological cultures defined in terms of pottery decoration) to social groups (e.g., ethnic groups) and social behaviors (e.g., trade) is fraught with methodological and ontological pitfalls (Conkey and Hastorf 1990; Hegmon 1992; Shennan and Wilkinson 2001).

While objects of similar styles often are broadly assumed to have similar sources, differentiating locally made copies from imports can be difficult even for something as well researched as Chinese bronze vessels (Allan 2007:466). Disentangling this distinction may be significant for understanding the dynamics of a political economy. The subtleties of variations and recognition of the potential for imitations can sometimes be sorted out via art historical reasoning. For example, an object different from a recognized norm might be an imitation if it is deemed to be more awkward or less functional than an artifact from the original source for that artifact type. On the other hand, less than stellar examples could be earlier prototypes, later perhaps degenerated versions of classic examples, or simply apprentice work (Crown 2001).

Ultimately stylistic and morphological analyses work best with carefully excavated and provenienced collections that provide a wide range of information on both archaeological and sociocultural contexts. The study by Sarah Allan (2007) is an example of integrating a stylistic understanding of early bronze vessels in China with a wide range of archaeological and regional evidence and certain theoretical perspectives to infer aspects of the ancient Chinese political economy. She provided evidence that regional examples of bronze vessels were not evidence of exchange but were locally produced copies of vessels made in dynastic centers. She argued that this demonstrates not political or economic domination of peripheral regions by dynastic centers, but that regional elites outside of state boundaries chose to emulate the material culture and cultural practices of early Shang political centers. Hence, she argued, the early dynastic histories are characterized more by cultural and not political hegemony.

Use of morphological and decorative style to address smaller-scale questions about exchange is an area that is being further developed in archaeology (Hegmon 1992). Ethnographic research has shown that stylistic variation can occur at many scales (individual to continental) and social groups (e.g., gender, ethnic, age group, technological province), and can signal many components of social dynamics and interactions that are of increasing interest to archaeologists. Examples include pottery styles that reflect intracommunity ethnic variation and political factions among Achuar and Quichua females in the Ecuadorian Amazon (Bowser 2000); spear styles whose circulation expresses relationships among sets of Loikop male age cohorts in Kenya (Larick 1987); and the co-occurrence of individual and group sources of stylistic variation in San iron projectile points and the distinct sociocultural roles the different style referents play in Botswana (Wiessner 1983).

In short, the recognition of style and the use of style to help disentangle evidence of a wide range of societal interactions, not limited to but including trade, exchange, or just contact, no longer simply means devising a typology and examining the geographic distribution of the types (even if that might be a useful starting point). Considerations of how the production of artifacts was organized, the sociopolitical and economic contexts of the style variants, the history of the technological tradition, the available means of transportation, and many other factors need to be brought into any discussion that seeks to relate style to various mechanisms of dispersal of objects of material culture.

Provenance Sourcing

To demonstrate that an artifact has moved in the past from its place of origin or manufacture, sourcing the raw materials from which the artifact was made represents a much more powerful argument than one based solely on final product morphology. Sourcing of raw materials, i.e., metals to particular ore deposits, is integral to understanding the economic geography of alloyed metals because raw materials, usually from more than one mineral source, had to be brought together often from substantial distances to make the alloy, and final and sometimes intermediate products (e.g., ingots) were subsequently distributed again, often over considerable distances. Sourcing data can provide concrete evidence not only for exchange but also for technological and social networks in prehistory.

Provenancing materials, however, has proven to be less than straightforward (Pernicka 2014). Wilson and Pollard (2001) set forth the six assumptions of the provenance hypothesis that all sourcing attempts must hew to in order to be both scientifically useful and archaeologically meaningful: (1) some characteristic, such as a trace element or isotope, of the raw geological material must be preserved in the finished material; (2) this characteristic must vary more between potential raw material sources than it does within one source; (3) the presence or amounts of this characteristic must be measurable enough to distinguish between sources; (4) raw materials from more than one source are not mixed, as in the recycling of metals; (5) the post-depositional environment does not alter the fingerprint of the characteristic; and (6) variations in the characteristic are interpretable in light of patterns of human behavior.

As a rule, provenancing has worked best with materials such as amber or obsidian, whose properties are not altered during working or deposition (Wilson and Pollard 2001). To source metal is more problematic. For decades there was hope that elemental patterns—the presence and amount of impurities or trace elements in the metal—would allow the metal of an artifact to be sourced to a particular orebody. Historically these attempts have been disappointing for several geological and technological reasons (reviewed by Pernicka [2014]; see also Pollard and Bray [2014:229–230]). The amounts of any trace element can vary widely even in a single orebody. The activities undertaken to win the metal, including degree of beneficiation, contaminants from the furnace walls and fluxes, the mixing of ores, and the mixing of metal from recycled artifacts can also affect the impurities within a particular piece of metal. Furthermore, elements such as lead, bismuth, and nickel can segregate differentially within a single batch of metal, leading to elemental inhomogeneity within the artifact (Henderson 2000:250–251). Although some correlations between regional ore sources and artifacts have been claimed on the basis of trace element studies (Chernykh 1992:17–23; Pernicka 2014; Rapp et al. 2000), attempts to link the metal of an artifact to a specific orebody based on the trace element content of artifacts have rarely been successful (Budd et al. 1992; Gale et al. 2003). Greater success has been claimed for the process of isolating regional or chronological groups based on trace elements to characterize metal stock (Pernicka 2014; Rohl and Needham 1998). For example, Coote (1990) has isolated two regional groups of tin impurity patterns in mainland Southeast Asia, but unfortunately no artifact sourcing work has followed from her article.

The most promising technique for sourcing metal used in artifacts, or at least being able to say with some confidence that the metal did *not* come from a particular ore deposit, has been lead-isotope analysis (LIA). Lead has four different isotopes (^{204}Pb, ^{206}Pb, ^{207}Pb, and ^{208}Pb) and individual ore deposits tend, in general, to have distinctive ratios of the four types (however, see Pernicka [1993]). Most pre-industrial metal artifacts, whether of copper, tin, or other metals, contain trace amounts of lead (Villa 2009). Claims have been made, particularly by some Mediterranean archaeologists, that artifacts can be sourced to a region of origin by analyzing the ratio of isotopes in silver, lead, and leaded bronze (Gale and Stos-Gale 1992), though this is a subject of vigorous debate (Gale and Stos-Gale 1989; Pernicka 1993; Sayre et al. 1992; and see Rohl and Needham [1998] for a case study).

Villa (2009) and Stos-Gale and Gale (2009) discussed the analytical challenges of lead isotopic analysis. In one example, Gale et al. (2003) compared lead isotope evidence from 141 copper-base artifacts from several Bulgarian eneolithic sites, including the cemetery at Varna and the cult site of Dolnoslav, to numerous copper ore sources in the region including ore deposits with evidence of ancient mines. Many copper-base artifacts could be traced to specific mines. It was also determined that the copper-base artifact assemblage from Varna came from many sources, and copper artifacts within individual graves often had different sources. On the whole, the lead isotope evidence from thousands of samples demonstrated that specific geographic relationships could be determined with clear implications for the variety of intraregional producer-consumer transactions at individual sites. In another example, Cattin et al. (2009) conducted a study tracing changes in settlement patterns in eneolithic to early bronze age western Switzerland, using 180 analyses of copper-base artifacts and a large-scale database of over 1200 ore and slag samples from Europe and Anatolia. Finally, Bode et al. (2009) were able to source the lead used in Roman military camps with 161 artifact samples and a large collection of published ore analyses. The rich lead isotope databases from central and southern Europe also allowed interesting results from analysis of a single artifact; lead isotope analysis of the Iceman's axe from ca. 3200 B.C. demonstrated that the metal originated in southern Tuscany, hundreds of kilometers from where the axe was found (Artioli et al. 2017).

Examinations of exchange and contacts that use particular culturally distinctive metal alloy compositions have also had success. This approach is based on the assumption that different cultures often used distinct recipes for making alloys; these recipes, added to the general similarity of ores and trace impurities from a particular region and typological and stylistic analysis of the artifacts, allow culturally distinctive metals and artifacts to be defined (Hamilton 1991; Northover 1980; Waterbolk and Butler 1965). For example, Northover (1982) traced distinctive metal circulation zones throughout the bronze age in Britain, Ireland, and northwest France by their patterns of arsenic, antimony, cobalt, nickel, zinc, bismuth, and silver impurities, as well as changes in technological habits such as the addition of lead. These characteristic elemental patterns have been well associated with the archaeological evidence (Henderson 2000:254) and this enabled Northover to trace the movements of metal "elemental recipes" from Ireland to Britain, and from Britain to northwest France (Henderson 2000:254–261; Northover 1982). The advantage of using the concept of "cultural metals" is that the recycling of metal within the culture would not affect the distinctive composition, and it does not require locating the specific spot of primary metal production. This combination of technique, raw materials, and cultural preferences in style can be referred to more broadly as part of a particular culture or time period's "technological style" (Lechtman 1977).

In eastern Asia, there have been few attempts to flesh out metal exchange in prehistory through sourcing studies or definition of cultural metals. Lead-isotope analysis has been applied with some success to Chinese bronzes (Barnes et al. 1988), and its application in Southeast Asia has begun (Pryce et al. 2014; see also Kuno et al. 1990; and TAM 2C, chapter 3). In the long run, attempts to define the existence and movement of culturally distinctive metals and the relationship of metals to ore sources will require a large number of elemental and lead isotope analyses conducted on a wide variety of artifacts and orebodies.

Mechanisms of Exchange

The presence of metal artifacts in a society indicates that exchange of some sort has taken place because producers of metal are unlikely to be the sole consumers and because of the geographically dispersed nature of the raw materials. But the socioeconomic context of exchange must be inferred from a variety of evidence. There is a sizable archaeological literature about documenting and modeling trade and exchange (e.g., Earle and Ericson 1977; Ericson and Earle 1982; Hårdh et al. 1988; Masson 2005).

Conventionally, three broad forms of exchange are recognized by anthropologists: *market exchange*; *controlled exchange*, sometimes called *redistribution*; and *reciprocity*, sometimes called *gift exchange* (Masson 2005:1139). Although any economic system may have a dominant or characteristic genre, the three basic forms of exchange often co-exist in any one society, sometimes with different products distributed via different forms of exchange (e.g., Masson 2005).

In market exchange, consumers at any level of society can interact with producers directly or through middlemen, and supply and demand regulate the value and distribution of goods. A physical market site is not necessary; goods can be exchanged via itinerant traders, roadside stands, consumer visits to production locations, and other methods.

In recent decades, there has been considerable focus by archaeologists on developing models of controlled exchange, relating production and exchange to the development of inequalities, including the development of elites, centers, and states (Cross 1993). As Masson (2005:1145) noted, it has been *assumed* that complex societies would have evidence of controlled and/or centralized production and exchange. Attached or patronized production is seen as providing specialized crafts that can be used to display elite status (prestige goods), or products whose distribution can be controlled and transformed into wealth and/or political power. A variety of political economy models have been proposed (e.g., Liu 2003; Stein 1996).

Reciprocal exchange has long been recognized by anthropologists as fundamental to non-state societies (e.g., Cross 1993; Mauss 1966; Sahlins 1972), and was likely a fundamental form of exchange for the non-state metal age of northern northeast Thailand. Reciprocal exchange is exchange among individuals or groups for the purpose of establishing and maintaining mutually beneficial social ties. Archaeologists have been slow to develop methods and theories that acknowledge the centrality of reciprocal exchange as the key exchange mechanism for non-state, middle-range societies. This exchange mechanism is common among many hunter-gatherers (Bettinger 1982; Cross 1993; Sassaman 1998:94), pastoralists (Larick 1987), and agricultural societies (Dalton 1977).

Economics of Middle-range Societies

> *it is important to redefine craft specialization in a way that is sensitive to the nature of production in nonstratified societies. In these societies, production is often characterized by low outputs and nonmarket distribution systems. (Cross 1993:62)*

The fundamental focus of middle-range society economies is subsistence, although the subsistence economies can be of various types, such as agriculturalists, pastoralists, complex hunting and gathering, horticulturalists, fishers, and so forth. However, production and exchange of utilitarian and non-utilitarian material objects, crafts, are important not only for distribution of needed goods, but also for creation and maintenance of social networks. In comparison to the archaeological attention paid to controlled trade and exchange in contexts with overtly hierarchical social systems, trade and exchange in less-stratified, acephalous middle-range contexts, particularly reciprocal exchange, appears to be undertheorized in recent archaeological literature (Cross 1993).

A variety of informal and formal variants of reciprocal exchange have been observed in ethnographic contexts (Earle 2002; Sahlins 1972; Welsch and Terrell 1998). In general terms, ethnographic evidence indicates that reciprocal exchange in non-state contexts and middle-range societies is structured around the creation and maintenance of networks of social ties and obligations on both collective and interpersonal levels (Cross 1993:80; Dalton 1977; Welsch and Terrell 1998). Exchange not only of valuables but also of basic foodstuffs figure prominently in these alliance networks and commonly encompass craft products including ornaments, implements, and articles of clothing. Livestock may store wealth and surplus that can then be transformed into "useful currencies such as social, economic, and political mutual help or debt relationships" (Hayden 2001a:579). Women also may be the focus in exchange transactions, and exchange networks with affinal ties could be particularly strong and enduring. The ties and obligations in turn serve a variety of purposes: giving access to resources outside the territory of the social unit, providing marriage partners, providing valuables for subsequent exchange transactions and ordinary goods, especially in times of need, and helping ensure group autonomy and relative peace between allied groups.

Exchange can co-occur with hostilities and "primitive warfare" (see Levi-Strauss 1943), but exchange alliances appear to suppress lethal fighting between major allies (Dalton 1977:200; Sahlins 1972). Groups seeking to strengthen their position in a regional alliance network can seek out valuables they would then employ strategically in reciprocal transactions (Dalton 1977:201). Individuals can also establish structural relationships via exchange. Cross (1993:66) referenced ethnographic examples from Africa and New Guinea where exchanges of weapons among males set up relationships and obligations among the men.

Ethnographic examples show that there are a wide range of occasions in which reciprocal exchange could take place, and unfortunately most of the details would be archaeologically invisible. Exchange occasions could occur at various collective levels (family, clan, village) and individual levels (marriage, friendship). They could be ceremonial (e.g., *kula* ring, Malinowski [1922]) or ad hoc (silent barter). The interdependence perpetuated by the exchange relationship militated against isolation and self-sufficiency. In the Trobriand Islands, "Peaceful expeditions went forth to visit and feast, conduct ceremonial exchange, trade, arrange marriages, and to use the natural resources of external groups with whom peaceful relationships existed, with, of course, their hosts' permission" (Dalton 1977:200). Market types of trade in utilitarian goods using barter would occur on expeditions as an adjunct activity to the ceremonial exchanges. Raiding occurred among the same groups, but victory in a raid did not entail permanent subjugation or political incorporation.

Inferring reciprocal exchange in prehistoric archaeological contexts will rest on marshaling a range of positive and negative indirect evidence such as the absence of indications for market places, administered economies, or attached workshops, and identifying and tracing out exchange networks in valuables as well as other commodities. Evidence from strontium isotopes from human remains may also eventually reveal past marital, mobility, and alliance relationships pertinent to exchange patterning (e.g., Bentley et al. 2005; Domett et al. 2016; King et al. 2013).

Wealth in Middle-range Societies

> *Wealth accumulation by itself does not result in social inequality or hierarchies. (Hayden 2001b:250)*

As reviewed in Chapter 3, though many middle-range societies give lip service to egalitarianism, social, political, and economic inequalities are commonly observed. In fact, middle-range societies provide the best examples for exploring variation in the development of differentiated societies as well as economies (Cross 1993; Hayden 2001a, 2001b, 2013).

Brian Hayden's (1995, 2001a, 2001b, 2013) discussions of what he calls "transegalitarian societies" have made contributions to understanding economics of non-state, middle-range societies, though many of Hayden's contributions on economics have been over-shadowed by his emphasis on the role of feasting and "aggrandizers" (acquisitive, aggressive, accumulative personalities) in fostering the development of socioeconomic inequality. In addition, his overly broad use of the word "prestige" (e.g., "prestige technologies," "prestige items," "prestige objects") blurs the useful distinction noted above between Earle's "prestige good" and Dalton's "valuable," confusing attempts to differentiate true exclusivity of elite access to special products from use of special products for socially, politically, and economically integrative purposes in gift exchange networks. Hayden's (2001b:255) statement on the function of prestige objects makes clear, however, that in his conception prestige objects function more like Dalton's integrating valuables than Earle's exclusive prestige goods.

Nonetheless, Hayden's explorations of evidence that differential wealth accumulations are common in middle-range, transegalitarian societies—in contrast to egalitarian bands where distribution of goods and services is based on shared resources and accumulation by individuals is discouraged—are very useful for understanding metal age Thailand, where wealth differentials in grave goods have long been documented (e.g., Bayard 1984a; Higham 1984b; White 1995). He argued that the key transition in human economies was from non-competitive sharing economies to "economically-based competitive" economies, and that transition enabled accumulation of surpluses by individuals, families, and groups. The social acceptability of accumulating surpluses enabled wealth differentials to develop, as variation in individual/family efforts and desires were allowed to be expressed in terms of acquisitions and accumulations, whether of food and/or material objects.

What was accumulated in subsistence economies was less likely to be rare craft goods than more useful, storable food surpluses such as domestic stock (Hayden 2001a), but valuables still had a role, especially in ceremonial exchanges (Dalton 1977). Such surpluses can be marshalled by individuals and families to serve various interests, from acquiring the most desirable mates through generous bride prices to establishing useful networks of exchange and debt relationships. Although Hayden focused on aggrandizer personalities as pushing acquisitive agendas out of self-interest in economic contexts with ample surpluses, scholars like Dalton focused more on the benefits of alliances for diversified social groups where ample well-distributed resources were perhaps less assured (e.g., protection/support in times of war or drought, access to goods from another territory, etc.). Alliances partially articulated via material exchanges could help sustain political autonomy (Dalton 1977:208). Presumably the degree of security of the subsistence base is a variable that can be assessed and studied on a case-by-case basis, and one might expect different trajectories of development depending in part on subsistence security.

Importantly, Hayden argued that transegalitarian societies contribute to understanding heterarchical political systems:

> heterarchies appear to characterize many transegalitarian groups...in particular, multiple

lineages or corporate groups (often internally organized as hierarchies) typically form multiple power centers within communities, interact as equals, or have unstable and changing relationships of power, wealth, social position and ritual roles. (Hayden 2001b:234)

He further commented (2001b:249) that "the emergence of heterarchies, in which as many emergent aggrandizers could be accommodated as possible, is expected from the surplus-based political model." An interesting outcome of a transegalitarian system is that the majority of the population will be well-to-do or at least self-supporting, and only a small minority poor—hence the inverse of a stratified system where wealth is concentrated in a small noble class and the vast majority of subsistence farmers are comparatively poor (Fig. 5.1; Hayden 2001b:257).

Furthermore, Hayden (2013) explicitly discussed a corporate economic model that he argued is a type of transegalitarian society directly pertinent to pre-metal and metal age societies of Southeast Asia. He argued that Southeast Asian societies of those periods likely had "residential corporate groups" (comparable to "house societies" as discussed in White and Eyre [2011]), whereby several families lived under the same roof and operated as an economic unit in varying ways. The degree of internal administrative centralization/autonomy among families in residential corporate groups varies (e.g., ranging from each family providing their own subsistence and having their own "wealth items" to centrally shared food resources with wealth items like cattle and pigs held in common; valuables like gongs and drums could be held by the lead administrator). Surpluses produced by constituent families beyond their needs were used for communal objectives, such as financing bride prices and funerals, acquisition of emergency food, and communal feasts that could serve a variety of integrative purposes. Residential corporate groups also provided defense against threats such as uneven agricultural yields and other social (e.g., litigations and conflicts), economic, and physical threats. An interesting point in Hayden's discussion is that within a residential corporate group, there are both well-off and poorer families who together had symbiotic social, political, and economic roles.

Regional Exchange Systems and Heterarchy

Middle-range transegalitarian dynamics can also be expressed at a regional scale where different communities interact with each other for economic advantage. A pertinent archaeological example of economic specialization of communities in a non-state socioeconomic context can be seen in Stephen Shennan's (1999) discussion of the *regional exchange system* that emerged in the early bronze age of central

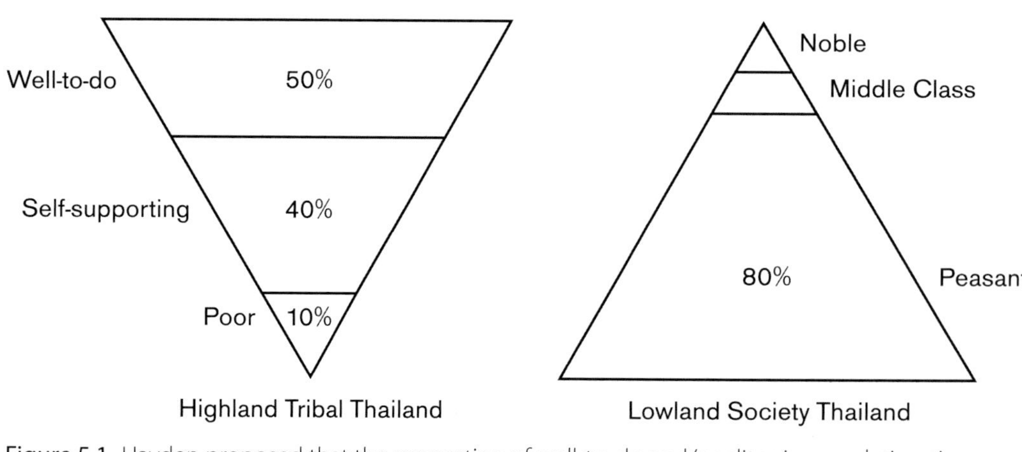

Figure 5.1 Hayden proposed that the proportion of well-to-do and/or elites in populations in transegalitarian societies is typically the inverse of stratified societies, based on archaeological research in British Columbia and the American Southwest and ethnographic research in Thailand. Figure adapted from Hayden (2001a:574, 2001b:257).

Europe. Applying "Ricardo's Law of Comparative Advantage," Shennan described a regional bronze age economic system whereby a region as a whole benefits and undergoes economic growth when constituent communities specialize and use their production not only for themselves but to exchange for products manufactured (or grown/collected/hunted) in neighboring communities. The economic research underlying this principle showed that in terms of the total labor expended, i.e., the cost/benefit ratio, it benefited communities to specialize and acquire other goods by exchange, rather than to maintain community economic self-sufficiency. The suitability of the law of comparative advantage for other historic and prehistoric contexts with regional exchange systems has been suggested by David (2012) and Phillips (1979, 2006; see also Phillips and Rathje 1977).

Significantly, such a system does not produce a uniformly egalitarian socioeconomic system, as different commodities and geographic locations were in effect weighted differently. Communities specializing in commodities that were of lower cultural value, where the capacity for improvements in efficiency was low, or that were geographically peripheral to the system as a whole, tended to be poorer, although they still benefited from being a part of the larger system. In the early bronze age of central Europe, communities near copper sources of the Mitterberg region were peripheral to the Alpine social world ca. 2000 B.C. They produced copper with inefficient methods but gained access to agricultural products from the lowlands. In the social system of this region, copper was neither a resource of central concern nor highly controlled and, thus, copper's role in the regional system was like a not-highly-valued valuable (rather than a prestige good) in exchange contexts.

The kinds of regional economies described by Welsch and Terrell (1998) for the Sepik Coast, by David (2012) for the northern Mandara Mountains of Cameroon in the last century, and by Shennan (1999) for the Alpine bronze age can be termed regional exchange systems. A key point made by Shennan is that a regional exchange system was not simply a matter of trading for a resource that was locally unavailable, but that communities exchanged for many items that they could have produced by themselves. The communities instead opted to specialize in a product (what Flad and Hruby [2007] would call "product specialization" as opposed to "producer specialization") and used their specialty to obtain through exchange the product they decided not to produce or to produce at levels insufficient for their full needs. No coercion was involved in the community's choice to specialize; monopolizing production of valuables like copper rib ingots did not provide unusual power or wealth; it provided access to a larger system. There was no value in letting violent conflicts jeopardize the system, so intercommunity lethal violence was low; and the cost/benefit valuations appeared to be to the advantage of the participants and communities in the system generally, in contrast perhaps to polarized systems where benefits concentrate in the higher social strata and costs are concentrated in the lower strata. Shennan also described how differential wealth concentration can occur (i.e., the system is differentiated but not egalitarian).

Regional exchange systems appear to provide a highly suitable economic foundation for regional heterarchical sociopolitical systems. Heterarchical systems are characterized by flexible hierarchies and lateral differentiation (White 1995) among other attributes. These qualities have been inferred as characteristic of metal age Thailand, especially for the bronze period (Eyre 2011; O'Reilly 2001, 2003; White 1995), but are also found in the iron period (Theunissen 2003; Tucci et al. 2014). Because of the range of stylistic, technological, functional, manufacturing, and consumption evidence that prehistoric metal assemblages can provide, evaluation of the full range of metals evidence has great potential for demonstrating specific networks of social ties and relationships in non-state societies. This study seeks to elucidate the role of metals in multiscalar exchange networks of prehistoric Thailand by examining the place of metals in social, economic, and political networks and by investigating the possible existence of a regional exchange system during the period prior to the appearance of states in the region (see TAM 2C, particularly chapter 6).

Summary

Tools available from technical analyses, concepts and methods of the New Paradigm, and thoroughly assessed archaeological evidence give modern archaeometallurgy unusual potential to shed light on prehistoric economies of both middle-range as well as state level societies. These tools facilitate elucidating the economic specifics of production, demand, consumption, and trade and exchange, including specific regional networks of resource distribution and circulation. The creation and distribution of goods and services are fundamental to every society, but many key goods and services have comparatively low visibility in the archaeological record. Metallurgy has unusual visibility in the archaeological record. Its complexity and richness as a source of data help it to be a proxy for investigating details of past social and economic networks, probably unparalleled by any other material.

6

Geomorphology of Metal Resources in Mainland Southeast Asia

Elizabeth G. Hamilton and Joyce C. White

Before one can reconstruct prehistoric technological or economic systems for manufacturing material goods, among the first questions to address are: Where did the required raw material resources exist in the landscape (e.g., nearby or long distance)? How were those resources structured (e.g., uniformly distributed, highly nucleated, dispersed)? And what implications did the circumstances of how the natural resource existed in the environment have for human acquisition (e.g., easily acquired surface finds, acquisition requiring hard labor or special equipment to exploit deep deposits)? The geologic and landform history of a region provides key evidence for understanding the provenance and accessibility of metal resources for occupants of that region. Considerable evidence of metal melting and metallic artifact production was discovered at Ban Chiang, Ban Phak Top, Ban Tong, and Don Klang, but the evidence for these manufacturing activities does not directly reveal the sources of the original primary metals found at these sites. The regional geographic and geologic background for metal resources is important in both investigating potential metal sources for these prehistoric societies and the economic networks that emerged linking consumers of metal products to primary metal producers. The geology of the Khorat Plateau demonstrates that no significant metallic minerals are found close to sites of the Ban Chiang cultural tradition. However, as a region, mainland Southeast Asia is unusually rich in metal ores (Khin Zaw et al. 2014a, 2014b).

A subsequent volume, TAM 2C, describes specific areas in Thailand and Laos where prehistoric mining for copper is currently known to have taken place. These sites offer only a partial answer for where copper resources for prehistoric societies in Southeast Asia could be found, and no archaeological data are available for the sources of other metals used by the region's prehistoric societies. Thus, the archaeological evidence is still quite incomplete.

This chapter describes and illustrates locations where modern surveys have discovered areas of copper, iron, tin, and lead mineralization that are rich enough for potential industrial exploitation. These areas include mining sites described in TAM 2C (chapters 2 and 3) but the description here places these metal extraction sites in their larger regional geological context, including areas that have yet to be examined archaeologically. An additional complication is that by no means do modern zones of metallurgical interest necessarily correspond to places of prehistoric mining activities, for several reasons. Deposits not large enough to exploit today could well have been useful in prehistory, when the quantities needed would have been far less. It is also likely that small, rich lodes or sources were completely worked out in earlier centuries, leaving little for the modern prospector to find. In addition, deposits worth exploiting in modern times, with modern mining technology, might have been inaccessible or too low-grade for those using simple smelting processes and tunneling. Killick (2014:13) noted that the comprehensive mining of the earth's copper over the past millennia, and particularly the switch from underground to open pit mining starting in the early 20th century, means that "many of the copper mines used in antiquity have been destroyed."

It should be remembered, then, that resources available in prehistory might have been different from those listed here and in modern surveys of mineral resources. It is notable, in fact, that most lists and maps of mineral resources published by governments and industry include few resources that appear to have been exploited in prehistoric times. The copper mines at Phu Lon and Sepon are exceptions. It should also be noted that it is difficult to determine the quality of the ore available to prehistoric miners even when known prehistoric mines are located, as the ore remains found in prehistoric smelting areas were likely discards (Hauptmann 2007:16).

All that geological survey can do is reveal known modern sources. Geological history can explain why those deposits exist in those particular areas, and both knowledge of modern sources and understanding of the geological events surrounding the mineralization of ores can suggest where archaeological survey in search of prehistoric mining sites should take place. Only on-the-ground archaeological fieldwork can show where prehistoric miners actually worked, rather than where they might have worked.

In this chapter, a summary of the geologic evidence for metal sources in mainland Southeast Asia is presented with a concentration on Thailand and Laos. The evidence is much more abundant for Thailand than for Laos. Other parts of Southeast Asia have limited data, though more information has been emerging about Vietnam in the last decade. The chapter begins with a description of the physical environment of the Khorat Plateau and its surrounding landforms. This plateau, a landscape favored by prehistoric agrarian societies, is a shallow, upland, gently rolling basin filled with sediments turned into rock. Other physical environments accessible to the people living on the Khorat Plateau include the mountainous regions to the north and east, the lowlands along the Mekong River to the southeast, the escarpment and hills of the Phetchabun Range to the west, and central Thailand's alluvial plain drained by the Chao Phraya River to the west and south of the Phetchabun Range. The latter subregion was settled by prehistoric metal producers and agriculturalists contemporaneous with the Ban Chiang cultural tradition. The chapter continues with a brief review of the extraordinarily complex geologic history of mainland Southeast Asia in order to locate tectonic activities that could lead to metallogenesis, or the creation of metal-bearing ores, including an introduction to how metal ores form. Known ore sources that occupants of the middle Mekong region could have accessed are then reviewed. This chapter shows that metallogenic belts with abundant exploitable copper, iron, tin, and lead resources were present only a couple hundred kilometers away from the peoples of the Ban Chiang cultural tradition.

Mainland Southeast Asian Geography

In order to situate potential metal resources in the landscape of prehistoric inhabitants of mainland Southeast Asia and particularly Thailand, a brief overview follows of key landforms and geographic features (adapted from Gupta 2005). The region is renowned for its variety of landforms, from steep, rugged mountains to extensive alluvial plains and deltas. The first part of this overview focuses on the landforms along the middle portions of the Mekong River, which crosses mainland Southeast Asia from northwest to southeast over a course of roughly 2390 kilometers. Within the Mekong Basin lie the sites of the Ban Chiang and other prehistoric cultural traditions, along with many possible ore sources used from early times.

The Mekong Basin

The Mekong River headwaters lie in the People's Republic of China (PRC) on the eastern edge of Tibet in the province of Qinghai. From this source the river extends about 4350 km to where its delta in southern Vietnam meets the South China Sea. The river is the twelfth longest in the world and has the ninth largest river discharge (Douglas 2005). The drainage basin of the Mekong includes parts of Yunnan Province in southwest PRC, a small portion of Burma/Myanmar, almost all of Laos, northeast Thailand, most of Cambodia, and a good part of southern Vietnam. The variegated landscapes within the basin—mountains, karst terrain, alluvial plains, lakes, and delta—support an extraordinary diversity of plants and animals. The climate is diverse as well, ranging from the cold, dry headwaters on the Tibetan Basin, with an annual rainfall of only 650 mm/year, to the tropical and subtropical areas farther south, where annual rainfall can average in some parts 3000 mm/year. In all areas of the drainage basin, the rainfall is seasonal; the summer (May–October) southwest monsoons result in considerable flooding (Douglas 2005). The river dominates life in the basin. The principal crop today is wet-season rice, either inundated rice in the lowlands or swidden (slash and burn) rice in the uplands, but a wide variety of plant foods are both grown and collected. Domestic animals are bred, but fish and other riverine products are major sources of food (Douglas 2005).

Present-day Landforms

The middle Mekong Basin (as defined by White and Bouasisengphaseuth [2008], drawing on Gupta and Liew [2007]) is contained within the present-day political boundaries of Laos and Thailand, extending from the Laos-China border to the Laos-Cambodia border. The principal landforms of the middle Mekong area are the Northern Mountainous Region, the Khorat Uplands, the Annamite Cordillera, and the Mekong Lowlands (Fig. 6.1; see Gupta [2005] for a more detailed description of all the landforms in Southeast Asia).

THE NORTHERN MOUNTAINOUS REGION

Following Gupta (2005), the Northern Mountainous Region is here defined as reaching from Myanmar east to northern Vietnam and includes most of northern Laos. For the purpose of this study, north and northwestern Thailand can also be considered part of the Northern Mountainous Region, although those mountains are drained primarily by the Chao Phraya and Salween Rivers. The landscape is marked by rapid changes in elevation, where mountain ranges are incised by steep-walled, V-shaped valleys. In northern Laos, the mountain ranges consist of a nearly parallel series of NE–SW-tending ridges and steep gorges; the topography is the consequence of differential erosion of sedimentary rocks folded and faulted in the early Paleozoic, mid–late Paleozoic, and the Triassic. The valley floors range from 600 to 1200 m lower than the mountain peaks. The only area in northern Laos without major dissection is the high basin of Tranninh including the Plain of Jars in Xieng Khouang, an archaeologically and metallurgically important province (Gupta 2005).

THE KHORAT UPLANDS

The Khorat Uplands include the Khorat Plateau, which lies mostly in northeast Thailand but also extends north across the Mekong as the flat plain around Vientiane in Laos (Gupta 2005). The Khorat Plateau comprises some 150,000 km^2 of undulating upland that slopes slightly to the south and east (Gupta 2005). This plateau, which is considered by geologists to be an unfaulted sag basin (Booth and Sattayarak 2011), makes up about one-third of Thailand (Fig. 6.2). It is bounded to the north by the southern foothills of the Northern Mountainous Region, on the east by the Mekong River, steeply to the west by the Phetchabun Mountains, and to the south by the Phnom Dong Rek Escarpment (Gupta 2005). The ridges along the edge of the Khorat Plateau can rise to over 1000 m above sea level (asl), though the average surface elevation of the plateau itself ranges from 150–200 m asl in the northwest to 50 m asl in the southeast. The escarpment to the south in some places rises to nearly 700 m asl (Gupta 2005). The landscape is gently deformed, with low hills and shallow lakes whose area varies with the season.

The Khorat Plateau has two shallow basins—the Khorat and the Sakon Nakhon—separated by the Phu Phan Hills. The southern, larger Khorat Basin is drained by the major Mekong tributaries, the Mun and Chi Rivers plus their contributing tributaries. The smaller, northern Sakon Nakhon Basin, where the Ban Chiang cultural tradition sites lie, is drained by several smaller tributaries to the Mekong, the Songkhram, Nam Luang, Nam Suai, Nam Kham, Nam Mong, and other small rivers. These two differing drainage patterns mean that the two basins are characterized by distinctive landscapes. The Sakon Nakhon Basin is characterized by a finely feathered riverine drainage system with numerous separated outlets connecting to the Mekong and limited incising of channels, resulting in an ecologically patchy undulating landscape. In contrast, the Khorat Basin's drainage aggregates into a single major outlet to

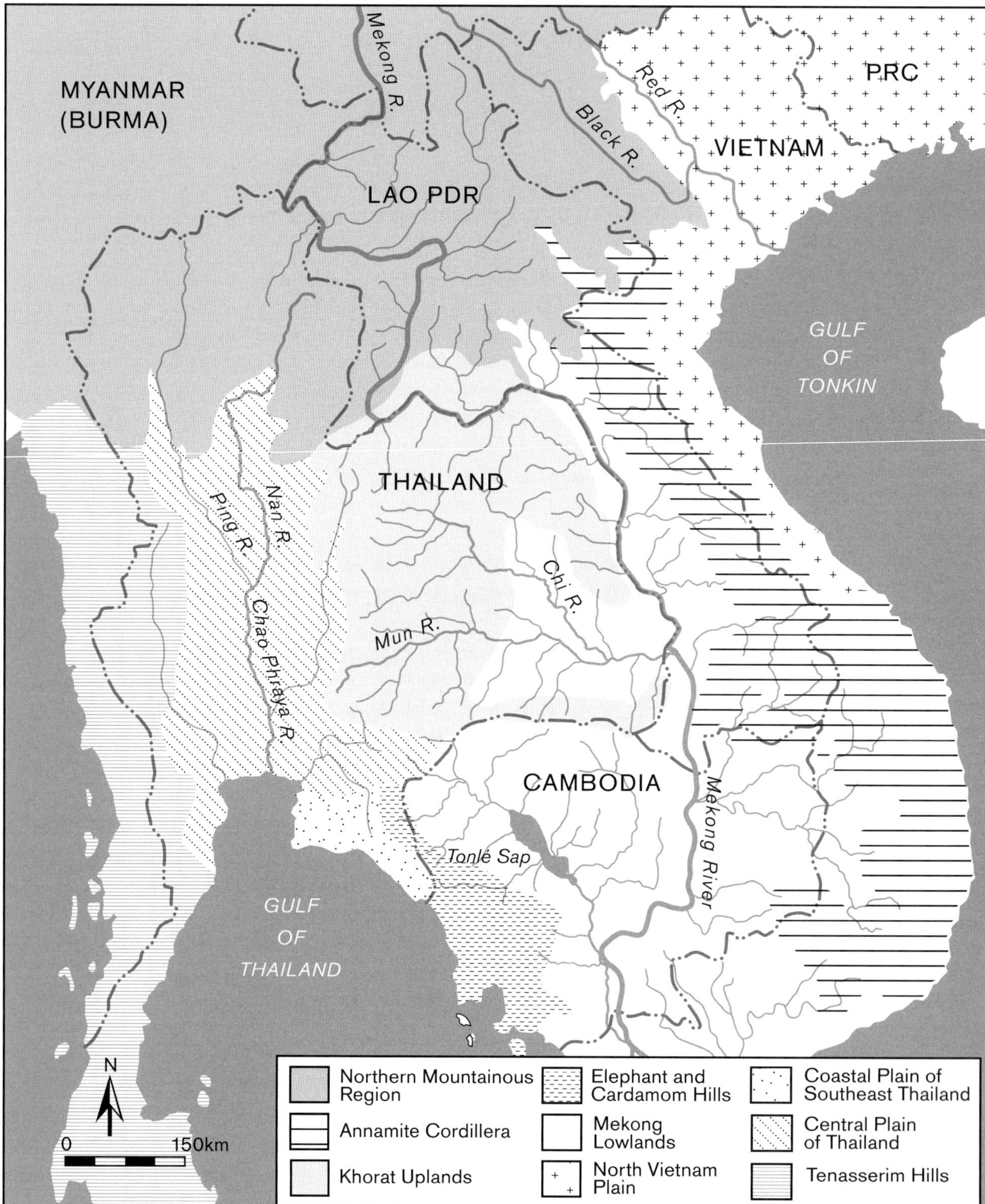

Figure 6.1 Major landforms of mainland Southeast Asia. Adapted from maps in Gupta (2005).

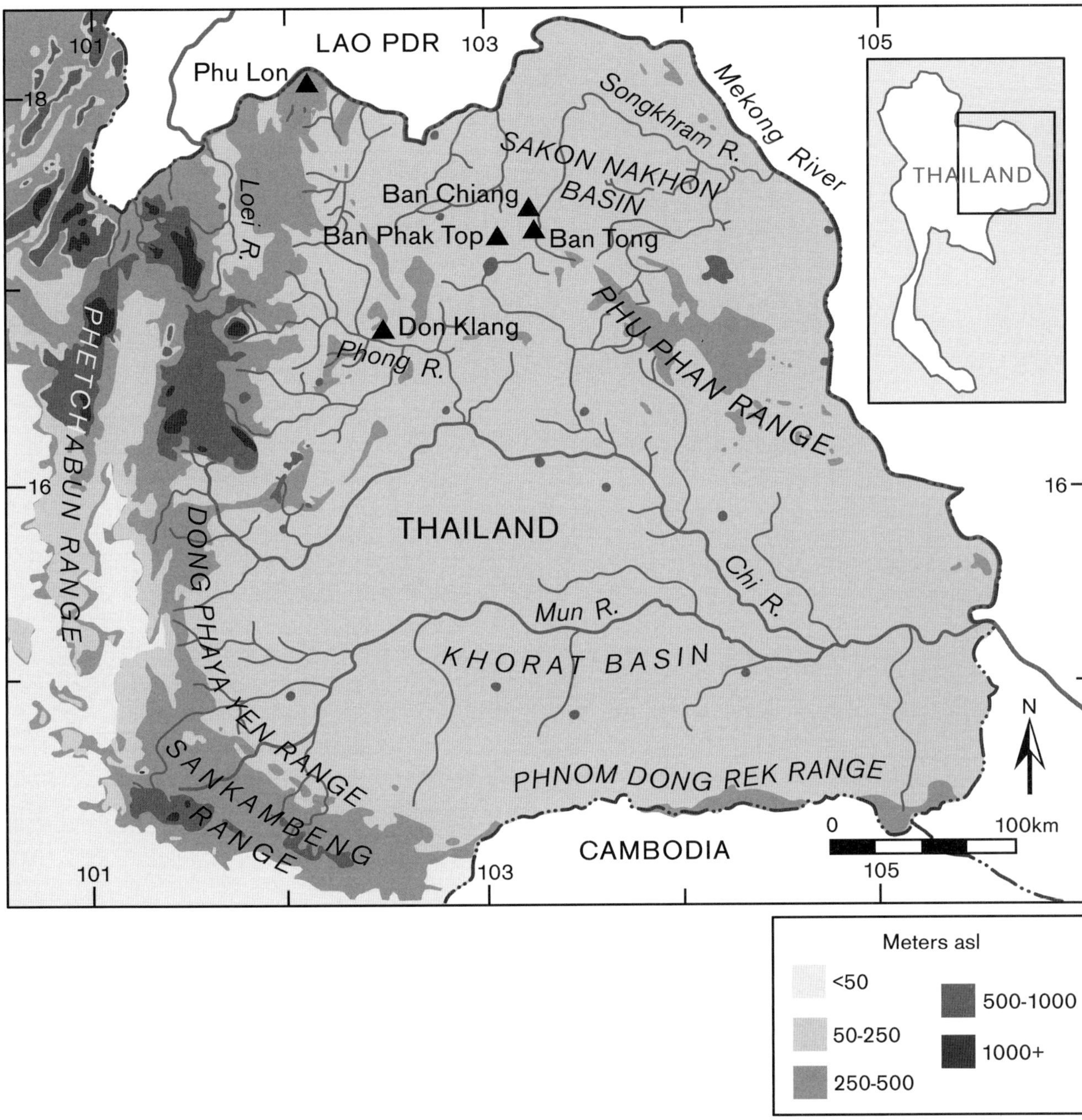

Figure 6.2 Map of northeast Thailand and the Khorat Plateau, after Damm (2008). Locations for the sites of Ban Chiang, Ban Phak Top, Ban Tong, and Don Klang are indicated.

the Mekong and has more deeply incised channels in wide valleys along the major Mun and Chi Rivers, creating more continuous parallel terraces across much of the basin.

The mountain ranges along the western and southern edges of the plateau act as partial rain barriers against the southwest monsoon, creating a rain shadow in the western portions of the Khorat Plateau. The southwest monsoons bring most of the annual rainfall between May and October. The lateritic layers under the topsoil promote waterlogging and flooding during the rainy season, and many water sources dry up during the dry northeast monsoon. Summers are extremely hot, with temperatures up to 40°C or more, but in the cool season, temperatures occasionally drop to 0°C at night (Rutnin 1988:52–53).

Large portions of the plateau are flooded in the wet season, but enough of this water evaporates during the dry season for the area to be classified as semi-arid (LaMoreaux et al. 1958). Erosion and deposition on the plateau during the Quaternary and preceding period resulted in a mixture of lowland and highland landforms that affect cropping regimes (Rutnin 1988:43–44); narrow alluvial plains and low terraces are suitable for inundated rice, middle and high terraces have upland crops, and hills and mountains are suitable for swidden agriculture.

THE ANNAMITE CORDILLERA

This is a series of mountains and plateaus that run from the Northern Mountainous Region along eastern Laos and western Vietnam to the southern portion of Vietnam. It forms the eastern boundary of the middle Mekong River Basin. The mountains are high and steep, many over 2000 m in height, with few passes. Their slope is steeper on the eastern side in Vietnam but with more gradual inclines along the western side in Laos. The cordillera includes the Boloven, Khammouane, Phouane, and Nakai Plateaus in Laos. Along most of the length of the eastern side of the Annamite Cordillera in Vietnam, lines of hills run steeply to the sea and enclose patches of coastal plain (Gupta 2005).

THE MEKONG LOWLANDS

These comprise the alluvial plain that lies principally in Cambodia and southern Vietnam, much of which floods during the rainy season. However, Gupta (2005) includes low flat terraces along middle portions of the Mekong and lower portions of the Mun and Chi Rivers, presumably because their topography and hydrology create similar flat alluvial conditions conducive to prolonged flooding, comparable to the terrain south of the Phnom Dong Rek Range.

Therefore, in Gupta's view (Gupta 2005), these lowlands begin at the southern end of the Northern Mountainous Region in Laos and stretch south to the sea. To the west of the northern portion is the Khorat Plateau and to the east is the Annamite Cordillera. The Mekong Lowland in Laos is narrow, marked both by flat areas where tributaries of the Mekong flow into the river, and by valleys bracketed by ranges of hills. In these narrow valleys the river flows over visible rocks.

Just north of the Laos-Cambodia border, the Mekong splits into numerous channels as it flows between rocky islands associated with the Khone Falls, and the river can broaden to 15 km in the rainy season. Once the Mekong passes south of the Phnom Dong Rek Range and into Cambodia, it flows over a broad alluvial lowland plain dotted with isolated hills (Gupta and Liew 2007). The western extension of this plain drains into the lake of Tonlé Sap in Cambodia, which then connects to the Mekong by the Tonlé Sap River (Gupta 2005). For the purposes of this volume, this portion of the Mekong drainage south of Laos, namely Cambodia and the southern tip of Vietnam is considered the "lower Mekong Basin." This physiographically distinct region is also characterized by somewhat distinctive trends in prehistory in comparison with the middle Mekong Basin.

Chao Phraya Basin

Although the four study sites of this volume all lie within the Mekong drainage in northeast Thailand, important metal ore sources and prehistoric metal producing sites are found in the adjacent Chao Phraya drainage basin to the west in central Thailand. These production sites are discussed in TAM 2C (chapters 2 and 3) and archaeology has conclusively demonstrated that metal age societies in the Chao Phraya and middle Mekong Basins interacted. A brief summary of the associated landforms follows.

The Chao Phraya's two main source rivers, the Ping and Nan and their tributaries, drain most of the Northern Mountainous Region that lies within Thailand's national borders. The rivers join in the archaeologically important province of Nakhon Sawan, and their confluence defines the beginning of the Chao Phraya that traverses the alluvial area known as the central plain of Thailand. The extensive lowland drained by the Chao Phraya River and its tributaries (Fig. 6.1) extends over 500 km north–south and almost 300 km east–west. Bordered to the north by the southern foothills of the Northern Mountainous Region, the plain is bordered in the east by the Phetchabun Mountains that mark the western edge of the Khorat Plateau, in the west by the Shan Highland and Tenasserim Hills, and in the south by the Gulf of Thailand. The transition zones from highlands to plain are marked by piedmont fans, isolated hills, and rocky spurs of limestone, granite, and quartzite, but the plain in the south is very low and flat, with a subsurface of marine alluvium and a surface delta of alluvium from the Chao Phraya and other rivers (Gupta 2005).

Plate Tectonics of Mainland Southeast Asia

> *The north-drifting journey of SE Asian component terranes was accompanied by several episodes of continental rifting, arc/backarc basin development, and collisional events. The associated magmatism, metamorphism, tectonic deformation and fluid flow have been the major driving forces behind the formation of major mineralized belts in the region and a diverse range of base and precious metal deposits in these SE Asian mineralized fold belts. (Khin Zaw et al. 2014b:9)*

The diverse landforms described above were created in part by a long and complex tectonic history. The details of the tectonic processes in turn provided contexts by which the region's metal ores formed. Metal ores often form in geologic fold belts created by collisions of tectonic plates, including smaller portions of plates called terranes. Terranes are tectonic units such as a block, massif, orogenic belt, or microplate, with each terrane having a distinctive tectonostratigraphic and magmatic history (Khin Zaw et al. 2014b:5). Therefore, a brief review of how the landforms discussed above relate to the region's tectonic history will shed light on potential metal ore resource availability. Although not all details are understood or agreed upon, the complexity of the geologic history of mainland Southeast Asia laid the foundation for the region's unusually rich metallogenesis (Khin Zaw et al. 2014a, 2014b).

Mainland Southeast Asia was formed by a collection of crustal terranes and fragments rifted at various disputed times from Gondwana and amalgamated into one landmass via a complex series of subductions, terrane-terrane collisions, and volcanic arc-terrane collisions and interactions. Gondwana was the southern supercontinent formed in Pre-Cambrian and Cambrian Eras. The various terranes that later made up Southeast Asia broke off the northern margins of Gondwana and drifted northward, progressively joining together as an accreted terrane annexed to the southern part of the growing Asian landmass during the interval from mid-Paleozoic to Triassic times (see Table 6.1 for geologic names and eras; for recent synthesis see Khin Zaw et al. [2014b]). How the terranes accreted is complex; it was not simply a matter of crustal terranes colliding one after the other into a larger stable landmass. Instead, many of the terrane-terrane interactions took place while the terranes were still drifting north. These "complex and multiple tectono-magmatic-metamorphic-hydrothermal interactions" (Khin Zaw et al. 2014b:6) led to the creation of numerous mineralized fold belts at the margins of the terranes, the most economically significant being the Truong Son and Loei Fold Belts (Fig. 6.3).

The stable nucleus or cratonic core of mainland Southeast Asia is composed of four main continental terranes (also called blocks) (Metcalfe 1990): South China, West Burma, Indochina, and Sibumasu (also called the Shan-Thai by Bunopas and Vella [1992] and others, and Sinoburmalaya by Hutchison [2005]). The two biggest terranes are the Indochina Terrane, which contains present-day northeastern Thailand, eastern peninsular Malaysia, Cambodia, and all but the northernmost portions of Laos and Vietnam, and the elongated Sibumasu Terrane, which contains present-day western Thailand, eastern Myanmar, and most of the Malay Peninsula (Buffetaut and Ingavat 1985). The Indochina Terrane consists of Pre-Cambrian basement overlain by Paleozoic marine and Mesozoic continental sediments, and is characterized, like the South China Terrane, by a tropical flora (Barber et al. 2011). The Sibumasu Terrane is composed of late Pre-Cambrian basement overlain by Paleozoic shallow marine deposits and Mesozoic shallow marine and continental deposits (Metcalfe 1990) and is characterized by a "Gondwanan" flora and fauna. The Sibumasu Terrane is bounded on the west by the West Burma Terrane along the Sagaing Fault and the Mogok

Metamorphic Belt. Other, smaller, terranes (Mount Victoria Land, Sikuleh, Natal, Semitau, and S.W. Borneo) were accreted to this four-terrane core during Late Mesozoic (Jurassic and Cretaceous) times to form "Sundaland" (Metcalfe 1990:626). A number of additional smaller blocks contributed to Myanmar, the tip of the Malaysian Peninsula, and the island arc (Hall 1998).

As Barber et al. (2011:507) noted, "Controversies still continue concerning the precise definition of the continental blocks, the timing of separation from Gondwana, the timing of collision with Asia and the identification and significance of the suture zones which represent sites of oceanic subduction and continental collision." Nonetheless, recent work on the tectonics and associated metallogeny of mainland Southeast Asia, along with the presentation of new models (Khin Zaw et al. 2014b; Sone and Metcalfe 2008), has provided a framework for understanding the major geological events. According to this model, during the Devonian through the Carboniferous Periods, the various major terranes that later comprised mainland Southeast Asia—South China, Indochina, Sibumasu, and West Burma—rifted from Gondwana and drifted in a vaguely northern direction. The rift between the terranes and Gondwana opened the eastern Paleotethys Ocean, along with, perhaps at the same time, in an eastern branch of the Paleotethys Ocean, the Ailaoshan-Song Ma Ocean, lying between the Indochina Terrane and the South China Terrane. During the Late Carboniferous and early Permian (ca. 300–270 mya), this Ailaoshan-Song Ma Ocean began to close by subduction of the ocean plate beneath the eastern Indochina Terrane; this process generated arc/back-arc basin magmatism (Lai et al. 2014). (Island arcs and back-arc basins are volcanic island arcs or ridges/basins made by rising magma melted by water released into the mantle overlying a subduction zone.) According to Khin Zaw et al. (2014b), there were at least two episodes of arc magmatism.

The timing of the collision between the South China Terrane and the Indochina Terrane, like so much else, is debated. Crow and Khin Zaw (2011), Hutchison (1989:151), Metcalfe (1998), Workman (1977), and Manaka et al. (2008) maintained that, during late Devonian and early Carboniferous, the Indochina Terrane collided with the South China Terrane to form the Cathaysian continental mass. Krobicki et al. (2008:102) stated that the collision occurred in the early Carboniferous or Silurian-Early Devonian. Lai et al. (2014), based on more recent work, suggested a late Carboniferous through Permian date. Lepvrier et al. (2004) argued that evidence recently acquired supports a later-Permian/early-Triassic date for this collision, and Cai et al. (2008) supported a late Triassic date. This gradual closing of the Ailaoshan-Song Ma Ocean and subsequent collision of Indochina and South China, along with the variably aged island arcs and back-arc basins in between, is marked by the Ailaoshan-Song Ma Suture Zone that runs generally along the line of the Song Ma Valley. The collision of the Indochina Terrane with South China Terrane produced mountains, faulting, and folds found in the Northern Mountainous Region in northern Vietnam and northern Laos (Lepvrier et al. 2004; Metcalfe 1990). The Truong Son Fold Belt may have formed as a result of this collision (Khin Zaw et al. 2014b; Sone and Metcalfe 2008), not only from the pressures of collision but also as the collision compressed the remnants of the continental and volcanic island-arc systems created by the subduction of the Ailaoshan-Song Ma Ocean Plate under the Indochina Plate (Lai et al. 2014).

A glance at Figure 6.3 shows that the modern mainland Southeast Asian core is composed of four major elongated and roughly parallel tectonic units, in addition to the South China Terrane on the northeast. Two of the units are the Indochina and Sibumasu Terranes already described; the other two are the Sukhothai Terrane or Sukhothai Arc, and the Inthanon Zone. The Nan-Uttardit Suture was traditionally regarded as the marker of where Sibumasu collided with Indochina, following the prolonged subduction of the main Devonian-Triassic Paleotethys Ocean Plate under the Indochina Terrane. This collision is part of the Indosinian Orogeny and is one of the two major geologic events that formed Southeast Asia; the land mass created is called Indosinia. Recent work has demonstrated a more complex series of events. According to Sone and Metcalfe's model (2008), the Paleotethys Ocean Plate began to subduct underneath Indochina at the start of the Permian. This subduction caused magmatism and created a series of back-arc basins and the Sukhothai island-arc system along the western edge of the Indochina Terrane; these collapsed under compression by the end of the Permian and the Sukhothai island-arc system, now the Sukhothai Arc Terrane, was folded and accreted to Indochina. The Sukhothai Terrane is bordered on the east by the Nan-Uttardit Suture and on the west by the Inthanon Zone. The mineral-rich Loei Fold Belt formed by eastward subduction of the Paleotethys Ocean Plate beneath the Indochina Terrane during the late Permian to Early Mesozoic. The resultant arc magmatism is believed to have contributed to the magmatic–hydrothermal fluids that gave rise to a variety of ore deposit types along the Loei Fold Belt (Kamvong et al. 2014).

The Inthanon Zone is regarded by Sone and Metcalfe (2008) as the major portion of the Palaothethys Suture Zone, where the ocean plate was finally destroyed by subduction underneath the edge of the Indochina Terrane and Sibumasu collided with the Sukhothai Arc. Above the subduction area developed an extensive accretionary complex made up of a mélange of uplifted ocean-floor materials and continental margin sediments; this complex, now termed the Inthanon Zone, was thrust westward over Sibumasu during the Triassic (Barber et al. 2011). The westernmost boundary of the Zone is not yet clear; the eastern boundary is the Sukhothai Arc Terrane, marked by the Chiang Rai Line (Barber et al. 2011). The zone is extremely complex, with many different elements, and is not fully understood. Barber et al. (2011) suggested that the definition of the Inthanon Zone be extended to include the southern Thai peninsula to the west of the Bentong-Raub Line, the main Paleotethys Suture of peninsular Malaysia. The Bentong-Raub Line is the continuation of the Nan-Uttardit Suture; it forms the boundary between Sibumasu and the East Malayan Terrane, an extension of the Indochina Terrane (Hutchison 2005). Sone and Metcalfe (2008) regarded the Nan-Uttardit Suture not as the main suture line of Indochina and Sibumasu, but as the result of the closure of back-arc basins created behind the Sukhothai Island Arc during the Permian.

The final major event in the tectonic history of Southeast Asia occurred during the Eocene of the Cenozoic Era, though Bunopas and Vella (1992) attributed it to the Cretaceous, the last period of the Mesozoic. At that time, a northeastward-moving India collided with Asia, but did not stop moving. The southeastern parts of Indosinia were down-faulted or tilted, producing the Gulf of Thailand and the South China Sea (Workman 1977). This Himalayan Orogeny produced broad regional folding and major strike-slip faults (Hall 1998; Workman 1977), a long subduction zone along the western edge of Southeast Asia, and uplift, erosion, and a sinistral shift along the major northwest-southeast faults (Ridd et al. 2011a). This orogeny is responsible for most of the configuration of the Southeast Asian mainland seen today, including the present course of the Mekong River (Gupta and Liew 2007).

In addition to these major events, the Cenozoic subduction of the Indian-Australian Plate at the Sunda or Java Trench produced a long archipelago of volcanic and non-volcanic islands that make up a great deal of island Southeast Asia. The complex faulting and subduction zones are made more complex still by the continuing westward pressure of the Philippine Plate (Hutchison 2005).

Many details of the tectonic history of both mainland and island Southeast Asia are still unclear (Barr and Charusiri 2011; Metcalfe 1990); this description is therefore vastly simplified. As Hutchison noted (1989:151), the late Triassic collision and orogeny has obscured the earlier events, particularly on the mainland. In addition, late Mesozoic and Cenozoic subduction and rifting processes and the collisions of India with Eurasia and Australia with Southeast Asia have further complicated the geological interpretation (Metcalfe 1990).

Metallogenesis

The above overview of mainland Southeast Asian tectonics provides a framework to start to assess the region's locations for potential ore sources. However, before potential orebodies can be evaluated, a better understanding of the geological processes that produce ores is needed. Not all ores were exploitable by pre-industrial methods.

Although metallic elements are widely distributed in the earth's crust, occasionally metallic elements are present in higher concentration as a result of specific geological processes, particularly tectonic processes. Ores are deposits of minerals containing metals in concentrations high enough for useful quantities to be extracted. A modern definition would include the phrase "extracted profitably," but this would not necessarily apply to prehistory. The minerals containing the metals, called ore minerals,

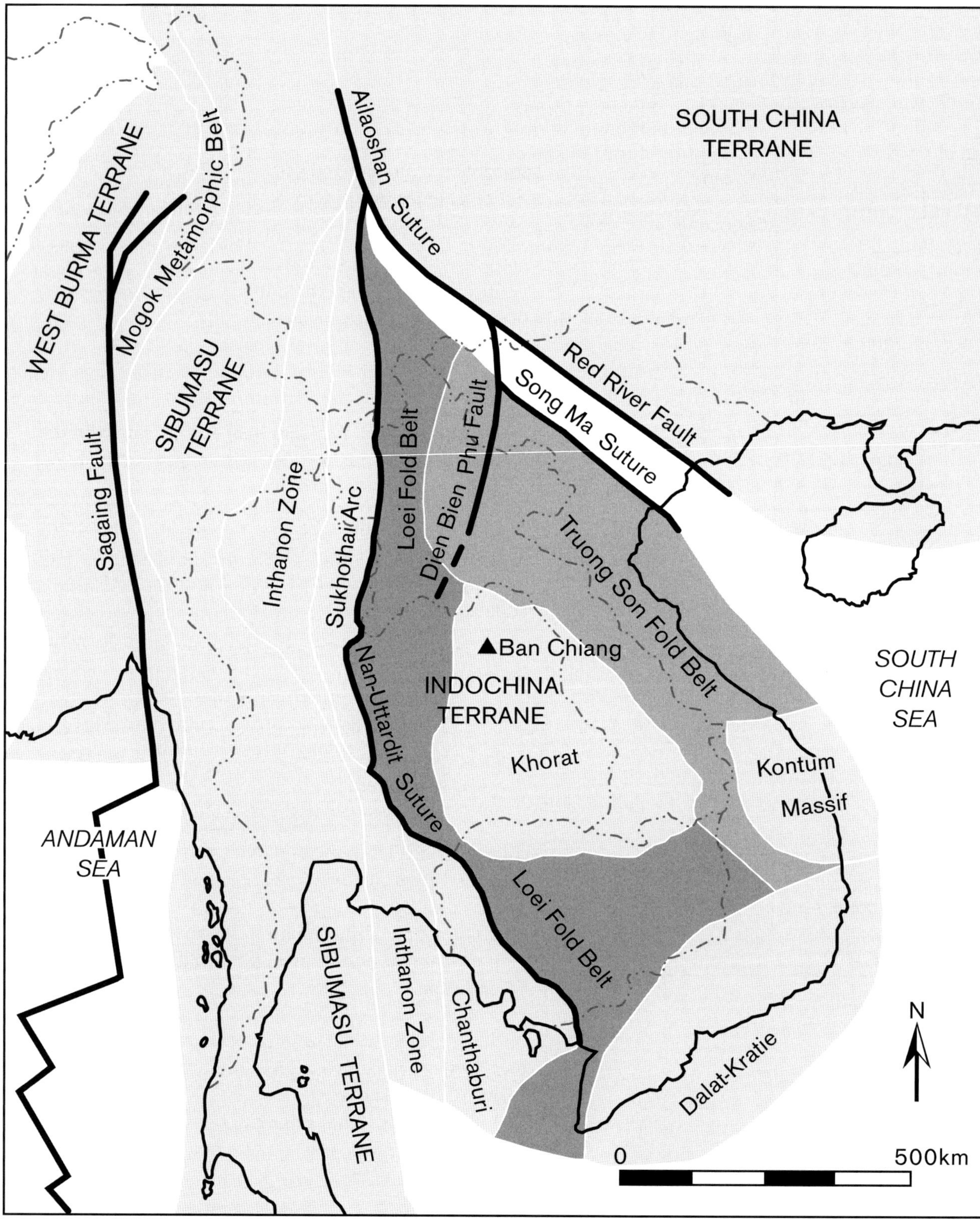

Figure 6.3 The principal tectonic subdivisions and fault belt systems of mainland Southeast Asia, after Sone and Metcalfe (2008) and Khin Zaw et al. (2014a, 2014b).

Table 6.1 Chronology of Major Geologic Events in Mainland Southeast Asia. Adapted from Hutchison (2005), Khin Zaw (2014b), and U. S. Geological Survey (2010), along with information from Gillieson (2005).

<table>
<tr><th>Era</th><th>Period</th><th>Epoch</th><th>Began (mya)</th><th colspan="2">Events</th></tr>
<tr><td rowspan="7">Cenozoic</td><td rowspan="2">Quaternary</td><td>Holocene</td><td>~11kya</td><td colspan="2" rowspan="3"></td></tr>
<tr><td>Pleistocene</td><td>2.6</td></tr>
<tr><td rowspan="5">Tertiary</td><td>Pliocene</td><td>5</td></tr>
<tr><td>Miocene</td><td>23</td><td rowspan="2">Laterization of granite and karstification of limestone outcrops.</td><td rowspan="3">Formation of Southeast Asian oil- and gas-bearing basins: rifting and filling by lacustrine deposits, followed by shallow marine deposits.</td></tr>
<tr><td>Oligocene</td><td>34</td></tr>
<tr><td>Eocene</td><td>56</td><td>Collision of India with Eurasia: Himalayan Orogeny and tilting of Southeast Asia to south and east.</td></tr>
<tr><td>Paleocene</td><td>66</td><td colspan="2">Indian Ocean spreading.</td></tr>
<tr><td rowspan="3">Mesozoic</td><td>Cretaceous</td><td></td><td>146</td><td colspan="2">Indian Ocean spreading. Collision of Burma Plate with the Shan Highland, producing tin-tungsten granites in Phuket and Myanmar.</td></tr>
<tr><td>Jurassic</td><td></td><td>200</td><td colspan="2">Indian Ocean spreading. Continental conditions in south China, Indochina, northern Thailand, and eastern peninsular Malaysia result in red bed deposition, e.g., the Khorat Plateau.</td></tr>
<tr><td>Triassic</td><td></td><td>251</td><td colspan="2">Indosinian Orogeny: Closure of the Paleotethys Ocean by collision of Sibumasu with Cathaysia (Indochina), with uplift, folding, and mountain building. Major tin-bearing granite emplacements from Belitung through the Malay-Thai peninsula to north Thailand. Deposits of limestone laid down, especially in northern Southeast Asia. Final closure of the Ailaoshan-Song Ma-Paleotethys Ocean.</td></tr>
<tr><td rowspan="6">Paleozoic</td><td>Permian</td><td></td><td>299</td><td colspan="2" rowspan="6">Closing of the Ailaoshan-Song Ma branch of the Paleotethys Ocean and accretion of Indochina to South China to form Cathaysia, descended from warm-weather portions of Gondwana. Formation of the Sibumasu Terrane (Sumatra, Phuket, eastern Myanmar, and Yunnan) from cold-weather portions of Gondwana.

Cambrian–Devonian: Deposition of shallow water marine strata widely over Southeast Asia, from the Yangzi Platform to peninsular Malaysia.
Thick sequences of limestone were deposited in the Ordovician through Triassic eras; these rocks were karstified in Cenozoic time.</td></tr>
<tr><td>Carboniferous</td><td></td><td>360</td></tr>
<tr><td>Devonian</td><td></td><td>416</td></tr>
<tr><td>Silurian</td><td></td><td>444</td></tr>
<tr><td>Ordovician</td><td></td><td>488</td></tr>
<tr><td>Cambrian</td><td></td><td>542</td></tr>
<tr><td>Pre-Cambrian</td><td></td><td></td><td>4600</td><td colspan="2">Formation of the Pre-Cambrian terrain of northeastern India and the Kontum Massif in central Vietnam.</td></tr>
</table>

Note: kya = thousand years ago; mya = million years ago.

include silicates, oxides, and sulfides, the last being the most economically important. These ore minerals are compounds of metallic elements with silicon oxide, oxygen, and sulfur, respectively (Grotzinger et al. 2007:69). For instance, hematite is an iron oxide (Fe_2O_3), and chalcopyrite, the most important copper ore, is a sulfide ($CuFeS_2$). A small number of metallic elements can be found in the native state, that is, pure and uncombined with other elements. Gold, copper, and silver can sometimes be found as native metals. Most metals, however, must be extracted from ores that are a mixture of valuable ore minerals and gangue, the unwanted rocky matrix.

As a rule, metallogenic ores are found preferentially in zones of tectonic activity, where plates meet or have met, where fold and fault belts exist, and where volcanos and mountains have arisen. Nonetheless, their occurrence is unpredictable, to the dismay of economic geologists (i.e., not all fold/fault belts have associated orebodies). Most, though not all, metallogenic (metal ore creation) activity is caused directly or indirectly by magmatic intrusions into the crust. Magma tends to form at plate boundaries, both divergent (creation of crust, rifting) and convergent (subduction and collision zones). Thus, plate boundaries offer some guide as to the possible location of orebodies.

The various ways that plates of different size, thickness, density, and temperature interact with each other produce distinctive geological processes and features. Collisions of continental plates in particular can create conditions suitable for metallogenesis. When continental plates collide, as was the case in the formation of Southeast Asia, and the continental plates are of equal density, neither plate will subduct (be forced downward), though one plate will often ride over the other. A good example is the collision of the Indian Plate with the Eurasian Plate. The Eurasian Plate is overriding the Indian Plate, but the Indian Plate is not being forced down into the asthenosphere; rather, both plates are floating on the asthenosphere and creating a double thickness of crust. The thicker crust, plus the compressive stress of the collision, created the Tibetan Plateau and the Himalayas. Many incidents of orogeny (mountain building) are due to similar continental-continental collisions (Grotzinger et al. 2007:28). The collision can create more than mountains; the lower portion of the crusts may partially liquefy, forming molten granites that rise and intrude into the crust above. Fold belts, where layers of rock are forced to bend as a result of compressive stress, and fault belts, where rock layers fractured in response to the stress, are products of orogenies (Grotzinger et al. 2007:158). Fold and fault belts are frequently marked by granitic intrusions with associated metallogenesis.

Three general points should be kept in mind about the geography of metallogenesis relative to plate boundary interactions. First, orogenic mountains, fold belts, and faulting can often be found hundreds of kilometers from the actual convergent boundaries. Second, plate-to-plate interactions are not one-time collision events. Rather, over the epochs and eras numerous multi-faceted plate boundary events can occur that produce suites and zones of contiguous and adjacent faults and folds. Therefore, metallogenesis that is manifested within a general geographic region can have been produced by many distinct tectonic and geological events, a fact important to the application of lead isotope analysis discussed in TAM 2C (chapter 3). And third, collisions involving several plates, blocks, and microblocks as well as different kinds of plate interactions—divergent, convergent, and transform (sliding)—can result in a variety of metallogenic processes as well as various metallic ores and ore combinations. The multi-terrane interactions summarized above for mainland Southeast Asia set the scene for the richness and diversity of the region's ore resources.

Ore Formation

Metallic elements can be concentrated by several geological mechanisms. These processes influence what ores form and if and how the metals can be exploited. Major metallogenic processes can be divided into three main groups: processes related to volcanoes, igneous intrusives, or sedimentation (Barnes 1988:43). Ancillary processes include hydrothermal (hot fluid) precipitation, metamorphic processes, magmatic processes, evaporation, ore mineral segregation, water-born sorting of eroded minerals, and other geochemical processes. Classification of the sources of ore deposits can be difficult, as many deposits are the result of more than one formation process (Hutchison 1983:7). In addition, ores within a single deposit may vary with exposure and depth;

sulfide minerals such as copper sulfides may oxidize near the surface or be enriched at greater depths by rainwater leaching dissolved minerals deeper into the orebody (Barnes 1988:13, 17).

Ore deposits fall into two general categories: syngenetic and epigenetic. Syngenetic ores were created at the same time as the host rock enclosing them; epigenetic ores were deposited later than the host rock, usually following cracks, fractures, or other weaknesses, and cut across strata (Barnes 1988:24).

Igneous Ore Deposit Formation

A number of ore formation processes are an indirect result of the epigenetic intrusions of igneous material into already existing host rock, largely as a result of plate tectonic movement.

HYDROTHERMAL PROCESSES

"Hydrothermal processes" is the collective term for a variety of mechanisms that are based on hot, aqueous (water-rich) fluids that flow through fractures and pores in rocks. Hydrothermal fluids have a particularly important role in the concentration and alteration of ore deposits (Hutchison 1983:5). Most hydrothermal fluids are brines and are widespread in the earth's crust; they form ore deposits only when the fluids meet certain fissures or reactive rocks.

Hydrothermal Vein Deposits

Hydrothermal solutions are created around bodies of molten rock that intrude into solid crust. The most common source of hydrothermal fluids is fluid released from cooling igneous intrusive bodies. In addition, hydrothermal solutions are formed when circulating groundwater comes into contact with the hot intrusion. The process is driven ultimately by magmatic or igneous intrusions. The heat drives the hot, mineral-rich liquid into fissures or cracks in the host rock (the rock into which the molten rock intrudes), or, in the case of geysers and hot springs, into the surface of the earth. The fluid cools quickly, resulting in rapid precipitation of the metal-rich minerals. The hydrothermal fluid can also leach more metals from the host rock, through which it is forced (Hutchison 1983:8), and deposit them in a process of alteration-mineralization. The ores are found in narrow fissures called veins or vein deposits. Frequently the ore minerals are precipitated along with quartz, calcite, or other vein minerals (Grotzinger et al. 2007:70). In areas where oceanic crust is being subducted underneath the continental crust, the oceanic crust can be melted to form igneous rocks. Major metallic deposits may be formed in these settings.

Prehistoric exploiters of copper often exploited vein deposits. Copper deposits at the prehistoric mining site of Phu Lon (see TAM 2C, chapter 2) included this type, where veins of carbonate copper ore were found in larger veins of quartz which could be mined in tunnels leading from the surface. In prehistory, the quartz would have been broken by stone mauls, freeing the lumps of ore.

Disseminated Hydrothermal Deposits

Another type of hydrothermal deposit is a disseminated deposit, a type of metallic mineral deposit where the desired minerals appear as scattered particles in the rock, but in sufficient quantity to make the deposit an ore of variable richness (U.S. Bureau of Mines 1996). Vein deposits are localized, but disseminated deposits can be spread over hundreds of kilometers. They are created from very large bodies of intrusive igneous rock, which are often the result of oceanic subduction that creates mountains at the edge of continental plates. For reasons not clearly understood, the intrusive igneous rock and the country rocks along the upper edges of the intrusive rock develop huge numbers of small fractures. Soluble metal-rich hydrothermal solutions penetrate and re-cement the rocks, precipitating ore minerals into vast networks of tiny cracks. The resulting orebodies are very large but low-grade. However, they can be exploited by modern industrial methods (Grotzinger et al. 2007:70–71). Stockwork deposits are an intermediary type of mineral deposit consisting of a three-dimensional network of veins spaced closely enough so that the entire body can be mined (U.S. Bureau of Mines 1996). Stockwork and disseminated deposits in most cases would be difficult if not impossible to mine using prehistoric techniques.

One of the most important types of disseminated hydrothermal deposits is porphyry copper. A porphyry rock is an igneous rock with a texture composed of large crystals surrounded by a fine crystalline matrix (Grotzinger et al. 2007:80). Porphyry copper deposits consist of disseminated copper minerals such as chalcopyrite found in veins that are relatively evenly distributed in large volumes of rock, forming very large but low- to moderate-grade (0.3–2.0% copper) ores. Copper ores can occur not just in the stockwork but also as larger veins and replacements, some of which may have been workable in prehistory. Over 60% of the annual world copper production today comes from porphyry copper deposits (John et al. 2010). Porphyry copper deposits are strongly associated with subduction zones along plate boundaries (Barnes 1988:60; Derry 1980:23). Due to their vast extent and low-quality ore, most porphyry copper deposits are worked in large open-pit mines (Barnes 1988:59).

METAMORPHIC PROCESSES

Metamorphic ore creation often goes along with hydrothermal processes. Sometimes minerals are found in host rock near the veins; the host rocks, especially carbonates (Barnes 1988:64), were changed as the hydrothermal solutions heated and penetrated the rock. These rock-altering processes are called contact metamorphism, which is the alteration of rocks near or next to magmas, and contact metasomatism (Murck 2001:254). Contact metasomatism is "a mass change in the composition of rocks in contact with an invading magma, from which fluid constituents are carried out to combine with some of the country rock constituents to form a new suite of minerals" (U.S. Bureau of Mines, 1996). These contact metamorphic deposits tend to be small but rich, containing both oxide ores such as magnetite and hematite (iron), cassiterite (tin), and others, and sulfide ores such as chalcopyrite (copper), galena (lead and silver), and sphalerite (zinc). The oxide ores occur closer to the igneous contact, the sulfides further away (Barnes 1988:64). Contact metasomatism can create skarn deposits when granite magma intrusions come into contact with limestone or other carbonate rocks. Hydrothermal fluids from the magma dissolve some carbonate and create the skarns, which are lime-bearing silicates. When they contain copper, iron, tin, and other ore minerals, they are called skarn deposits. The mineral deposits at Phu Lon (TAM 2C, chapter 2) include skarn deposits. Regional metamorphism, which is when large areas are altered because of plate collisions, can segregate minerals in distinct and concentrated bands, though these are primarily non-metallic minerals such as mica, asbestos, and graphite (Murck 2001:254).

MAGMATIC PROCESSES

The largest deposits of ores found in igneous rocks are located near the bottom of intrusions. Some silicate magmas can hold in solution up to 10% metallic sulfides. Layered deposits of ore minerals form when dense mineral-rich liquids crystallize from molten magma and settle, because of their greater density, on the bottom of a magma chamber. Although many such deposits lie so deeply in the crust that they are inaccessible to humans, in cases where ocean floor has been thrust up over continental crust, ophiolite complexes occur, and some, such as those in Oman and Cyprus, were important copper sources for the bronze periods of Mesopotamia and the eastern Mediterranean (Killick 2014:18–19). Such ophiolite complexes occur near the Ailaoshan-Song Ma and Nan-Uttardit Sutures (Lai et al. 2014; Sone and Metcalfe 2008).

Mechanical, Surface, and Sedimentary Processes

Water motion and weathering create placer deposits, which are secondary deposits of syngenetic ores. Placers are surface or near-surface deposits formed by the depositional concentration of mineral grains that have been eroded from primary bedrock locations and transported by surface water. The usual concentration process is fluvial, or by river flow, but it can also be by marine, eolian (wind), lacustrine (lake), or glacial action (Hutchison 1983:256–257). As flowing water moves over and sifts the sediments, the denser ore minerals sink and become concentrated. The mineral fragments can be transported very far from the source (Suthakorn 1992). Commercially valuable placer deposits concentrate minerals with a high specific gravity, such as elemental gold, cassiterite, or uranium. Tectonic uplift and rapid erosion by heavy seasonal rainfall contribute to placer formation (Hutchison 1983:256). Placer deposits of the tin ore cassiterite are assumed to have been important in prehistory, although Killick (2014:23) noted, "Depressingly little work has been done on the early mining and metallurgy of tin." See the section on tin deposits of Southeast Asia below for a further discussion of types of placer deposits.

Iron, salt, and phosphorus deposits can form by sedimentary evaporation or precipitation, as minerals carried in solution or suspension. Indeed, most iron produced is from sedimentary ores (Barnes 1988:32). Simple evaporation leads to salt, gypsum, and potassium salts, but biochemical reactions lead to the precipitation of iron in banded iron formations (BIFs). These banded iron formations, the most important source of iron today, are all more than 1.8 billion years old (Murck 2001:256) and were formed at a time when there was little oxygen in the earth's atmosphere (Murck 2001:155). Killick (2014:22) argued BIFs were little used in prehistory due to their hardness unless heavily weathered. Softer iron ores like hematite and limonite, along with northern bog iron ores, were probably preferred in prehistory; these ores were formed in shallow waters by precipitation.

Leaching can lead to enrichment of ore deposits. After an ore deposit has been uplifted as a result of tectonic activity, rain can dissolve soluble minerals and deposit them underneath the ground water layer. Laterite, a hard subsoil that is sometimes mined in tropical zones for iron and nickel, is a result of leaching (Murck 2001:257) and is the most abundant iron ore in Southeast Asia. Alternating wet and dry seasons dissolve elements in the soil except for insoluble minerals including iron. "Iron and aluminum ions are slowly moved downwards during the rains but reprecipitate as oxides and hydroxides between the high and low annual levels of the water table during the dry season" (Killick 2014:22). Laterites can be relatively rich (15%–30% weight percent) in iron (Pryce and Natapintu 2010). Laterites were smelted in Mauritania during prehistory (Killick 2014:22). Smelting of laterite for iron in prehistoric Southeast Asia has been both proposed and questioned (Cawte and Boyd 2010; Pryce and Natapintu 2010). Cawte and Boyd (2010) have observed variability in the iron content in the region's laterite, and some may have been sufficiently high grade to be smeltable.

Enriched copper sulfide ores can also result from leaching. The oxidized, leached cap of a sulfide ore deposit is called a gossan (Hutchison 1983:9). Gossans are mostly hydrated iron oxides, and beneath them can sometimes be found oxide and carbonate ores of copper, if the deposit is of copper sulfide. (See TAM 2C, fig. 2.5 for a schematic drawing of the leached copper deposits at the mining site of Phu Lon.) Gossan formations were probably important signals to pre-industrial copper prospectors (Killick 2014:19).

Metallogenesis in Mainland Southeast Asia

Plate Collisions and Fold Belts

The geological overview so far presented, which focuses on positioning northeast Thailand in Southeast Asian tectonic history, is a necessary simplification of the extremely complex and detailed evidence for innumerable events and tectonic features discussed in the primary research literature. Although geologists in a synthetic work may refer to, for example, "a Loei Fold Belt" or one of its various aliases, as a belt extending from Yunnan to western Cambodia, investigation of the primary research publications shows that within the larger belt are multiple separately named and shorter sutures, faults, fold belts, and mineral belts. As noted above, not every fold belt has metallic deposits, and by no means do all "mineral belts" have metallic deposits. On the other hand, at a macro scale, zones of metallogenesis that more or less align with some larger-scale fold zones that relate in most cases to terrane interactions can be recognized.

The collision of the Indochina Terrane with the South China Terrane produced a Red River Fault Zone and metamorphic belt north of the Song Ma Suture in northern Vietnam, along with the Truong Son Fold Belt southwest of the Song Ma that runs northwest–southeast from north of Vientiane down the Laos-Vietnam border as far south as Savannakhet and the Kontum Massif of Vietnam (Fig. 6.3). Metallogeny, including of copper ores, is associated with this plate interaction to the north of the Song Ma Suture, and more specifically north of the Song Hong or Red River (Tran Trong Hoa et al. 2008). The Red River Metamorphic Belt in South China, near the Red River and in northernmost Vietnam, is rich in iron, hydrothermal tin, tungsten, lead, zinc, iron, antimony, and residual bauxite, plus copper and gold (Tran Trong Hoa et al. 2008). An additional zone of folding related to the Triassic Orogeny can be found in southern Laos east of Salavan (United Nations 1990).

The Truong Son Fold Belt meets the Loei Fold Belt north of Vientiane; the junction area includes Xieng Khouang Province, where the Plain of Jars is located. The Truong Son Fold Belt appears to result from the closure of the Ailaoshan-Song Ma branch of the Paleotethys Ocean, along with the extensive continental and oceanic island arc created in between South China and Indochina (Khin Zaw et al. 2014b). The Truong Son Fold Belt, roughly congruent with the northern portion of the Annamite Cordillera, includes the known prehistoric mining site of Sepon, Laos (see Fig. 6.4 and TAM 2C, chapters 2 and 3). The Devonian and Pre-Devonian sedimentary rocks of the Truong Son Fold Belt are highly deformed, especially to the north (Hutchison 1989:150). The Devonian strata are extensively intruded into by volcano-plutonic granitoids and diorites. These highly deformed geologic deposits are rich in many mineral resources, including copper, iron, tin, and gold.

The complex folding and mountain building where the Sibumasu and Indochina Terranes met created excellent conditions for metallogenesis. The collision where Indosinia overthrusts the southern Asia Terrane caused intense deformation along the eastern edge of the Sibumasu Terrane. Lower Paleozoic rocks were extensively penetrated by Triassic and Cenozoic granitic and granodioritic metallogenic intrusions (Crow and Khin Zaw 2011). One of the long belts of tin-bearing granites that runs from Yunnan to West Thailand and through to northwest Indonesia was extruded into country rock during the Sibumasu and Indochina collision (Bunopas and Vella 1992). The Eurasian Continental Plate that resulted from this collision currently extends as a peninsula as far south as Indonesia.

In Laos and Thailand, the collision of the Sibumasu-Indochina Terranes and the smaller arc/complexes in between produced two north–south trending parallel fold belts on either side of the Nan-Uttaradit Suture Line. The Sukhothai Fold Belt is on the western side of the suture. The Loei Fold Belt (called Loei-Phetchabun by several authors, the Loei-Prachinburi Mineral Belt by Crow and Khin Zaw [2011], and the Phetchabun Volcanic Belt by Barr and Charusiri [2011]), or the Loei volcanic arc, lies on the western edge of the Indochina Terrane, bounded on the west by the Nan-Uttardit Suture and on the east by the escarpment of the Khorat Plateau. The Loei Fold Belt resulted from the subduction of the Sukhothai microterrane under the Indochina Terrane (Lepvrier et al. 2004). The Loei Fold Belt extends south from the mountainous area north of Vientiane, Laos, and runs along the western edge of the Khorat Plateau and down through the Khao Wong Prachan Valley before it turns southeast and runs into western Cambodia, bisecting Thailand. To the north, it extends into northern Laos, where it either takes an abrupt turn to the southeast (Gatinsky et al. 1978; Sitthithaworn and Wasuwanich 1992) or meets the Truong Son Fold Belt (Manaka et al. 2008; United Nations 1990), depending on how it is labeled. The fold belt extends further north into Yunnan (United Nations 1990). The belt is characterized by complex subduction, collision, accretion, and deformation (Kamvong and Khin Zaw 2009), with significant magmatism, and with evidence for several cycles of arc-related and volcanic events, separated by erosion, with volcanism from the Devonian through the Cenozoic. The fold belt contains volcanic rocks that are predominantly rhyolite and andesite-dacite, with subordinate basalt. Many areas have granites and granodiorites that were intruded into limestones and other sedimentary rocks. Major metallogenic zones are associated both with volcanic rocks and granite, granodiorite, and diorite intrusions (Jungyusuk and Khositanont 1992). The Loei Fold Belt hosts several economic ore deposits of base and precious metals, including the copper-gold skarn at Phu Lon (TAM 2C, chapter 2). Copper and iron deposits can be found along the margins of the intrusions. Copper ores in Loei Province are particularly rich, though good deposits exist as far south as Lopburi (TAM 2C, chapter 2). Deposits in this belt in western Cambodia have yet to be investigated by archaeologists. Modern literature sources (e.g., Crow and Khin Zaw 2011) also list gold, zinc, lead, iron, and silver. Ores of note include chalcopyrite and malachite, along with magnetite, hematite, and electrum. Ore types include copper and copper-gold skarns, bedded and vein systems, and porphyry copper (Crow and Khin Zaw 2011). The Sukhothai Fold Belt runs parallel to the Loei Belt in the Sukhothai Terrane on the other side of the Nan-Uttardit Suture; most economic geologists include this belt in the Loei Belt.

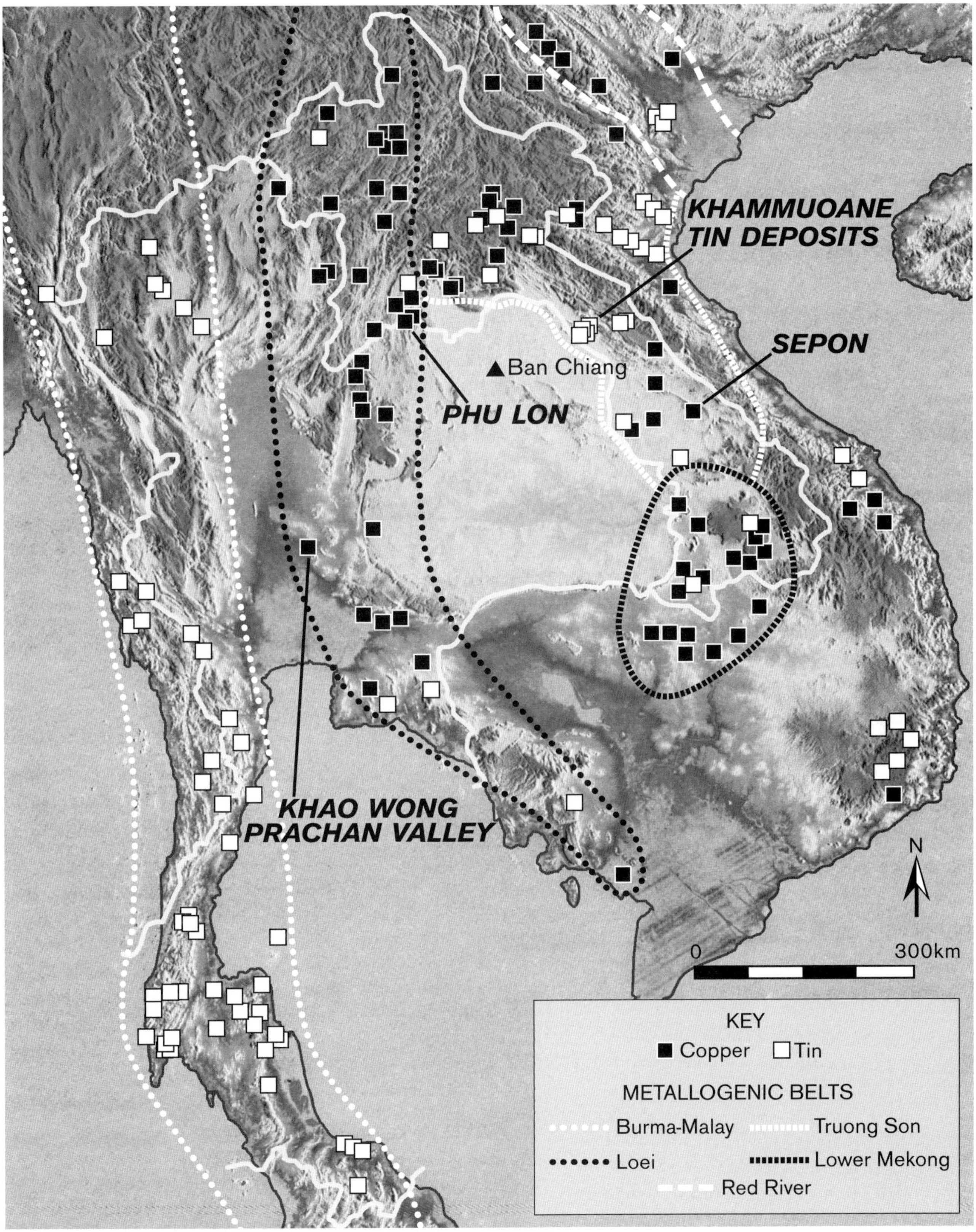

Figure 6.4 Known modern sources of copper and tin in Thailand, Laos, Cambodia, Vietnam, and southeast Myanmar in relation to metallogenic belts. Map based on relief map in Ridd et al. (2011b), with metallogenic data from U.S. Geological Survey (2005), Smith et al. (2012), and Khin Zaw et al. (2014b). Locations are only of those ore sources where copper ore or tin ore is the primary metal ore found.

The Truong Son and Loei Fold Belts rim the Khorat Plateau on its east, north, and west sides. Excavations at prehistoric metal age copper extraction sites located in these two fold belts are discussed in TAM 2C (chapter 2). The Loei and Truong Son Fold Belts are the most relevant for the purposes of locating ore resources near Ban Chiang, but additional fold belts with significant metallogenesis are (Gatinsky et al. 1978):

- Burmese-Malayan Belt (or Thai-Malaysian Belt [Smith et al. 2012]), containing over 60% of the world's tin reserve, most associated with late Paleozoic and Mesozoic granites. This belt is comprised of two parallel tin-rich belts running from the Malaysian Peninsula through western Thailand and up to Yunnan; the belts also contain concentrations of copper, tungsten, lead, silver, antimony, gold, iron, and zinc ores.

- West-Burma-Sunda Belt, associated with the collision of the Indian Plate with Asia, running south from Myanmar down through the Indonesian islands. This belt is distinguished by Cretaceous-Cenozoic hydrothermal lead-zinc-silver, gold, porphyry copper, and iron.

- Philippines Belt, also containing Sulawesi, contains large quantities of copper ore.

Metallogenic Belts

For the purposes of archaeologists, it is very useful to recognize and name these usually elongate zones, which in the case of Southeast Asia approximately conform to fold belts, or perhaps more descriptively, "fold belt systems." Rundquist's definition (1984:28) seems very apt for our purposes: "metallogenic belts…are composite, linear, elongate fold systems, which can be traced for many hundreds or even thousands of kilometers."

Viewed more broadly, as a general rule, the metalliferous minerals found in Southeast Asia including western Thailand and southern central Laos are associated with the north–south tectonic zones and fold belt systems produced by many geological events during the Indosinian Orogeny, including the Permian-Triassic collisions of the South China, Sibumasu, and Indochina Terranes described above (Crow and Khin Zaw 2011). In addition to these Indosinian mineralizations, a late Cretaceous–Eocene episode of mineralization along the Thailand-Myanmar border is related to the granitic plutonism resulting from the collision of an island arc with the western edge of the Sibumasu Terrane. An effect of the Himalayan Orogeny in the Eocene was the production of a basement dome in the Inthanon Zone in northern Thailand along the Sukhothai Fold Belt. The core and flanks of this dome are areas of significant mineralization, with inclusions of tin, tungsten, gold, arsenic, and iron (Crow and Khin Zaw 2011).

Stuart Smith (Smith et al. 2012) has recently broadly delineated five "mineral provinces" in mainland Southeast Asia, which are useful for conceptualizing and describing orebody zones to which prehistoric inhabitants of the region could have had access. Smith's five zones are adopted and adapted here but with terminological modifications. Although Smith employed the term "province," the shapes tend to be elongated and belt-like as they generally are related to fold belts from plate boundary interactions discussed above. Therefore, the term "belt" is used here, following Rundquist (1984:28; see also Gatinsky et al. 1978 for use of "metallogenic belts" in Southeast Asia). Secondly "mineral" is less apt than "metallogenic" as the depicted belts refer to orebodies but not mineral deposits such as salt. It is important to note that the heuristically useful metallogenic belts may in some cases encompass more than one geologically defined fold belt system. For example, the ores in the adjacent Sukhothai and Loei Fold Belts would be subsumed, following Smith, into the Loei Metallogenic Belt, both being products of the Sibumasu-Indochina Terrane interactions. Thus, there is not a simple isomorphism between metallogenic belts and individual geological fold belts discussed in the primary literature. One final terminological adjustment to the Smith system is that some belts are assigned variant names that more closely identify their geographic positions.

The five metallogenic belts following Smith's system are here termed, starting from the west: (1) the Burma-Malay Metallogenic Belt, extending from central Myanmar to the tip of the Malay Peninsula; (2) the Loei Metallogenic Belt that generally follows the well-recognized Loei Fold Belt from northern Laos to southeast Cambodia and, which for purposes here, includes the Sukhothai Fold Belt; (3) the Truong Son

Metallogenic Belt, generally aligning with the Truong Son Fold Belt system; (4) the Red River Metallogenic Belt, lying north of the Song Ma Suture; and (5) the newest metallogenic zone, which is here called the Lower Mekong Metallogenic Belt, replacing Smith's Khorat orebody zone. As explained above, the Khorat Plateau in Thailand proper has few metallic resources. Smith's delineated zone with notable copper resources actually extends from the southern edge of Laos in Champasack and Attapeu Provinces to northeast and north central Cambodia (Fig. 6.4). This area is immediately to the west of the Kontum Massif and is possibly related to its tectonic interactions within the Indochina Terrane. Recent exploration there has revealed evidence of ore-bearing intrusive deposits that had been overlooked or minimized in earlier geological explorations (McDougall 2009).

Metallic Resources in Mainland Southeast Asia

As noted, many of the metallic resources of mainland Southeast Asia are found in geologic fold belts created by terrane collisions. Many of these metallogenic deposits were produced when molten magma intruded into rock and solidified into granite or granodiorite. The rich Southeast Asian deposits of copper, gold, lead, antimony, and zinc are commonly found in veins, stockworks, and disseminations near these granitic or granodioritic intrusions, and result from hydrothermal and/or volcanic activity during the Permian to late Triassic orogenies. Quartz veins in this area often include pyrite, which, in some areas, contains copper and gold. Chalcopyrite, chalcocite, molybdenite, galena (lead sulfide), sphalerite (zinc sulfide), and stibnite (an ore of antimony) are also found in such veins and in zones of fracturing, brecciation, and alteration of intrusive rocks. There are also deposits containing lead, zinc, copper, and other sulfides, and some contact metasomatic magnetite-hematite-pyrite deposits (United Nations 1990).

Middle Mekong Basin Metal Resources

Despite a general association with fold belts, metal ores that are exploitable, especially by pre-industrial methods, tend to be highly localized. A review follows of known and potential orebodies that could have been of interest to prehistoric occupants of the middle Mekong Basin.

Khorat Plateau

At the heart of the middle Mekong Basin, with ample evidence for metal age occupation, lies the Khorat Plateau, whose geological history has endowed it with limited metal resources. The geological surface of the Khorat Plateau is predominantly composed of Jurassic to mid-Cretaceous red beds, so-called because they are composed of iron-rich sediments. They are a "rather monotonous thick series of predominantly red, tan, gray, and brown fine-grained sandstone and shale beds" (LaMoreaux et al. 1958:16). The iron-rich sediments eroding from newly raised mountains were deposited on floodplains and lake bottoms, resulting in layers of sandstones, clays, and lacustrine (lake) limestones that are termed by geologists the "Khorat Group." The deposits, often rich in Mesozoic fossils, can in some areas be as much as several thousand meters thick (Buffetaut and Ingavat 1985; Buffetaut and Suteethorn 1998). Despite this depth of deposit, the Khorat Group Basin is merely the youngest of a series of superimposed basins of shales, sandstones, and limestone, the oldest of which go back to the middle Paleozoic (Booth and Sattayarak 2011). The underlying Carboniferous through Triassic deposits were laid down in marine conditions, providing the source for percolated salt deposits still mined today. Quaternary alluvial silt, clay, sand, and gravel also overlay the area (LaMoreaux et al. 1958:1).

However, the plateau has undergone a variety of tectonic events, primarily around its margins. The oldest was an orogeny in early Carboniferous time, followed by rifting in the late Carboniferous. The Indosinian Orogeny produced volcanism and considerable granitic intrusions, notably in Loei Province, along the Loei Fold Belt at the western margin of the plateau. At the end of this orogeny, the Khorat Plateau began subsiding, particularly in the middle, leading to the massive deposition of sedimentary red beds. Finally, during early Tertiary time, there was another episode of orogeny along the suture lines, again along the western margin of the Khorat Plateau (Kozar et al. 1992). The Phu Phan uplift that separates the Sakon Nakhon from the Khorat Basin occurred during the Tertiary.

The only copper mineralization discovered so far on the plateau itself is at Phu Wiang on the western edge of the plateau, where modern records note occurrences of malachite-azurite, along with uranium mineralization. Two lenses of relatively low-grade ore, with a copper content of 0.8%–3%, have been exploited in modern times (Crow and Khin Zaw 2011). Nevertheless, the proximity of that source to the metal age site of Non Nok Tha deserves exploration.

The sedimentary basins of the Khorat Plateau are more likely to contain oil deposits than minerals of economic interest (Crow and Khin Zaw 2011), but the Khorat Plateau is surrounded by rich metalliferous deposits. Many of these are of the low-grade disseminated orebody type exploitable mainly by modern methods. However, hydrothermal vein deposits and skarn deposits that were or could have been exploited by pre-industrial techniques were accessible within 150 km.

Metallic Mineral Deposits

Many of the ore deposits used by early miners and metallurgists no longer exist... (Killick 2014:23)

Khin Zaw et al. (2014b) summarized some of the important ore deposits (listing 58 deposits for copper, gold, lead, zinc, and silver) for mainland Southeast Asia. Tin placer and ore deposits are innumerable and widespread, with a particular concentration along the Burma-Malay Metallogenic Belt. The abbreviated discussion that follows is based on modern geologic surveys that have assessed mineral deposits of potential industrial value, bearing in mind Killick's caution that the earliest deposits may have been destroyed. The description in this summary focuses mostly but not exclusively on copper, tin, and lead resources located near the Khorat Plateau.

Laos and Thailand are rich in metallic mineral resources (see Figs. 6.4–6.5 for major deposits of copper, tin, lead, and iron). Inspection of the 1990 ESCAP geological map (for Laos) and the United Nations Atlas (1968) reveals as expected that, as noted above, the Khorat Plateau itself has few metallic deposits. Within an estimated 100–200 km of Ban Chiang, however, are numerous deposits of gold, iron, lead, copper, zinc, and antimony, as well as tin found in placer deposits between the Mekong and Theum Rivers. For dwellers of northeast Thailand, accessible deposits are especially rich in Loei Province, with occurrences of copper, iron, gold, and lead rich enough for modern exploitation. The Loei Fold Belt also extends north into Laos. In Laos, the Plain of Jars lies at the convergence of the Loei and Truong Son Fold Belts and has many granitic intrusions. It is well supplied with a particularly wide range of ores, including ores of copper, iron, gold, tin, and lead. Two rivers, the Ngum and the Nhiep, flow from the Plain of Jars area to the northern Khorat Plateau, offering access to copper for ancient inhabitants of the Ban Chiang area.

As Bronson (1992) noted, the great majority of the prehistoric sites in mainland Southeast Asia with distinct bronze ages (as opposed to later metal-age sites that demonstrated concurrent use of bronze and iron) were within 200 km of known sources of copper and tin ores.

LOEI METALLOGENIC BELT

The Loei Fold Belt is distinguished by Paleozoic and Early Mesozoic iron and copper skarns, porphyry copper, vein lead, zinc, antimony, and, in Laos, vein tin (Gatinsky et al. 1978). Sites in this fold belt that have been the subject of modern day economic geological survey include the prehistoric copper mining site of Phu Lon, with its early Triassic mineralization (Crow and Khin Zaw 2011). The deposit found here is primarily gold-copper, with 2.54% copper in the ore. The orebody is described as a calcic skarn, with disseminated sulfides and quartz sulfide stockwork (Crow and Khin Zaw 2011). In 1985, following a survey and core-drilling program at the site by the Thai Department of Mineral Resources, this weathered copper sulfide orebody was surveyed by Thailand Archaeometallurgy Project geologist William W. Vernon (e.g., Vernon 1996–1997). He noted that the surface is an oxidized zone (gossan) containing veins of malachite, a primary ore of copper that is easy to smelt and is easily identified with the naked eye. This zone also contains some chrysocolla, limonite, hematite, and residual chalcopyrite and pyrite (Pigott and Weisgerber 1998). This upper zone of the deposit has significant evidence of prehistoric mining (e.g., Natapintu 1988; Pigott and Weisgerber 1998). The primary deeper-lying zone of the ore deposit contains sulfidic

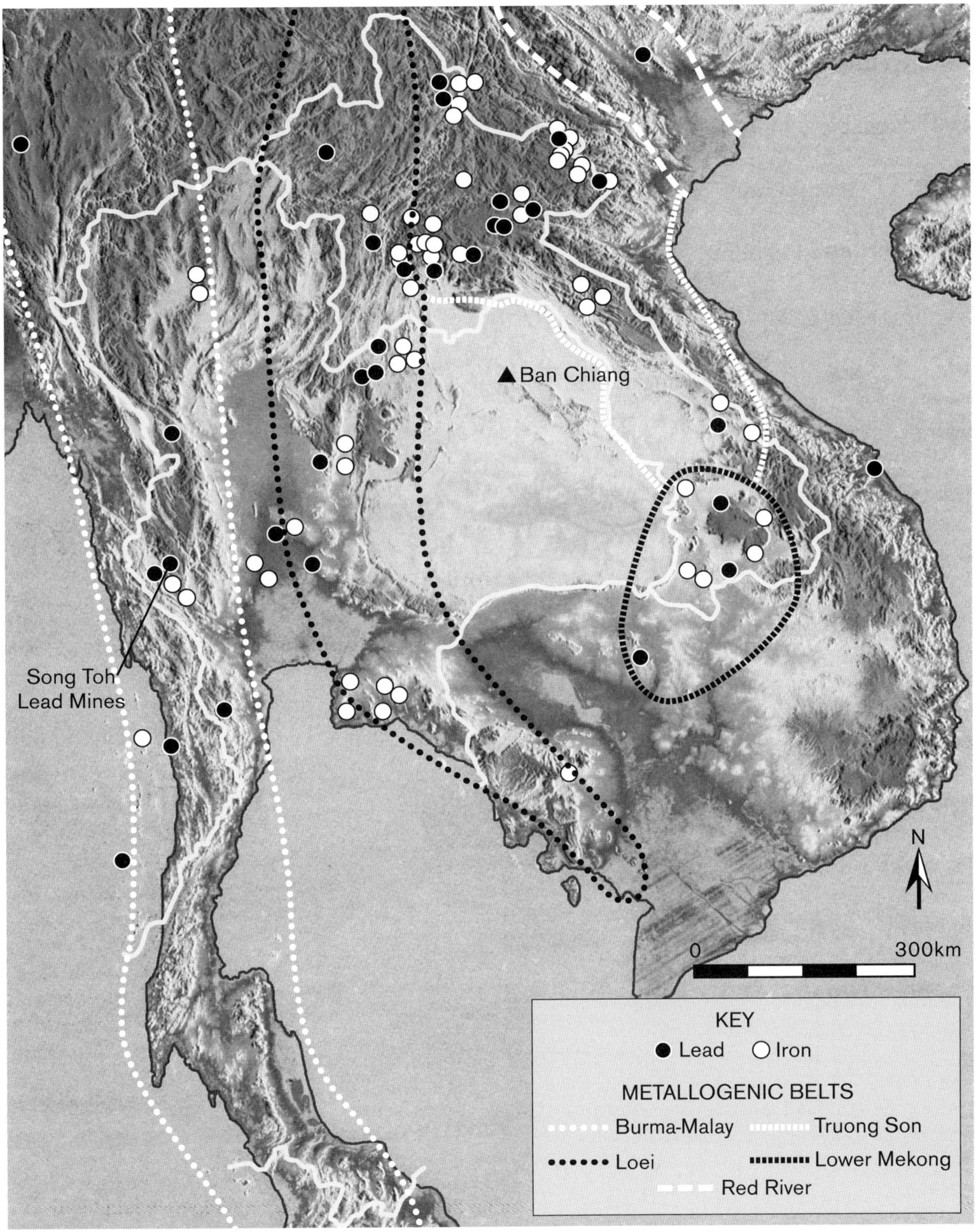

Figure 6.5 Known modern sources of lead and iron in Thailand, Laos, and surrounding areas. Map based on relief map in Ridd et al. (2011b), with metallogenic data from U.S. Geological Survey (2005) and Smith et al. (2012). Locations are only of those ore sources where lead ore or iron ore is the primary metal ore found.

ores like chalcopyrite, pyrite, and also the iron ore magnetite, all of which are embedded in a quartz-rich skarn matrix. (See TAM 2C, chapter 2 for more on Phu Lon and its metallurgical and archaeological deposits.)

Modern industrially worked deposits south of Phu Lon include Puthep, which consists of two combined deposits that were mineralized during Permian–early Triassic time. The deposits are low-grade (0.53% Cu) copper and copper-gold. The orebodies are copper-iron porphyry skarn in veins and fractures, gossans, and pockets of malachite. East of Bangkok, where the Loei Fold Belt turns east, is French Mine (or Frenchman Mine), with copper-gold minerals; the gold is low-grade. The orebody is a copper-gold skarn associated with other minerals; mineralization occurred during late Permian–early Triassic time (Crow and Khin Zaw 2011; Khin Zaw et al. 2014b).

In other areas of Loei Province, the copper orebodies are lower in grade, with many outcrops and boulders of gossan. Both oxide and sulfide ores are present, mostly the latter with chalcopyrite and also bornite and covellite. The oxidized zone contains native copper, cuprite, malachite, and chalcocite, which can appear as a secondary mineral (Workman 1972:37). The ore found in disseminations is of a low grade but widely distributed; the rarer quartz veins are rich in copper (Workman 1972:38). The copper deposits are also associated with gold, which is now receiving some industrial interest (Crow and Khin Zaw 2011).

Deposits also occur in Phetchabun Province, along the Huai Phi stream and in the Pa Daeng region. The copper minerals there include chalcocite, cuprite, malachite, and azurite located in quartz veins. The deposits are not of economic interest today (Workman 1972:39). Deposits at Khok Kathiam, Lopburi Province, consist of chalcopyrite in brecciated zones in the granite. Deposits at Phanhom Sarakham, Chachoengsao Province, are very unusual for Thailand, because they contain native copper and other copper minerals in mica schist with no association with igneous rocks (Tantisukrit 1978).

To the north of Chachoengsao Province, deposits in the Khao Wong Prachan Valley in central Thailand have been investigated by the Thailand Archaeometallurgy Project because of the extensive evidence for prehistoric copper smelting in that valley (e.g., Natapintu 1988:115, table 2; see also TAM 2C, chapter 2). There are three areas of copper mineralization in the valley complex: Khao Tab Kwai, Khao Phu Kha, and Khao Pa Daeng; the latter two have evidence of pre-modern though undated exploitation (Pryce et al. 2011). The deposits there do not appear to be economically important enough to be featured on modern geologic and metal-resource maps (see Fig. 6.5 and Crow and Khin Zaw 2011), although they were exploited on a massive scale in prehistory, another example of the disjunction between industrially important resources and resources used in antiquity. The geology of the Khao Wong Prachan Valley consists of primarily Permian sediments intruded into by andesite, diorite, granite, and granodiorite. The contact zones between the intrusive rocks and the sedimentary rocks are characterized by malachite, chalcopyrite, hematite, and magnetite (Vernon 1988 in Pryce et al. 2010).

TRUONG SON METALLOGENIC BELT

In the north of Laos and just into southern China, the Loei, Truong Son and Song Ma Suture mineral belts converge. Laos contains numerous deposits of copper ore, usually disseminated or in veins near granodioritic intrusions (see Fig. 6.4). The closest deposit to Ban Chiang is the Ban Nam Thong field in Vientiane Province (19°00'N, 103°15'E), a skarn deposit. Copper-gold deposits are located at the junction of the Loei Fold Belt and the Truong Son Fold Belt (running to the east of northeast Thailand) about 100 km north of Vientiane (Manaka et al. 2008). Deposits of malachite and chrysocolla, mostly of stratiform type with an origin unrelated to intrusive granites, can be found in the south of the country. In the Truong Son Fold Belt itself are some of the largest metalliferous deposits in the region: Phu Kham, a porphyry-related oxidized skarn deposit with enriched copper-gold mineralization; and Sepon, a sediment-hosted gold deposit with additional porphyry-related copper skarn deposits (Khin Zaw et al. 2014b). Sepon in Savannakhet Province in Laos (16°57'N, 106°00'E) is a known site of modern and prehistoric copper ore exploitation (see TAM 2C, chapter 2). Khin Zaw et al. (2014b: table 1) list additional deposits in the Truong Son Fold Belt with copper mineralizations.

OTHER COPPER RESOURCES

There are other noteworthy copper ore deposits within reach of dwellers on the Khorat Plateau, if only by trade, although they are not strictly within the Loei or Truong Son Fold Belts. These include copper resources on the north edge of the Kontum Massif in south central Vietnam (Khin Zaw et al. 2014b:16) and directly to its west in southern Laos and northern Cambodia (McDougall 2009) (i.e., the Lower Mekong Metallogenic Belt; see Fig. 6.4). Several copper deposits are noted north of the Song Ma Suture (i.e., associated with the Red River Metallogenic Belt) parallel with the southern edge of the South China Terrane on either side of the Red River Fault. The copper deposits are sometimes combined with gold or nickel (Tran Trong Hoa et al. 2008:488). There are other scattered deposits reported; of particular note is the big Monywa deposit in central Myanmar.

TIN

Southeast Asia is one of the richest areas for tin of the world. Much of the tin output of present-day Malaysia, Thailand, and Indonesia is from placer deposits in the Burma-Malay Metallogenic Belt (see Fig. 6.4). The richest source of tin for placer deposits in Southeast Asia is the granites of the Cretaceous fold belt, followed by the granites of the Triassic fold belt, which are frequently associated with tungsten (Suthakorn 1992). Most of the tin production in Southeast Asia is from late Cenozoic alluvial placers, or mineral-rich sand and gravel deposits created by flowing water. Most of the tin was deposited in braided streams, piedmont fans, and at the base of cliffs, though some tin remains near the parent ore as eluvial placers (Hutchison 1983:259–260). Related to alluvial placers, an eluvial deposit is created by the disintegration of mineral-bearing rock in place. Eluvial deposits, typically less than 3 m in depth, are rare, but rich, with the primary tin vein found underneath. The Mae Boh Kaew deposit in Chiang Mai is an eluvial deposit (Suthakorn 1992).

The tin found today in placers came from several different types of deposits. It is probable that the tin deposits worked in antiquity were of the placer type, but many modern tin mines directly exploit the larger deposits from which the placer ore was eroded. The most abundant type of in situ tin deposit in Thailand is hydrothermal quartz veins developed in the zones close to granite batholiths (Hutchison 2005). These deposits occur when superheated steam from a siliceous fluid rich in tin, tungsten, copper, arsenic, iron, and other minerals is ejected from an intrusive granite body into fractures in the country rock. Minerals are precipitated within these fractures as the steam condenses (Robert Giegengack, pers. comm. 2012).

Cassiterite, which precipitates at 350–400°C, crystallizes in the intrusive body and in veins near the body; copper ores, which crystallize at between 240°C and 400°C, overlap the zone of tin precipitation but also extend farther away from the granite intrusions (Barnes 1988:67; Hutchison 1983:134–135). Further still, in a zoning pattern based on the precipitation temperature, is silver-rich zinc-lead, and farther out still are zinc and lead sulfides (Barnes 1988:68–69). The tin ore was concentrated into rich placer deposits when the granites were exposed in the Miocene (Hutchison 2005).

The second richest in situ tin source is pegmatites, the very coarse-grained rock precipitated in veins, pipes, or lenses, usually granitic in composition and found near granitoid intrusions (Suthakorn 1992). Pegmatites represent the last portion of a magma intrusion to cool, and tend, for that reason, to be exceptionally rich in minerals. Like hydrothermal quartz veins, their mineral composition is zoned. In Thailand, they are also associated with tin-bearing aplites or fine-grained pipes or dikes (Suthakorn 1992). Tin is also found in disseminated form as small grains in the marginal zones of intrusive granite altered by gaseous emanations derived from solidifying magma (U.S. Bureau of Mines 1996). Disseminated tin is also found in the country rock near granite intrusions. Relatively rare elsewhere in the world, disseminated tin is found in many tin fields in Thailand (Suthakorn 1992).

More commonly a source of copper, contact metasomatic deposits can also be a source of tin. When granite intrudes into limestone or other carbonate rock, a metamorphic "cloud" develops that can range from a few meters to scores of meters wide. The limestone becomes skarn. Close to the intrusion is a zone of oxidized mineral deposits, which can include ores such as magnetite, hematite, cassiterite, and copper oxide ores. Farther away are sulfide ores such as chalcopyrite and silver-rich galena (Barnes

1988:64). The massive Pinyok Mine in Yala Province in southern Thailand is an example of a tin-rich contact metasomatic deposit (Suthakorn 1992).

Location of Tin Deposits

Peninsular and northwestern Thailand forms a large part of the tin- and tungsten-rich Burma-Malay Metallogenic Belt that extends south to Indonesia and to the north even to southwest China. The belt is the world's richest, but it is not continuous and is concentrated in discrete areas of rich ore (Hummel and Phawandon 1967). During the 20th century, over 70% of the world's tin came from this belt in Thailand, Malaysia, and Indonesia (Hutchison and Taylor 1978), though many of the mines have since closed because of lack of world demand (Crow and Khin Zaw 2011). The tin is usually in the form of cassiterite. In other places, such as eastern Laos, the cassiterite may be in placer deposits in alluvium, or concentrated in laterite by groundwater circulation and weathering (United Nations 1990:7).

The major tin zones in Thailand are in northern Thailand (Chiang Rai, Chiang Mai, Lamphun, Lampang, and Tak Provinces), central Thailand west of the Chao Phraya (Uthai Tham, Suphanburi, and Kanchanaburi Provinces), the Thai peninsula along the west (Phetchaburi, Ranong, Phangnga, Trang, and Phuket Provinces), and the eastern peninsula (Prachuap Khiri Khan, Surat Thani, Nakhon Si Thammarat, Songkhla, and Yala Provinces) (Suthakorn 1992). Some small operators, using minimal equipment, pan or mine tin in northern Thailand—mostly between Chiang Mai and Chiang Rai. Many traces of cassiterite have been found as placers in river gravels in the north, so little machinery is needed to extract tin ore (Workman 1972:27–29).

Tin has also been discovered in southeastern Thailand, in Chanthaburi Province, mostly in the form of placer deposits of fine-grained cassiterite. Like most such deposits in Southeast Asia, these deposits are associated with granite intrusions and associated mineralization. Beach placer deposits of cassiterite along the Gulf of Thailand have also been found in Rayong Province (Tantisukrit 1978). However, the sources closest to Ban Chiang would have been in Laos, associated with the Truong Son Metallogenic Belt.

In Laos, Khammouane Province, due east of the Sakon Nakhon Basin and north of Savannakhet Province where the Sepon copper deposits are located, has several areas with exploitable tin. One major tin field is located along the Nam Pathene Valley some 210 km southeast of Vientiane near the Mekong River. The principal industrially worked ore is cassiterite in laterite, a surface enrichment of iron oxide with quartz and needles of cassiterite. The ore contains 0.3%–1.3% tin. On the north side are microgranite veins containing cassiterite, some of which is enclosed by pyrite. There have been some investigations of a possible large area of alluvial placer tin near this field that may extend into the extreme corner of northeastern Thailand (Workman 1972:25). Barthelemy (1938) gave an ethnographic account of how low-technology tin panning and smelting was carried out by villagers on the edge of the Nam Pathene Valley (see Fig. 7.1 for an illustration of their methods.). The source of the tin in the Nam Pathene Valley was eluvial deposits resulting from the decomposition of pyrite veins (Barthelemy 1938). Currently mined areas in a small basin a few kilometers north of the Nam Pathene location include Nongsum, Phon Tiou, and Bo Neng. Other tin deposits are found some 50 km west-north-west of Vientiane (United Nations 1990), and in the mountainous Ngum River valley about 70 km directly north of Vientiane. Workman (1972:26–27) also reported that the geological continuity of northwest Laos with the tin-bearing zone of northern Thailand and the important tin deposits in Yunnan suggested the good possibility of tin finds there.

Coote (1990) has pointed out that tin deposits vary widely in their mineral associations; for example, tin-iron and tin-copper are found only in the Malaysian East Coast tin belt, and tin-calcium and tin-copper-lead-antimony are more characteristic of the areas along the Thai border and fields in the Kinta Valley and Kuala Lumpur. She suggested that this variation could be used, with caution, as a way to begin to locate the source of the tin used in prehistoric bronze. This intriguing suggestion has yet to be followed up.

LEAD

Thailand contains numerous hydrothermal deposits bearing lead-zinc minerals. They are particularly abundant in formations intruded by Triassic granites and granodiorites associated with the Loei

Metallogenic Belt. In northeastern Thailand, lead-zinc deposits of economic interest can be found only in Loei Province, for example at Phu Khum 1 and 2 (10–15 km southeast of the town of Loei) (Workman 1972:46), in zones of sulfide mineralization at the contact between limestone and intrusive granodiorite (Tantisukrit 1978). Workman noted that pre-modern workings (which were not described) could be seen at Phu Khum 1, near to where granodiorite intrusions abut a Permian limestone hill (Workman 1972:46). At the two Phu Khum sites, there are what Workman (1972:46) described as "large overgrown pits of unknown age…near the contact between a small intrusion of porphyritic microgranodiorite and the enclosing Permian limestones." These sites were also surveyed by the Thailand Archaeometallurgy Project in 1984. A wide area pocked with numerous large open pits dug into the lead/zinc surface deposit was found in association with slag-scattered smelting sites. One opencast pit, for example, was ca. 300 m long and 70 m wide. Given that lead and zinc (for brass making) were utilized on a large scale in the Historic Period, it is presumed that the workings and smelting sites date to this time frame (Vincent Pigott, pers. comm. 2014). Phetchabun Province also contains reported deposits of lead minerals. In western Thailand, deposits are documented in Kanchanaburi and Phetchaburi Provinces (Tantisukrit 1978). Most lead ore in Thailand occurs as galena and most zinc as sphalerite (Rachdawong 1959:35).

In Laos, concentrations of galena (lead sulfide) are widespread within the Truong Son Metallogenic Belt. These lodes usually also contain sphalerite (zinc sulfide), stibnite (antimony sulfide), and arsenopyrite (arsenic sulfide), along with silver. Major deposits are in north central Laos (Xieng Khouang and the Plain of Jars), northern Laos at Pha Louang, 130 km north of Vientiane, and at Phou Van San, near the Thai border in northwest Laos, which is also associated with silver. Other deposits are located in Savannakhet Province at Tchepone, some 13 km (as mapped by Google Earth) from the Sepon copper-gold mines. (See TAM 2C, chapter 2 for discussion of the evidence for prehistoric copper mining at Sepon.) Galena deposits also can be found in southern Laos at Phou Tapak, which was exploited by local inhabitants in the 19th century (Workman 1972:45).

ANTIMONY

Currently, Thailand is an important producer of antimony. Most of the sources are in the west, north of the peninsula, in Chiang Mai, Lampang, Phrae, Chiang Rai, Tak, Sukhothai, and Mae Hong Son Provinces, along the same zone as the sources of tin and tungsten. Antimony is also mined in the central and eastern parts of the country, in Ratchaburi, Kanchanaburi, Chonburi, Rayong, and Chanthaburi Provinces, and in the south in Surat Thani, Nakhon Si Thammarat, Chumphon, and Krabi Provinces (Tantisukrit 1978). All the antimony ore in Thailand and Laos is stibnite or the secondary mineral stibiconite. The ore appears in both disseminations and veins. The ore in veins is usually associated with quartz (Workman 1972:50).

Some prehistoric artifacts in the Ban Chiang collection have registered several percent of antimony, so antimony-rich copper ores must have been smelted. Bronson (1992) also reported that according to Gardner (1967), high-grade antimony ores that are easily smeltable were exploited in the northern Thai kingdoms; similar ores are found in southern Thailand and in Myanmar. Crow and Khin Zaw (2011) report numerous deposits of antimony, particularly in the Chiang Mai Belt. No pure antimony artifacts have been reported from prehistoric excavations, but one Early Period amorphous artifact from Ban Tong contained over 9% antimony.

In Laos, stibnite is usually found in sulfide lodes along with galena and sphalerite. Numerous veins of purer stibnite are found in Luang Prabang Province. In Vietnam, sources of antimony and gold-antimony ores are found in the north, to the northeast of the Song Ma Suture Line and the Red River (Tran Trong Hoa et al. 2008).

IRON

The cataloged reserves of iron ore in the middle and lower Mekong Basin are not large by industrial standards but would have been sufficient for pre-industrial use. The deposits themselves are numerous, though low-grade by modern standards. Most are "contact metamorphic-replacement type deposits…hematite and magnetite are the principal minerals. Most deposits are amenable to open-cut mining methods" (United Nations 1968:41). Other deposits are lateritic, such as those in Amphoe Long,

Phrae Province. It is unclear how many of these deposits would have been smeltable in pre-industrial bloomery furnaces. The average grade of the ore from three prospects located in Loei Province and assayed by Workman (1972:7) varied from 48.5% to 56.3% iron oxide minerals, though it is possible that ancient smelters had already depleted the deposits with the richest iron contents (Pryce and Natapintu 2010).

Major deposits of rich iron ore lie in Loei Province, not far from the Mekong River. Most of the ores are contact-metasomatic magnetite deposits from Triassic granites intruding into limestone. Other deposits known or presumed to have been used for bloomery smelting in the historic period are found in the Lopburi region (Suchitta 1983, 1992), including in the Khao Wong Prachan Valley (Pryce and Natapintu 2010; see also discussion of evidence for iron production at Non Mak La in this valley in TAM 2C, chapter 2).

Workable deposits of magnetite and hematite are found at Amphoe Thoen, Lampang Province. In Phetchabun Province, deposits of magnetite and hematite are found together with lenses, veins, and pockets of copper minerals in porphyritic rocks. A large and relatively rich (68% iron) magnetite deposit associated with intrusive granites was found in eastern Thailand in Chonburi Province. This ore could have been enriched by judicious beneficiation. Lenticular seams of magnetite were found in Chachoengsao Province and magnetite and hematite bands in Rayong Province (Tantisukrit 1978).

Laterite is a limonite-hematite ore created in iron-rich tropical soils by the action of rainfall washing other soil components to a deeper level. Laterites are frequently reported in Southeast Asia. In some parts of the world, including Southeast Asia, laterites have been used as a source of iron, but the assumption by some (e.g., Bronson 1992) that, most parts of Southeast Asia would have had access to smeltable, if low-quality, iron ore appears to be unjustified. Cawte and Boyd (2010) noted that the frequent co-occurrence of slag and laterite at archaeological sites is no sure sign of smelting; most of the laterite appears to have been deposited naturally. Pryce and Natapintu (2010) have argued that a successful bloomery smelt needs an ore containing at least 75% iron oxide. An ore could contain that naturally or through beneficiation (Cawte and Boyd 2010). Pryce and Natapintu (2010) assessed the iron content of several samples of laterite from Thailand and suggested that very little of it would have been smeltable. Cawte and Boyd (2010) assayed three laterite samples after benefication; two were still unsmeltable, but one, a laterite brick, might well have been usable. Their results demonstrated that the iron content of laterite on the Khorat Plateau varies considerably, and that the automatic assumption that production of iron metal could be conducted at a localized village level may be unwarranted; it is more likely that the numerous slag finds at many sites were the result of smithing iron blooms produced elsewhere.

Nitta (1991, 1997) excavated 17 iron-smelting furnaces at Ban Don Phlong in the Mun Valley. He speculated, based on finds at the site and the distance to any known source of hematite or other more standard iron ore, that low-grade iron sands and laterites, available on the Khorat Plateau, were used for smelting. From radiocarbon determinations and relative dating of ceramics, the iron-smelting activities took place between the 3rd and the 1st centuries B.C. In Lopburi Province, Suchitta (1983) excavated the 6th century A.D. iron-smelting site of Ban Di Lung. The considerable remains of iron smelting include chunks of hematite ore, probably derived from nearby Khao Tab Kwai Mountain, still mined for iron ore today (and possibly mined for copper ore in prehistory, see TAM 2C, chapter 2 and TAM 2C, fig. 2.14). A number of laterite pebbles were found at Ban Di Lung; analysis of the numerous tuyere fragments demonstrated that bits of laterite were used as temper in the tuyeres.

The minimal evidence available does not answer the question of what were the iron sources for the prehistoric smelting furnaces on the Khorat Plateau such as those at Ban Don Phlong. Until other sources of iron ore with evidence for prehistoric exploitation on the plateau are located, the presence of iron artifacts in most sites on the Khorat Plateau would seem to imply the existence of trade and exchange with the still unidentified primary iron producers (Pryce and Natapintu 2010).

According to Workman (1972:5), the only areas of Laos to contain industrially valuable deposits of iron ore are in Xieng Khouang Province, northeast of Vientiane, though wide-ranging investigations have been limited. These ores were of high grade by industrial standards—up to 70% iron—and consisted mostly of gray magnetite-hematite with some bands of red hematite. In some places, the ore was weathered to limonite (Workman 1972:5). The 1990 ESCAP survey confirmed the presence of iron ore in this area, though the extent of the reserves was still not known. In addition, the ESCAP survey reported occurrences of iron ore near the source of the Like River and in the tin field area in Khammouane Province along the Nam Pathene Valley (United Nations 1990).

ARSENIC

No information is available about commercial occurrences of arsenic, arsenic ore, or arsenic-rich copper ore, though, as Bronson (1992) points out, arsenic-rich ores must exist in large enough quantity to account for the occasional presence of arsenic in compositional analyses of archaeological copper-base artifacts. (See TAM 2B, appendix A, table A.3; three of the 42 artifacts from Ban Chiang, Ban Tong, and Don Klang that were analyzed by PIXE spectroscopy contained over 1.5% arsenic.) Bronson (1992) reported, based on an account in Högbom (1913–14), that an iron ore with 47.5 percent arsenic oxide was being smelted for bullets in northern Thailand in 1910.

Conclusions

The Ban Chiang culture area sites are in the Sakon Nakhon Basin of the northern Khorat Plateau, a basin, because of its geological history, almost devoid of metallic mineral resources. However, only a few hundred kilometers away, including upland areas immediately bordering the plateau, lie numerous sources of copper, tin, and lead ores. The northern Khorat Plateau is tucked up under the intersection of the most metallogenically rich fold belts of mainland Southeast Asia, the Loei and Truong Son.

For copper, the closest source would be the various deposits in the Loei Fold Belt that runs along the western edge of the Khorat Plateau. The known prehistoric mining sites of Phu Lon located roughly 150 km distant and, further away, the sites of the Khao Wong Prachan Valley, both discussed in TAM 2C (chapter 2), are located in this belt. Another source, demonstrated by lead isotope analysis conducted by Pryce (TAM 2C, chapter 3) to match the profiles of some Ban Chiang artifacts, is Sepon, Laos, in the Truong Son Belt, some 200–300 km east of Ban Chiang.

However, lead isotope analysis also showed that most of the small number of Ban Chiang and Don Klang artifacts so far tested did not match the lead isotope signatures of the known prehistoric mining sites, suggesting that much more archaeological survey of copper sources is needed. Particularly important for archaeological investigation would be the Phu Kham copper-gold deposit, called by Khin Zaw et al. (2014b:9) "one of the most important...porphyry-related oxidized skarn deposits in the Truong Son Fold Belt." This deposit, valued for both copper and gold, has recently been exploited on a large scale (see http://www.panaust.com.au/phu-kham-copper-gold-operation; accessed May 4, 2016). Phu Kham, located about 200 km north of Ban Chiang in the northern part of Vientiane Province, as well as several copper-gold skarns a little further north near Phonsavan in the archaeologically important Xieng Khouang Province, should be investigated as soon as possible before modern exploitation destroys any remnants of prehistoric mining activities, if it has not already done so.

Few published works have dealt with the potential sources of prehistoric tin in Southeast Asia, though see Coote (1990) for suggestions about tin ore sourcing in that area. Given the low level of technology needed to acquire and smelt placer tin, archaeological traces would be much harder to find than those resulting from copper mining and smelting. The closest known source of tin for Ban Chiang-related sites would be the large deposit in the Nam Pathene Valley in Laos, which is still today an industrial source, and which Barthelemy (1938) recorded as a tin source for village-level tin smelting even in the 1930s. This valley should be of highest priority for archaeological exploration as a possible source for tin from Ban Chiang cultural tradition sites.

Further afield, Khin Zaw et al. (2014b) provided numerous leads for other areas that could have provided metal raw material to prehistoric societies of Thailand as well as for the larger region. For copper, numerous localities along the Loei Metallogenic Belt were mentioned. Much farther away, the

Monywa copper gold deposit in central Myanmar near Mandalay was described as "one of the largest copper resources…in mainland SE Asia" (Khin Zaw 2014b:19). The copper resources in northern Vietnam north of the Song Ma suture (Tran Trong Hoa et al. 2008:488) should also be explored for prehistoric activity. The extraordinarily rich tin deposits of the Burma-Malay Metallogenic Belt, running from the Malay Peninsula through western Thailand into Myanmar and up to Yunnan, are considerably further away, but still accessible by trade. Though some of these resources were very distant from the study sites of this volume, the technological similarities of the metal artifacts produced throughout prehistoric mainland Southeast Asia indicates the network of shared metallurgical knowledge spanned these areas, indicating connectivity of some sort, whether by trade in finished products, raw materials, and/or by movement of craftspersons. Only systematic study will provide the details.

This brief review demonstrates the importance of understanding the geological history of mainland Southeast Asia—the complex sequence of orogeny, folding, and metallic mineralizations—as a guide to the location of possible prehistoric mining and smelting areas, and, perhaps of equal importance, where they will definitely not be found. The geological history also confirms that sources for the metals used in prehistory were not far away and gives both prehistoric and modern prospectors guidelines for where to search. Unlike much of the world, making bronze in Southeast Asia did not require vast long-distance trade networks to obtain tin, lead, or copper; obtaining these materials was within the reach of small communities without elite organization. This propinquity, it is hypothesized, had a profound effect on the social organization and the creation and maintenance of technological systems in the northern Khorat Plateau and possibly across mainland Southeast Asia as a whole.

7

Chaînes Opératoires and the Study of Archaeological Metals

Elizabeth G. Hamilton

This chapter, drawn from the discipline of archaeometallurgy, seeks to provide a basic background on how metals excavated from prehistoric sites in Thailand were exploited, though most of the descriptions of techniques apply worldwide, not just to Southeast Asia. Archaeometallurgists have studied metallurgical processes for many decades, using a variety of techniques to learn how primary metal production (i.e., the extraction of metal from ores) and secondary production, (i.e., the manufacturing of metal products) took place in pre-industrial contexts. They perform laboratory studies of artifacts, excavate mining and processing sites, analyze slag, ceramic and metal remains, and conduct experimental replication studies, all to understand not only the techniques and the wider technological system, but also the social, economic, and ideological contexts—the socio-technical systems (Pfaffenberger 1992)—in which the production and use of metals were embedded (Craddock 1995; Pigott 1999a; Roberts and Thornton 2014; Scott 1991; Smith 1981; Tylecote 1974, 1992; Young et al. 1999).

To understand how particular technologies, be those technologies agricultural, metallurgical, or other, are used in any culture, archaeologists must reconstruct what they can of subsistence and settlement patterns, trade, and social and political structures. They must also acquire a detailed knowledge of the properties of the objects employed in the technologies used in the ancient culture, such as the properties of copper and the ceramics used in high-temperature metals processing. In addition, they must understand the basic technical aspects of what is involved in acquiring and manipulating the relevant raw materials in the specific context being studied. For example, using an example from agricultural technologies, growing rice differs from growing manioc, and growing rice in paddies in alluvial basins in Thailand requires different techniques and strategies than growing rice in swiddens in the hills of Laos, or growing rice in the flatlands of California. The archaeologist needs basic knowledge of rice biology and techniques of rice cultivation, as well as a detailed understanding of the environmental context, particularly since many important properties of prehistoric rice and rice agriculture cannot be studied directly through excavation or excavated materials but can be inferred from other evidence.

The same can be said for metals. The technical requirements of exploiting small, scattered surface deposits of copper carbonate and oxide ores differ from those of exploiting deep veins of copper sulfide ore, for example. To reconstruct a mining technological system requires, as the first step, the knowledge of exactly what resources were available to the prehistoric miners. The archaeologist also needs familiarity with what is physically involved in smelting, casting, and working metals. Only a detailed knowledge of the specific *chaîne opératoire*—"the selection of the correct raw materials, the sequences and timings of actions, the creation and identification of the right conditions, and the addition of substances" (Roberts et al. 2009:1018)—used in any particular technological system will allow interpretation of the remains of that system left behind in the archaeological record. In addition, only a detailed knowledge of the technology allows one to determine what are the irreducible technical *requirements*—e.g., the melting point of pure copper—and what are technical *choices*—e.g., the decision to use crucibles instead of furnaces with superstructures to smelt copper—that vary with the cultural context.

Archaeological Evidence

The two distinct production stages in metallurgy—raw material extraction and object manufacturing—can leave considerable evidence behind for the archaeologist to recover (Table 7.1). Metal extraction activities include prospecting for ore sources, mining, ore preparation, fuel production, and smelting. Byproducts of these activities in the archaeological record include waste rock from mining, ore fragments, charcoal, furnaces, ceramic crucibles, tuyeres (thick clay tubes that protected bellows ends in the fire), molds, and slag. Artifact fabrication is the process of shaping the raw metal into artifacts. The products of object-manufacturing activities include not only the finished artifacts, but also casting debris or metal splashes, charcoal, crucibles, metalworking hearths, slag, tuyeres, molds, sprues, anvils, hammers, and other technological deposits. Copper-base metal and iron processing may leave different traces in the archaeological record.

Copper and Bronze

Copper appears to have been the first metal worked by humans in the Old and New Worlds (Craddock 1995:97; more recently Roberts et al. 2009), though lead may have been the first smelted (Craddock 1995:205; Roberts et al. 2009). In nature, copper appears in two forms: native and combined with other elements in the form of ores. Native metals appear in a mostly pure state in veins that may be chiseled out of the rock or as nodules that may be picked up. Except for certain areas in North America, native copper is relatively rare. Most copper deposits are in the form of ore, which is copper bonded to other minerals, and must be smelted—thermochemically reduced—to liberate the copper.

To alloy a metal means deliberately to mix it with varying quantities of other metals under heated conditions in order to enhance desired performance characteristics, such as hardenability, ease of casting, or a particular color. Copper can be alloyed with numerous metals, such as tin, zinc, lead, antimony, arsenic, silver, and gold. The various copper alloys differ somewhat in their material properties. Pure copper is quite soft, ductile, easy to work, but difficult to cast. There are two aspects of molten copper that challenge the caster. Unalloyed molten copper tends to absorb gases, which are emitted during solidification and result in blowholes in the metal. In addition, molten copper tends to be relatively viscous and does not always fill in small details in a mold wall.

Even unalloyed copper has its functional uses; the combination of ductility and toughness of copper means that it functions better as an awl than wood, bone, or stone. Notably, awls or punches are among the earliest copper artifacts found in many early metal assemblages including those from Southwest Asia at Çayönü Tepesi in Turkey (Maddin et al. 1999), pre-Harappan South Asia (Agrawal 2000), Hungary (Gömöri 1988:80), and Portugal (Müller et al. 2007:16). Unalloyed copper chisels and hammer-axes from 4th millennium B.C. in southeast Europe show evidence of extensive use (Glumac 1991:352), and experimental reconstruction and testing of copper and bronze axes have shown that axes made of pure copper are fairly efficient at the task of felling trees (Kienlin and Ottaway 1998) and more efficient than stone (Glumac 1991:353). Heeb and Ottaway (2014:183) record that experimentally a copper axe was serviceable for up to 30 minutes in cutting wood and stripping bark before requiring straightening and resharpening. The copper axe carried by the 4th millennium Iceman found in the Alps was work hardened and clearly showed signs of wear (Taylor 1999). "Certainly as a weapon for killing animals, human or otherwise, the hammered copper ax would have been perfectly practicable; for many agricultural uses it would have been tougher and more suitable than stone" (Charles 1980:168).

In addition, the infinite recyclability of copper tools—they could be straightened or resharpened quickly by hammering the cutting edge or reshaped into an entirely new form by casting—may in many contexts have been considered a great advantage over polished stone tools. It has been suggested that, in

Table 7.1 Archaeological Evidence for Copper-base Primary Metal Production and Finished Artifact Manufacture. Adapted from Northover (1989b) and Miller (2007: fig. 4.12).

Process/activity	Archaeological Evidence	Notes
Prospecting	None.	
Mining	Bits or piles of ore fragments; pits, shafts, and/or adits cut into rock; lenses or piles of finely crushed rock.	
Techniques		
Battering with stone mauls	Elongated stones, often with pecked ends and grooves for hafting. Peck marks on mine walls.	
Fire-setting	Smooth curved mine walls; layers of ash, charcoal, and burned rock.	
Metal picks	Long thin scratches, often parallel, on mine walls.	
Antler picks	Broken or worn antler picks in mines.	
Ore preparation		
Techniques		
Beneficiation	Bits of discarded ore; thick deposits of uniformly sized, finely crushed host rock; small hand-held hammerstones with pecking; shallow depressions in flat rocks used as anvils.	Crushed rock should be examined to determine efficiency of separation.
Ore roasting	Possible remains of enclosures, pits, or bits of porous ore.	
Smelting	Large amounts of slag, bits of ore, flux, charcoal.	
Techniques		
Crucible	Fragments of slagged refractory crucibles; bits of ore; crucibles sometimes vitrified; slag often broken up to retrieve prills of metal. Co-smelting of carbonate and sulfidic ores produces more slag than smelting of carbonate ores. Charcoal in soil.	Little slag produced if ore was rich in carbonate or oxide.
Furnace (slagging)	Large quantities of dense, dark-colored slag; bowl-shaped or hole-like burned depressions; remains of clay tuyeres; pieces of thick ceramic furnace pieces, sometimes vitrified and slagged. Charcoal in soil.	Metal produced from slagging furnaces has a much higher iron content than crucible metal.
Melting/Refining	Crucibles with slag, dross. Light, vesicular slag.	
Alloying	Bits of metal or ores of the alloying element; prills in melting crucibles.	
Casting	Mold fragments; metal drippings; casting flashes on finished products.	

fact, reusability was metal's main utilitarian advantage over stone and bone (Castro et al. 1998, as cited by Chapman [2003]).

Despite some evidence for the use of pure copper for tools, the widespread use of metal for tools and weapons did not occur until the discovery of arsenic-rich copper ores and the generally later innovation of deliberate alloying, in particular the alloying of copper with tin to form bronze. Bronze has a number of useful properties: it is harder than pure copper (a well-hammered 10–12% tin bronze is harder than pure iron [Scott 1991:82]); it is easier to cast than copper, both because it is more fluid when molten and so can more easily flow into narrow spaces in the mold and because the addition of the alloying element, arsenic, tin, or lead, reduces the amount of gases the molten material absorbs (Knauth 1974:56). Its melting temperature is lower than that of copper, requiring less fuel to melt. The addition of tin or arsenic can change the color of the metal, making it paler, brighter, and shinier. A bronze with approximately 10% tin is considered optimal for many uses. In one experiment, a copper axe could chop wood for 30 minutes, but a 10% tin bronze lasted 320 minutes before needing to be resharpened (Heeb and Ottaway 2014: 198). Bronze with over 10–13% tin results in a metal that is more brittle, so that although the metal becomes harder if hammered, it is also more likely to crack. Bronze can be cast more easily than pure copper, but the casting qualities are much improved by adding at least 3% lead; not only is the fluidity of the molten metal increased, but the melting point is lowered even further. However, the addition of lead also softens the metal, so depending on the desired performance characteristics in the final product, a balance must be struck between improved casting quality and hardness. Lead does not dissolve in copper, and when added, it remains scattered as globules of free lead. Craddock (1979) maintained, though, that any percentage of lead over 2% offers no advantage in casting. This is borne out, according to Craddock (1979), by the consistently erratic levels of lead found in ancient artifacts, a sign that the founders were unconcerned about achieving an exact quantity of lead in the metal.

It is important to remember, however, that the qualities of the alloy desired by the makers (i.e., its desired performance characteristics) did not necessarily include those important to purely utilitarian purposes. Many qualities, such as sound, color, or mode of manufacture, could be valued above what persons in an industrial culture might consider necessary. Northover (1989b:225) noted that many early bronze period cultures, including those of the Levant and Italy, preferred an alloy of copper and arsenic for prestige items, but used softer unalloyed copper for tools. The arsenical copper would have had a silvery surface that tarnished to gold and, clearly, this was valued.

A copper alloy can result from the deliberate addition of other metals, or from impurities naturally in the copper ore. Some copper ores (e.g., stannite, Cu_2FeSnS_4) contain tin naturally, and when smelted, the resulting copper can have considerable tin (see Radivojević et al. [2013] for analyses of Chalcolithic Balkan artifacts, probably from stannite ore). For archaeological purposes, it is necessary to define what a deliberate alloy is, and what place, if any, in a *chaîne opératoire* the alloying step has. Because the addition of at least 2% tin has noticeable effects on the color and working properties of the metal, in this volume, following Stech (1999), a deliberate tin bronze is defined as having over 2% tin. (As noted above, in rare cases a natural alloy can be derived from smelting stannite, but no deposits of stannite are noted in Southeast Asia, though see Pigott [2012b] for stannite sources in the eastern Urals.) This is admittedly somewhat arbitrary; percentages of alloying elements can be altered by scrap recycling and accidental choice of ores (Miller 2007:157). Because of the rarity of copper ores with appreciable tin contents, though, the 2% dividing line works reasonably well for tin bronze. The same applies to leaded copper or bronze; a metal with at least 2% lead is considered to be a deliberate alloy.

The *Chaînes Opératoires* of Making Copper Products

Determining the *chaîne opératoire* of any technological system must begin with understanding the sources of raw materials used in the system and how to acquire those resources in a given environment.

COPPER EXTRACTION

To produce raw copper metal from ore requires a complex series of steps and activities. These steps (adapted from Ottaway 2001) are simplified below and do not refer exclusively to practices in prehistoric Thailand.

Prospecting

Good ores can often be located by the color of visible minerals. Malachite and azurite, bright green and bright blue respectively, are apt to be visible at the surface of weathered copper sulfide orebodies, as is the red of ocher. These minerals were collected for ornaments and pigments long before the inception of metallurgy (Miller 2007:148), so this prospecting and collecting was not an innovation. The presence of ore can also be revealed by indicator plants (Craddock 1995:31).

Mining

Collection of ores can vary in difficulty depending on qualities of the ore deposit. Most copper ores start as compounds of copper, iron, and sulfur. Typically, when the ore is exposed to air and rain, most of the sulfur, the iron, and some of the copper leach down into the deeper orebody (see TAM 2C, fig. 2.5). The top layer of the orebody weathers into a zone that can contain native copper and some easily smelted carbonate, oxide, and silicate ores, such as malachite, azurite, cuprite, and chrysocolla. This weathered layer is called the *gossan*, or "iron hat," from the distinctive, colorful, mostly red and purple, iron oxides also present in the weathered layer. Much of the copper originally in the top layer, however, has leached down into the "secondary enrichment zone," a layer under the gossan with high copper and sulfide concentrations. This layer also frequently contains arsenic, antimony, iron, and other elements. Beneath this enriched layer is the primary copper sulfide ore, with considerably less copper than the higher secondary ore. (It should be noted that the above description refers to a geological schematic; specific individual ore deposits may not have this idealized and predictable layering.)

Activities involved in mining may vary in arduousness. Ore could be collected from outcrops, dug from open pits, or excavated from adits (horizontal tunnels) or shafts (vertical tunnels) that follow a vein of ore through a host rock. For reasons discussed in Chapter 6, deposits worked in prehistory would largely have been vein deposits, created by hydrothermal, contact metamorphic, or, rarely, magmatic processes, not the more dispersed and poorer disseminated deposits worked today. Stone hammers (mauls), picks of antler or bone, or, later, metal picks were used to pry ore from the host rock. Each of these kinds of tool leaves a distinctive mark on the wall of the tunnel (Craddock 1995:10). Sometimes fire-setting, building a fire against an ore seam to heat the water in the rock and make the rock crack, was used to break chunks of ore away from the parent rock. Fire-setting leaves a very smooth surface to the tunnel (Craddock 1995:36), unlike the jagged surface expected from the use of picks and hammers. The assumption that water needed to be thrown on the heated surface of the rock to break off slabs of rock was, after experimental replication, shown to be false; the superheating of water already in the rock was sufficient to weaken the rock face (Timberlake 2007). Archaeological remains of mining can include adits and shafts (if not obscured by later workings), picks, hammers, chunks of unprocessed ore, and waste rock from the process of beneficiation.

Timberlake (2007) also reported the results of experiments with stone mining tools. Unmodified (ungrooved) river cobbles very similar to those found at mining sites in Thailand were used, first held by hand, then swung by ropes against the fire-weakened rock faces. The most successful use of the mauls came when they were hafted with sticks and twisted withies. The characteristic spalling on the ends seen in both Timberlake's experiments and the Thai examples was produced when the mauls struck the rock face that had not been fire weakened (Timberlake 2007).

Beneficiation

The activity of separating the ore from the lumps of host rock recovered by the mining is called beneficiation. The rocks are crushed to the size of peas with stone hammers often on the surface of stone anvils, and the richer ore is either picked out by hand or separated by gravity and water or wind. The process leaves behind large amounts of tailings (low-grade ore and waste rock) and fines (crushed or powdered waste), worn hammerstones, and often flat anvil stones with depressions worn in them (Craddock 1995:159; also see TAM 2C, figs. 2.10–2.12 for prehistoric Thai examples of beneficiation remains and an anvil). Skillful beneficiation will considerably decrease the amount of fuel needed to smelt the metal

(Merkel 1985). Beneficiation areas are sometimes found near mines, suggesting that the process was carried out by the miners, but evidence of beneficiation is sometimes found in settlements as well.

Roasting

Some ores require additional activities to prepare them to be smelted successfully. Carbonate and sulfidic copper ores can be improved by being roasted below 800°C on a wood fire in the open air prior to smelting. Although it is possible to successfully smelt mixtures of oxide and sulfide ores without roasting them first (Rostoker et al. 1989), to smelt pure sulfidic ore requires roasting (Budd et al. 1992). During roasting, water is driven off, carbonate ores are converted to copper oxides, the iron sulfides in sulfidic ores are converted to iron oxides, and the sulfur is driven off as a gas. The ore pieces are rendered more friable, with more cracks to allow carbon monoxide to enter during the smelt (Craddock 1995:167). This would also allow for improved benefication, as the ore chunks would have been rendered easier to crush. Archaeological evidence of roasting, however, has been difficult to find, especially at prehistoric sites, though it might be inferred from the evidence of shallow postholes for low-walled enclosures too large to be furnaces or from pits dug into hillsides. Ethnographic and historical data show that both have been used for roasting iron ore (Craddock 1995:168; Rostoker and Bronson 1990:52).

Charcoal Preparation

Charcoal, being chemically pure, was often preferred as a fuel for smelting and smithing in antiquity, as it has a high heat capacity (calorific value) and will not add unwanted elements to the metal. Charcoal is almost pure carbon and is the result of the incomplete combustion of wood. Other fuels such as wood, dung, and grasses had some metallurgical applications, but mostly for roasting and preheating the furnaces (Rostoker and Bronson 1990:61–62). In an open fire, charcoal can burn at 900°C; with an air blast it can achieve much higher temperatures (Henderson 2000:229). It has disadvantages that will affect the *chaîne opératoire*; making charcoal requires a large amount of timber, it is fragile, and it does not transport well, because the motion causes the particles to abrade into a fine powder that is useless for furnace work. As a consequence, charcoal preparation should preferentially take place not too distant from where it is going to be used; even up to the 18th century in England the preferred distance for carrying charcoal was no more than 5 miles (Rehder 2000:61).

Charcoal can be produced in a number of ways. For example, seasoned wood sticks, preferably hardwood, can be laid in a spiral or cartwheel pattern, continuing until a stack is formed; an open shaft is left in the middle of the stack. Earth, sod, or other covering is spread over the stack and tamped down so that no air can enter. The stack is lit via air inlets at the bottom or side, which are then closed off, as is the central shaft. The rate of combustion can be controlled by opening or closing the vents. Charcoal can also be made in a pit. The combustion process can take one to two weeks; several days are also needed for the heap or pit to cool (Rehder 2000:56). (See Rehder [2000] for details of the chemical reactions involved.) In modern rural Thailand, charcoal can be made by covering sticks with a mound of dirt or dirt mixed with straw and rice husks (Fig. 7.1), or, more casually, by covering a mound of sticks with a thick layer of rice husks (Fig. 7.2) (Joyce White, pers. comm. 2016).

The effort and time required to produce good quality charcoal are underappreciated aspects in reconstruction of prehistoric smelting. The quantities required could be considerable; in prehistoric contexts, Merkel (1990) found that it took 20–50 kg of charcoal to produce 1 kg of copper, based on experimental reconstruction of copper furnaces at Timna in southern Israel (Henderson 2000:230).

Smelting

Many of the details of the earliest copper smelting processes are still unclear (Killick 2014:39). The archaeological remains can be either absent or difficult to interpret, and both experimental replication and slag analysis are needed to understand the physical processes and the nature of the surviving evidence (Killick 2014; Timberlake 2007).

To extract copper from an ore requires a high temperature in a reducing atmosphere; i.e., an atmosphere rich in carbon monoxide that will bond with the oxygen in the ore and leave the metal behind. A refractory container is required to hold the ore and charcoal during the smelt. Three separate though related issues need to be dealt with by any smelter,

Figure 7.1 Thai women sawing up logs to be made into charcoal. The charcoal domes, covered with mud and straw, are in the background. (Photograph courtesy of Joyce C. White.)

Figure 7.2 Remains of a charcoal-making mound made by covering sticks with rice husks. (Photograph courtesy of Joyce C. White.)

or in any discussion of prehistoric copper smelting: the structure of the refractory container, the nature of the ore, and whether the process was slagging or non-slagging.

Slagging versus Non-slagging: The earliest smelting process seems to have been non-slagging, or a simple process in a crucible or clay-lined hole with draft provided by human lungs through blowpipes, for example. From the evidence provided by slag analysis, most of these smelts were not hot enough or reducing enough to produce a free-flowing slag, i.e., where the unwanted gangue separated in a liquid form from the reduced metal. This is referred to as a non-slagging process; the end result was usually a mass of partially reduced ore with prills of copper inside. The mass of partially reduced ore would need to be broken apart to pick out the metal. The quantity of metal obtained was only a few hundred grams (Craddock 1995:135). A slagging process, where the gangue became liquid in the form of slag, allowed a higher recovery of metal from lower-grade ore and greater heat efficiency. The reduced metal could coalesce more easily to form a larger mass, and the quantity of metal extracted would have been considerably greater (Craddock 1995:146).

Smelting Structures: Smelting of copper ore can be conducted, with varying levels of efficiency, in several different types of structures that provide the needed heat, air, and reducing atmosphere. Tylecote (1992:9) described three stages (his term) of copper smelting:

(1) crucible smelting, with heat provided from above and outside of the vessel. In much of the world—for example in Peru, Egypt, and bronze age Europe (Craddock 1995:178)—air could be supplied by human lung power through blowpipes.

(2) smelting in a non-slag-tapping furnace, such as a bowl furnace (see Fig. 7.3 for an example of an iron-smelting bowl furnace), or a short-shaft furnace.

(3) smelting in a slag-tapping furnace where the slag could be drained away during the smelt, usually in a shaft furnace (see Fig. 7.4 for an example of an iron-smelting shaft furnace).

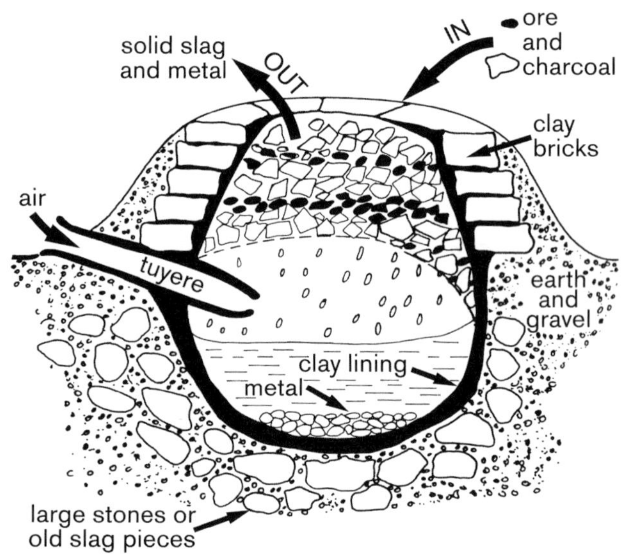

Figure 7.3 Generalized sketch of a bowl furnace, usable for both copper and iron smelting. They are easy to build, but few have any provision for tapping the slag, meaning that the smelt can only be conducted for a few hours before the rising level of slag chokes the tuyere hole (adapted from Rostoker and Bronson 1990:29).

Though crucible smelting has been assumed to be the earliest stage of copper smelting (Craddock 1995; Tylecote 1992), recent work in four Vinča sites (ca. 5000–4400 B.C.) in Serbia and Bosnia have revealed the earliest known remains of copper smelting in the world (Radivojević and Rehren 2016). The evidence—small bits of slag, black and green copper minerals, and slagged sherds—suggested that the process took place in a hole lined with broken sherds, with no crucible. Slag analysis indicated that the atmosphere during the smelt was moderately reducing and temperatures reached slightly above 1083°C; the use of richer ores resulted in only a small amount of slag being produced. Radivojević and Rehren noted that although there is no evidence for crucibles in Vinča sites, they must have existed in order to melt and cast the numerous copper artifacts produced at this time.

Although there are many variations, in crucible smelting, a charge of ore and charcoal is placed in a crucible, and the crucible placed in a pit or area surrounded by stone or short clay walls. More charcoal may be piled on top, and air is supplied perhaps through human-powered blowpipes thrust into or above the crucible; thus, the crucible charge is

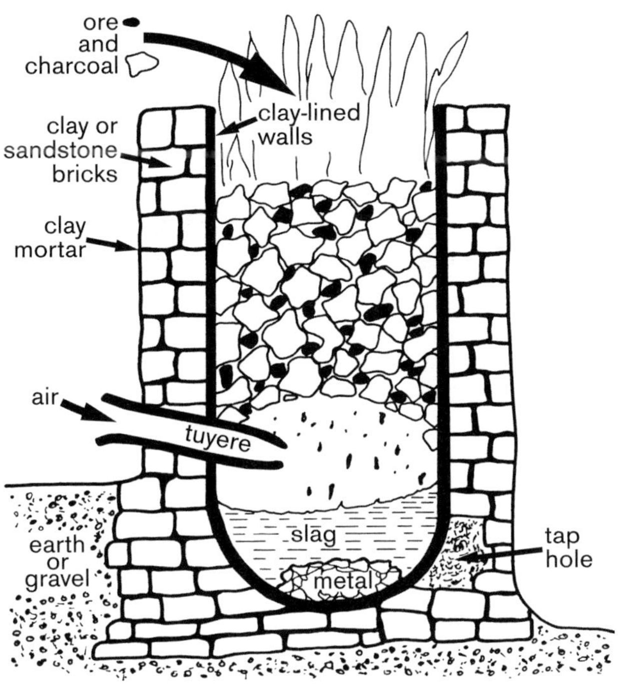

Figure 7.4 Generalized shaft furnace, usable for both copper and iron smelting. The height of the shaft allows the charge to be preheated before it reaches the combustion zone near the tuyere hole. This furnace shows a slag-tapping hole. The ability to drain the slag during the smelt allows for a much longer smelt time (adapted from Rostoker and Bronson 1990:29).

heated internally. There is some speculation that in areas with strong prevailing winds, such as central Thailand during the dry season, crucible smelting in antiquity may have been driven by wind power. The crucibles tend to be spouted vessels, and usually small enough to be easily portable. A wide variety of crucible shapes are possible (see Tylecote [1992:24] for an illustration). Because crucible smelting takes place in small hearth-like features with no superstructure, in this volume these features are called "hearths." An installation with a superstructure intended for a slagging process is called a furnace. This use of these terms follows Craddock (1995, 2000).

Bowl furnaces are more elaborate than a crucible smelting installation; they are dug more deeply into the ground and have some superstructure of clay or bricks (Fig. 7.3). The bowl would accommodate a larger ore-and-charcoal charge than a crucible, and more charge could be added during the process, unlike during crucible smelting. The larger volume and better insulation would retain heat, allowing for slagging. Eventually, though, the buildup of gangue and slag would choke the tuyere and force an end to the process, because unless the bowl is cut into a slope there is no way to drain the slag.

A shaft furnace has a taller superstructure, with a larger capacity for a charge, room for several tuyeres, and the ability to remove the slag by tapping, allowing the smelt to continue for as long as the metalworker wishes and the furnace holds out. Furnaces extract a greater quantity of metallic copper from the ore than crucible smelting does: a much larger charge of ore and charcoal can be smelted, more charge can be added during the smelt, the larger size makes for much greater heat retention and allows for more charcoal to be used and, thus, a more strongly reducing atmosphere to be created. Bellows can be used for air supply, and gangue and impurities can be turned to molten slag and drained away, enabling the use of lower-grade ores. The cost, of course, lies in the greater quantities of ore that must be mined to make a processing episode worthwhile, the larger the amount of charcoal that must be made per processing episode, the greater effort that must go into furnace construction, the greater time spent on the smelt, and the greater the degree of specialized knowledge that is required.

Tylecote (1992:9) recognized that the divisions between these types of smelting structure are not firm. If a tuyere or blowpipe is inserted deep into a large crucible filled with mixed ore and charcoal, the crucible becomes a small, internally heated bowl furnace. A bowl furnace that consists of a pit dug in the ground can become slag-tapping if it is dug into the slope of a hill and a channel dug to the bottom of the pit to allow slag to escape. It should also be noted that although Tylecote used the term "stages," with the implication that technical development proceeded along a fixed track from simple crucible smelting to full-fledged shaft furnaces, there is increasing evidence that more than one of these processes could be found at the same site at the same time (Thornton et al. 2010).

Oxide and Sulfide Ores: Copper oxide ores such as malachite and azurite can be easily reduced to copper in a single step, but it is likely that the majority of the world's copper ores used by prehistoric miners would have been oxides mixed with copper sulfides. Weathered copper sulfide bodies would have been capped by oxide-rich gossans. As seen in TAM 2C, fig 2.5, copper ore deposits often consist of a copper sulfide body with a zone of weathered oxide/carbonate enrichment at the top. These different ore types may require different preparations before smelting. The primary copper sulfide ore is usually chalcopyrite ($CuFeS_2$). Although it was possible to obtain nearly pure oxide or carbonate ore by careful beneficiation, examination of mines in the Near East showed that in reality a wide variety of minerals, including sulfides, were present in the near-surface zones. The evidence suggests that in prehistory whatever kind of ore was encountered, including sulfide ores, was smelted from early times (Craddock 1995:135), without the miners necessarily being aware of the distinction between ore types. How the early smelting of sulfide ores was accomplished is debated (Bourgarit 2007; Killick 2014; Timberlake 2007).

Many very early smelting sites in the world contain very little slag, and the assumption has been that only the richest oxide and carbonate ores were smelted. Such ores, smelted even under the simplest conditions, would leave little or no slag (Killick 2014). However, the majority of the world's copper ores are sulfidic—mixtures of copper, sulfide, iron, and other minerals, and even when miners sought oxide-carbonate ores, there probably would have been some mixing of oxide and sulfide ores. Bourgarit (2007) has examined slags from several Chalcolithic sites, such as Al Claus and La Capitelle in France, Dolnoslav in Greece, and Murgal in Anatolia, and he determined that several samples were the result of the multistep crucible smelting of sulfidic ores. Furthermore, he suggested that very few early copper processes were truly non-slagging. He argued that the lack of ore at early sites is the result of the crushing of gangue and partially reduced ore/slag to such fineness that Bourgarit (2007) called it slag sand.

In historic times, smelting roasted sulfidic copper ore was done in at least two steps in a process called matte smelting. The roasted and then beneficiated ore was smelted in a furnace in a reducing atmosphere. This step created two products: (1) a mixture of iron oxides, which were removed by slagging with a quartz flux; and (2) copper sulfide with some iron sulfide, called matte. The matte was then cooled, pounded into small pieces, and roasted to convert the remaining iron sulfide to iron oxide. This roasted ore was then resmelted with quartz flux to produce blister copper (so-called because the remaining sulfide gases produced blowholes), which was refined in a further step [Craddock 1995:149–150]).

This complicated multistep procedure is considered to require a considerable technical advance on the part of metalworkers. It has been shown, however, that "co-smelting," or smelting oxide ore together with sulfide ore (or primary sulfide ore together with well-roasted sulfide ore) in either a crucible or a bowl furnace can be performed in one step, without the necessity of a strongly reducing atmosphere, as the smelt is driven by sulfur gases and not carbon-rich gases (Rostoker et al. 1989). The experiments were conducted by placing the ores in crucibles of refractory clay, covering them in charcoal, and heating them to approximately 1250°C or higher. Adding crushed quartz or a layer of crushed slag on top of the charge to promote slagging greatly increased the efficiency (i.e., yield) of the process (Rostoker and Sadowski 1980; Rostoker et al. 1989). The process also produced large quantities of slag, as the smelting of pure oxide ores did not. Rostoker et al. suggested that because early ores were often mixed oxides/sulfides, and because the crucible smelting of these mixed ores was relatively easy, the process of learning to deal with sulfide ores in the Old World was a gradual one rather than the result of a dramatic discovery of matte smelting. Matte smelting, though more complicated and tedious, likely was adopted both because it greatly increased the amount of copper that could be extracted from the ore (Rostoker et al. 1989), and because easily accessible oxide ore sources may have become exhausted. Crucible co-smelting of copper carbonate and sulfidic ores has been suggested for several Old World sites (Craddock 1995; Pigott 1999b), though only in the Khao Wong Prachan Valley in Thailand (see TAM 2C, chapter 2), has it been empirically demonstrated (Pryce et al. 2010).

Further laboratory experiments of crucible smelting of sulfide ores were conducted by Zwicker et al. (1985), who smelted an ore of 29% copper, 27% iron, and 31% sulfur in a crucible with charcoal and 3–6 blowpipes. The ore was roasted first in the crucible itself at about 800°C to drive off the sulfur as sulfur dioxide. The progress of the roast could be tracked by smelling for sulfur dioxide. Then this roasted ore, mixed with charcoal, was smelted in the crucible at 1100°C to form an impure copper. Whether slag was produced, or if the smelted result needed to be broken up to extract the metal, is not mentioned. These were both laboratory simulations; attempts to experimentally replicate the non-slagging copper sulfide processes that archaeological evidence, at least in the British Isles, has suggested, so far have led to inconclusive results (Craddock et al. 2007).

Evidence for crucible smelting of copper ore in open hearths has come from the 5th millennium B.C. site of Abu Matar in the southern Levant (Golden 2014), sites in southeast Spain dating to the 3rd millennium B.C. (Craddock 1995:127, 132–134), Tal-i Ibris (6th–4th millennium B.C.) and Tepe Gabristan (late 5th millennium B.C.), both from Iran (Pigott 1999b:74–77), the mid- to late-5th millennium sites of Chatalka and Dolnoslay in Bulgaria (Radivojević and Rehren 2016), and debatably, some 7th millennium crucible fragments from Çatalhöyük in Anatolia (Roberts et al. 2009), along with 2nd–1st millennia B.C. prehistoric Thailand (see TAM 2C, chapter 2). The evidence at these sites demonstrating a non-slagging process was recovered from small amounts of residual vitrified material adhering to the crucible fragments. Analysis showed that the slag had never fully melted and consisted mostly of partially reduced ore and gangue. Slag of this composition is a sign that the melting temperature of the gangue was never achieved and so the molten gangue/slag could not separate from the reduced metal. No tuyeres were found in these sites, suggesting that air may have been supplied through lung-powered blowpipes. Experiments have shown that crucible smelting of high quality oxide or carbonate ores produces very little slag and gives up to a 90% recovery of metal; even the crucible smelting of well-roasted sulfide ore recovered 50% of the metal in the ore and produced only small amounts of slag (Craddock 1995:135; Zwicker et al. 1985). Nonetheless, the small crucibles had a limited space for the charge (ore plus charcoal); the amount of metal produced was not large and was often in the form of prills that had to be broken out. Because the reducing conditions were poor, iron was not reduced and could not enter the copper. As a result, copper produced by this non-slagging process contains very low levels of iron (Craddock 1995:138). Thus, elemental analysis of early copper objects can suggest whether the process used to smelt the copper was done in a crucible or a furnace. Non-slagging smelting would leave little trace in the archaeological record; only the crucibles would remain, and it is difficult to distinguish crucibles used for smelting and those used for melting.

Furnaces with Slagging: Although the technological system for copper extraction employed in prehistoric Thailand focused overall on crucibles, further comparison with extraction processes using stationary furnaces illuminates the technological options and choices made by the prehistoric metalworkers in the Old World. An installation with a superstructure intended for a slagging process during which all or most impurities are removed as slag is called a furnace. Furnaces are distinguished from metalworking hearths, following Craddock (1995, 2000). In the bronze period of Southwest Asia and in many other areas such as Europe, copper was smelted in relatively efficient furnaces using fluxes or flux-rich ores in a slagging process, i.e., slag (molten gangue) was deliberately created to encourage the separation of gangue and metal (Craddock 2000). Slagging process furnaces took place in installations with a clay or stone superstructure and bellows and tuyeres for air supply. These furnaces are usually identifiable in the field by the presence of large quantities of slag, vitrified or slagged pieces of the furnace superstructure, in situ furnace bottoms, and/or slag-filled pits.

The development of the slagging process allowed the use of lower-grade ore. Slagging smelting took place in furnaces, either a shaped hole in the ground with a modest superstructure, as in Peru at Batan Grande (Shimada et al. 1982), or, more commonly, with a small refractory clay- or stone-lined hollow with a clay or stone superstructure. No evidence of copper-smelting furnaces with a fixed superstructure has been found in prehistoric Thailand (see TAM 2C, chapter 2), but the remains of shaft furnaces used for smelting iron and dating to the late 1st millennium B.C. have been excavated (Nitta 1991, 1997), so some metalworkers at least were familiar with the idea. The shape of a furnace shaft can seldom be reconstructed, and usually archaeologists only find the furnace base or fragments of the furnace walls scattered amid the slag heaps. As a rule, however, most shaft structures seem to have been round so that heating would be uniform. The clay of the shafts would normally be heavily tempered and coarse to increase the heat resistance, but because the furnaces were subjected to such stress—holding up to several hundred kilos of solid and molten charge and having air pumped in with bellows, along with slag that eroded the walls—few lasted more than one smelt without having to be refettled (repairing the lining) or, more likely, completely rebuilt (Craddock 1995:147, 172–173). Merkel's experimental crucible and furnace smelts, based on evidence from bronze period Timna in the Levant, suggested that a single smelt, including the necessary preheat of the furnace, took between 5–10 hours (Merkel 1990).

The slagging process relied upon the use of fluxes, materials that would combine with the gangue to reduce the melting temperature of the gangue and allow it to form a liquid slag at temperatures around 1300°C. The fluxes used varied with the composition of the ore. If the ore was rich in iron, as in pyrite ores, then crushed quartz was added. In many cases, the clay of the walls contributed to the fluxing of the slag, and was a part of the separation process. If the ore was rich in silica, as when the ore came from a quartz dyke, then hematite or other iron oxide was added. Sometimes the gangue plus ore contained the right ratio of metal oxides and silicates to form a free-flowing slag without addition of fluxes—these are known as self-fluxing ores (Craddock 1989). Most of the slags produced by pre-industrial processes are silicates or iron with admixtures of calcium and other metal oxides and inclusions of ore and metal.

Slag forms the primary archaeological evidence of smelting. It is largely indestructible and unusable for other purposes, so long-term smelting locations are often marked by large heaps of slag mixed with waste rock, ore, and remnants of furnaces and tuyeres. Craddock (1989) calculated that the typical beneficiated copper ore contained one part copper to three parts gangue. Merkel's experimental furnace smelts used a furnace charge of one part fuel to one part flux and ore (Craddock 1989). At the end of the smelt, the ratio of slag to copper would have been 8:1, i.e., eight times as much slag as metal. Thus, even limited smelting could, over time, result in large quantities of slag accumulation. Macroscopic examination of slags seldom reveals what metal was being smelted, as all metallurgical slags are mostly iron and silicates. Occasionally prills of copper will be visible to show that the slag resulted from copper smelting. In other cases, semi-quantitative analyses by X-ray fluorescence of slags will answer this question.

Air Supply: Smelting requires large, constant, and controlled supplies of air. Crucible smelting is feasible with lung-powered blowpipes tipped with a thin layer of clay; from experimental work and Egyptian tomb paintings, the number of blowpipes varied from two to six (Craddock 1995:177; Timberlake 2007). Slagging furnaces, on the other hand, require bellows tipped with tuyeres, to sustain their higher temperatures. (One exception to this was in Central and South America, where blowpipes were used in small slagging furnaces, possibly, at least in the Andes, because of the cultural requirement that human breath be involved in the process [Epstein 1993]. Only prills of metal were produced.) Blowpipe clay tips have been recorded not only in the Americas, but also frequently in chalcolithic and bronze period sites in Europe (Craddock 1995:178). Interestingly, they do not seem to have been an Asian phenomenon (Vincent Pigott, pers. comm. 2016).

Much more sustainable air flows could come from a variety of bellows: bag bellows, pot bellows, piston bellows, and concertina bellows. The temperature of the furnace depended on the amount of air delivered. (For more physical and chemical details of air supply in charcoal burning hearths and furnaces, see Rehder [2000, chapter 8].) The simplest are bag bellows, but experiments have shown that these can deliver 150 liters of air per minute, compared to about 20–40 liters for human lungs. Pot bellows of ceramic, first developed in the early 2nd millennium Near East (Davey 1979) and still found today in Africa and southern Indian furnaces, were more efficient; concertina bellows, which first appeared in the eastern Mediterranean in the 1st millennium B.C. and spread to India, could attain between 400–500 liters of air per minute. East Asian bellows included the piston bellows (e.g., using bamboo trunks) and fan bellows. Pot bellows would leave behind ceramics of distinctive shape, but the other bellows' designs would leave only their tuyeres in the record. Merkel's experimental furnace smelts used three tuyeres at once but found that copper could be smelted with a minimum of two (Craddock 1989). Fewer than two risked the sole tuyere becoming blocked by slag, but too many tuyeres risked a hampering degree of overcrowding at the furnace. The usual number appears to have been between two and six (Craddock 1989). No tuyeres have yet been excavated from prehistoric copper-producing sites in Thailand, but they have been found at the 1st-millennium B.C. iron-smelting site of Ban Don Phlong (Nitta 1991) and at Non Mak La in central Thailand, but the latter are of uncertain date (Vincent Pigott, pers. comm. 2017).

Smelting locations: Particularly in pre-industrial contexts, metal, both copper and iron, may be smelted at or near the mine because of the labor and logistical challenges of transporting large quantities of heavy ore. (Tin may be an exception, since the quantities of tin ore required to make a bronze artifact are much less than copper.) Any smelting operation even using crucibles likely involved a workgroup of several individuals who may have had differing areas of expertise. The need for large quantities of fuel can also constrain where smelting operations were carried out. Expert knowledge would be required to fashion crucibles, build the furnaces, place the blowpipes or tuyeres, choose the ratio of fuel to ore, decide how long to smelt, and numerous other technical decisions. Smelted metal was often cast into *ingots*, or small, relatively standardized shapes that could be transported and traded.

COPPER AND BRONZE FINISHED OBJECTS

Once the metal raw material is prepared, several more steps are completed to create a finished object. Metal may be further modified by refining and alloying, and metal that will be cast must be melted. The heat-resistant crucibles and molds for melting and casting must be created. Post-casting treatments to further harden, shape, or decorate the final product may be undertaken.

Refining the Copper

Smelted copper, whether produced in a crucible or a furnace, contains many impurities, especially metals such as iron, arsenic, antimony, bismuth, or nickel. Copper produced by the slagging process will contain iron from the ore or the flux. Copper smelted by simple processes used in small-scale operations often retains metallic oxide contaminants that render the copper brittle (Craddock 1995:202). In many cases, the smelted copper was usable only after it was refined. This activity could be accomplished by melting the metal in an open crucible and skimming off the oxidized impurities that float to the top, thereby reducing the quantities of these contaminants. If the copper contained a great deal of iron—as the copper smelted in a slagging furnace usually does—scattering sand on top would produce crucible slags that could be skimmed off. Experiments have shown that the iron content of copper can be reduced from several percent down to 0.5% by melting and skimming. Oxidation of the molten copper could be prevented by stirring the molten metal with green sticks, a process called poling (Craddock 1989; 1995:202). These crucible slags, often found on metalworking sites, tend to be much lighter in weight and more porous than smelting slags, with a heterogeneous distribution of inclusions, but can also show the smooth tapping surface created by being poured out on the ground from the crucibles (Miller 2007:155). Refining slags tend to be high in wood ash and copper (or whatever other metal was being refined) and low in iron.

As a rule, refining appears to have been done elsewhere than smelting. Survey of mining and metalworking sites in the Near East (Berthoud et al. 1980) has shown that refining slags with tin, lead, bismuth, and other impurities are found not at smelting sites but at urban metalworking sites, suggesting that refining the copper was the task of the metalsmith rather than the person who smelted the ore. The smelter, after all, as Craddock (1989:209) pointed out, had "little incentive to purify his product and thereby reduce the quantity of saleable metal, and the only real guarantee the metal smith had of the quality of this raw material was to purify it himself."

Alloying

As noted above, copper can be combined with other metals to form alloys whose properties and performance characteristics differ from unalloyed copper. The proportions of the metals in the alloy need to be within certain ranges for the desirable material qualities to manifest. Alloying a copper with over 2% tin lowers the melting point, improves the quality of a cast by improving fluidity and reducing gas porosity, allows for much greater hardness if hammered, and changes the color. Deliberate or accidental alloying with arsenic or antimony will have the same effects, while alloying with lead will improve the castability while making the metal softer.

Alloying can be done three ways: (1) by mixing two or more separately produced metals; (2) by smelting together (co-smelting) two different ores; or (3) by mixing an ore and a metal. A related operation is to deliberately select ores for smelting that are known to have enhanced properties as a result of naturally occurring mixtures of two or more metallic elements, even if the specific reason for the enhanced properties was not known to the smelters. Alloys can be accidentally produced; however, because tin is relatively rare and its ores seldom occur together with copper ores, most tin bronzes are deliberately produced. Mixed ores of copper, tin, and iron, such as stannite, do exist. As noted above, following Stech (1999) and Stech Wheeler and Maddin (1976), in this monograph a deliberate tin bronze alloy is defined as containing 2% or more of tin. The 2% proportion for defining an alloy is used with leaded copper/bronze and copper-arsenical alloys as well, though defining an arsenical copper as a deliberate alloy is problematic. (See discussion of arsenical and antimonial copper below.)

Archaeological signs that alloying took place in a particular location would include nodules of the ore of the alloying elements, such as cassiterite for tin or galena for lead, or even more clearly, the presence of lead or tin casting debris among the finds. Analysis of the metal adhering to crucibles can also suggest whether alloying took place at that spot. Elemental analysis of artifacts will reveal the presence of alloying elements, though not where the alloying took place.

So far, no identifiable pieces of cassiterite ore have been found in prehistoric Southeast Asian sites, and there is no evidence yet to suggest the mode of alloying employed during the bronze period. Analyses conducted on materials excavated from the mid- to late-1st-millennium B.C. site of Khao Sam Kaeo on the upper Thai-Malay peninsula, though, suggested that a cassiterite cementation process (cassiterite ore smelted together with copper) was used to produce high-tin bronze ingots (Murillo-Barros et al. 2010). Given the relative consistency of tin percentages in the bronze artifacts, especially in the later periods, it seems likely that the metalsmiths over time developed the ability to control the content of their alloyed metal which, despite the findings at Khao Sam Kaeo, suggests that tin was smelted separately and not co-smelted with copper ore. No direct evidence has emerged to confirm this, though with a melting temperature of 232°C, tin is far easier to smelt than copper, easily done in a village setting. Figure 7.5 illustrates how tin smelting was carried out in a remote Laotian village in the 1930s, using methods that may well date back to antiquity.

Casting and Working

Metal products can be fashioned in basically two ways: casting and working. Melting and casting is the process of pouring molten metal into a mold. Working is the process of hammering the metal into the desired form; in iron smithing, "working" is called "forging." (Some workers in copper and bronze also refer to working as forging.) The use of one method does not preclude the other, since a cast object may have the final finishing and shaping done by hammering.

Working is either *cold working*, i.e., working carried out with the metal at room temperature, or *hot working*, carried out with the metal at a minimum temperature of 400°C–500°C. Working is carried out by hammering, raising, turning, or drawing an object to the desired shape, using a variety of hammers, anvils, and lathes. The process of working an artifact can be challenging. The ductility of metal is due to the ease with which the atomic planes slip over each other. As copper or bronze is hammered, the metal grows harder as defects caused by the working pile up in the atomic planes. These defects, called dislocations, accumulate and prevent the planes from slipping over each other. The accumulation of defects can increase the hardness of metal by two times or more. With the increase in hardness, though, comes an increase in brittleness; eventually, with too much hammering, the metal will crack. To ease the brittleness, the smith *anneals* the metal, by heating it up to 400°C–500°C, not hot enough to melt the metal (pure copper has a melting temperature of 1083°C), but hot enough to allow the molecules to rearrange themselves and remove the dislocations in the atomic planes that create hardness and brittleness. Once annealed, the metal is again ductile, while still retaining its worked shape, and further working can take place. Often several cycles

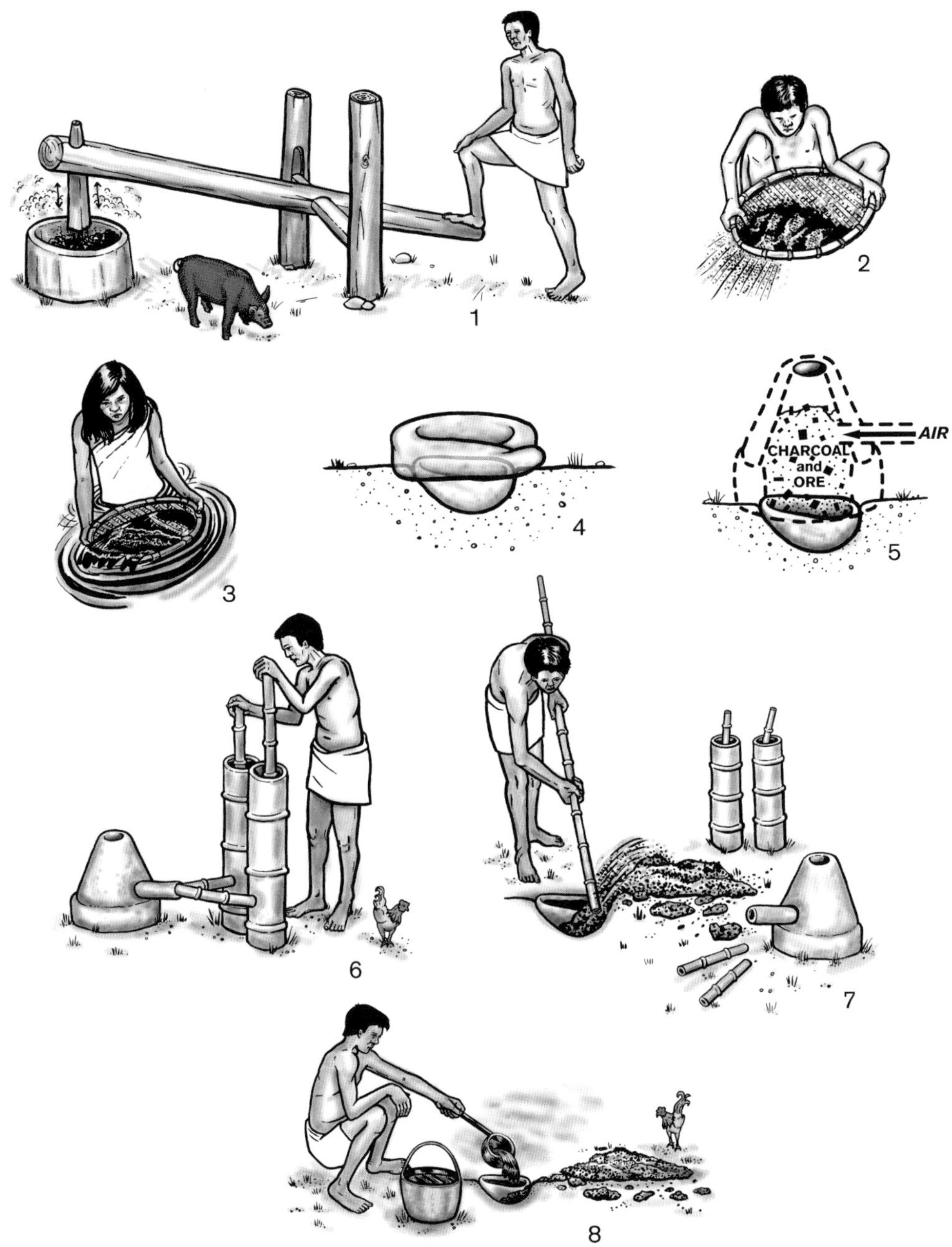

Figure 7.5 Tin smelting in a remote Laotian village in the 1930s. (1) The ore, picked up from an eroded hillside, is crushed in a wooden mortar and stamp. (2) The crushed ore is sifted, with the oversized pieces returned to the mortar for further pounding. (3) Reduced to the desired size, the crushed ore is washed and concentrated in large wooden pans. The concentrate is separated out by hand. (4) A small pit about 15 cm in diameter is dug in dry ground. The crucible is placed inside the pit, which is then ringed by a collar of wet clay. (5) A baked clay oven with an open top and a side tuyere hole is placed on top of the collar and sealed. A burning ember is placed inside and the oven-furnace charged with layers of ore and charcoal. (6) During smelting, air is provided by two bamboo piston bellows linked to a single clay tuyere. (7) When the smelting is finished, the furnace and blowers are dismantled and the embers and molten tin in the crucible are stirred with a wooden rod to clear dirt and ashes from the tin. (8) The spongy tin ingot is cooled by water and removed from the crucible. (Drawings and descriptions derived from Barthelemy 1938.) According to Barthelemy, the tin ore averaged about 9% tin; washing and concentrating increased this to 57% and the overall recovery of metal was about 20%.

of working and annealing are necessary to complete the formation of the piece. Whether the final stage was annealing or working can give a clue to the intended function of the piece and the knowledge the smith had of the properties of metal. If the artifact was left annealed, then it would be relatively soft, which is not an optimal condition for an axe blade intended for chopping wood, for instance. If it was left worked, it would be hard. If the artifact was a bangle, then working it would be unnecessary in that hardness will not improve its function, although the smith could have wanted to shape the bangle and so bangles may be found in a worked state.

Casting, by far the most common technique in prehistoric Thailand, has other difficulties. Casting involves heating the metal until molten, pouring it into a mold, and allowing it to solidify. Stone, clay, sand, and metal can all be used to make molds. Molds can be single sided, i.e., consisting of a depression in stone, clay, or other material into which the molted metal is poured so that one side of the cast is flat; two sided or bivalve, able to create more complex three-dimensional shapes; or multipart, piece molds (used in early dynastic China) that can make elaborate shapes such as ritual wine vessels. Tools made in bivalve molds can have a socket fashioned by insertion of a plug in the hafting element portion of the mold. Single-sided and bivalve molds made of stone or even clay can often be reused, as demonstrated by molds found at sites in the Khao Wong Prachan Valley (Vincent Pigott, pers. comm.), but often clay molds are broken while extracting the object.

Lost wax casting is another method that can make very intricate or elaborate artifacts, although it can also be used for simple shapes such as rings. The lost wax method involves shaping a model of the desired object in wax, and then encasing the wax model in clay, with outlet and inlet holes provided. If the clay and wax are then heated, the clay is hardened and the wax melts and runs out of the outlet holes, leaving a hollow mold into which molten metal can be poured. The melting of the wax and the pouring of the molten metal can be combined into one step, as well. As in other methods of casting, during cooling the metal emits gases that, if not allowed to escape, can cause bubbles to appear in the finished artifact. The metal can also shrink, especially if cooling is rapid. The production of a fine cast piece requires considerable skill.

The archaeological evidence that casting activities took place on site most commonly includes the small bits of casting debris, the solidified remains of the molten metal that accidentally splashed out of the crucible or mold during the process of heating and pouring. Crucibles, whole or fragmentary, are also common finds. Other remains can include stone or ceramic molds, mold plugs, and pieces of low-iron crucible slag. Also occasionally found are sprues, little cones of metal formed when the molten metal is poured into the mold and the excess metal solidifies in the pouring hole (which can also be called a sprue). Evidence that a specific artifact was cast can be found macroscopically and microscopically. An artifact may retain a slight ridge of metal, called a casting flash, left where the two pieces of a bivalve or two-piece mold met, which also signals that the artifact was cast. Casting is also clearly demonstrated when the artifact is sampled and the microstructure examined under a microscope (see discussion in TAM 2B, chapter 4), which indicates the object was left in the as-cast condition.

Other Post-casting Treatments

A cast artifact may need further treatment to fulfill its social purpose. The casting flash must be hammered or ground off if a bivalve mold was used. The area of the artifact by the pouring hole of the mold must be cleaned and smoothed. The artifact can be polished with sandstone or water and sand, engravings or decoration made, and, if desired, the edge of a cutting tool hardened by hammering and honed with a whetstone. Whether cast or worked, copper and copper alloy work in antiquity was often carefully finished by planishing, or lightly hammering the surface to remove working traces, and then filing and polishing for a fine finish (Hodges 1976:76). Polishing can sometimes be seen if the artifact is inspected under a microscope, and edge hardening and sometimes final shaping can be detected by metallography, but otherwise few of these finishing touches will leave macroscopic traces for the archaeologist.

SPECIAL ALLOYS

Although tin bronze and leaded tin bronze were the most common alloys of copper in prehistory, several other alloys were known to have been used in many parts of the world.

Arsenical and Antimonial Copper

Whether arsenical copper (some scholars use the term arsenical bronze [Lechtman 2014]) was deliberately produced in any one area can be difficult to determine, as many copper ores contain arsenic naturally (Craddock 1995:285). Experimental smelts have shown that under normal reducing conditions, in such ores the arsenic tends to be retained in the copper, so that it is quite possible to produce an arsenical copper by accident (Lechtman 1999; Rostoker et al. 1989). Arsenic has the same hardening effect on copper as tin, especially after working, but the inhomogeneous distribution of arsenic in an arsenical copper (made from copper ores with arsenic content) makes it very difficult to predict the properties of the metal from its bulk composition alone (Craddock 1995:286). Establishing whether an arsenical copper was the product of a deliberate attempt at alloying is not easy. Craddock (1995:287) listed three ways: (1) compare the composition of the metal with the ore, if known; (2) compare the arsenic percentages with other impurities, as a high percentage of elements such as antimony would suggest that the ore used was naturally arsenic-rich; and (3) look for correlations between level of arsenic and artifact type. Arsenic has a substantial hardening effect and also produces a silvery surface color; therefore, was an arsenic-rich alloy used preferentially for cutting tools or for ornaments? For example, during the European bronze period, in many areas blades in particular tended to be made of copper-base metal with higher levels of arsenic, i.e., up to 4% (Craddock 1995:289). Northover (1989a, 1989b) also noted that high arsenic levels are in some areas linked to specific classes of metalwork, especially high-prestige items such as daggers or ornaments, probably because of their silvery color. The most well-known example of arsenic-rich copper used for high-status ornamental items comes from the famous Chalcolithic period Nahal Mishmal hoard of the southern Levant, known also for antimonial copper artifacts (Golden 2014).

Pigott et al. (2003) described three methods of making arsenical copper, (with particular reference to the Iranian Plateau during the 4th and 3rd millennia). The first was to add native arsenic to molten native or smelted copper, or to melt naturally arsenic-rich native copper. This would be a simple procedure, requiring temperatures of only ca. 900°C, but arsenic-rich native copper deposits are rare in the world, though some are easily accessible in Iran. The second method involves the smelting of weathered copper sulfide orebodies, especially *fahlerz* ores, or gray-black copper sulfides rich in arsenic, antimony, silver, and bismuth. To smelt these ores would require either a preliminary roasting step to drive off the sulfides or co-smelting the arsenic-rich fahlerz with oxide ores in a process that required a temperature of at least 1250°C and produced a great deal of slag. The third method is to smelt arsenates, arsenic- and antimony-rich minerals from the oxidized zone of copper orebodies. Such arsenates are colorful and resemble distinctive copper oxide ores such as malachite. Such a process could be conducted in a simple bowl furnace at a temperature as low as 700°C and produce little or no slag. The arsenates smell like garlic when hammered, providing a clear sensory method to distinguish them from other ores. As with the first outlined method, the level of arsenic in the resulting copper will be below 5% (Pigott et al. 2003).

Higher levels of arsenic, however, could have indicated a fourth method, which might have been to add pure smelted arsenic to the molten copper. Arsenic is so volatile that special equipment would be needed to reduce it to the metal. Arsenic minerals such as realgar could also have been added; they are often found associated with copper ores.

The simplest way to produce arsenical copper would have been to smelt naturally arsenic-rich copper ores if they were present; these are distinctive in appearance and odor when heated (Craddock 1995:289). This deliberate choice of arsenic-rich ores is also defined as alloying (Stech 1999). For example, in Mesopotamia (Stech 1999), the Levant (Golden 2014), Anatolia (Lehner and Yener 2014), Iran (Thornton 2014), Iberia (Roberts 2014), and the Balkans (Kienlin 2014), smelted copper or smelted arsenical copper appeared in the archaeological record well before tin bronze. The number of artifacts made of arsenical coppers declined in frequency during the early bronze periods of Southwest Asia and Europe, and during this period in these areas it is not unusual to find artifacts with low levels of both tin and arsenic (Tylecote 1992:19–20). The analyses performed for this study also found two artifacts with low levels of tin (2–5%) and over 2% arsenic, both from the Early Period or Early-Middle Period at Ban Chiang. It is rare to find artifacts with arsenic as the major alloying ingredient in any copper alloy-using culture in the Old World later than the earliest period of copper alloy use, often termed the "Chalcolithic" in various culture areas.

Other elements such as antimony also have a hardening effect when added to copper. High-antimony (5–15%) artifacts are not uncommon in the early bronze periods of the Near East and Europe (Tylecote 1992:11), though they are infrequent after that time.

High-tin Bronze

High-tin bronze is a bronze that contains 20% or more tin. The addition of 20–30% tin reduces the melting temperature of the copper alloy to below 900°C (see TAM 2B, fig. 4.7a,b). This alloy is extremely brittle (i.e., no longer ductile) when treated like a binary 10% tin bronze, i.e., cast and allowed to cool slowly to room temperature. The casting can only be shaped further by hot working, which is hammering the metal while it is between 550°C–798°C (Srinivasan and Glover 1995). There are two variants of this alloy: high-tin with over 2% lead and high-tin without lead. Leaded high-tin bronze was used to make Roman, Islamic, and Chinese mirrors (Craddock 1979; Meeks 1993a; Zhu and He 1993), Ptolemaic period rings, early medieval European buckles and belt fittings (Meeks 1993a), and Late Iron Age Celtic "potin" coins (Meeks 1993b:250), among other items. High-tin bronze without lead appears in early Islamic Persian vessels (Allan 1979; Melikian-Chirvani 1974; Pigott 1989), medieval Korean vessels (Voce 1975), and Indian mirrors and Indian vessels (Srinivasan and Glover 1995). Unleaded high-tin wires are found at Don Klang and Ban Chiang (see TAM 2B, chapter 4); so far, these wires appear to be unique to these sites. The same alloy is found in vessels in late 1st millennium B.C. Ban Don Ta Phet in central Thailand (Rajpitak and Seeley 1979) and 4th century B.C. Khao Sam Kaeo (Murillo-Barroso et al. 2010). Both sites show evidence of interactions between the peninsula and South Asia. This technology appears now to have originated in India (Glover and Bennett 2012).

Because of its sonority, this unleaded high-tin alloy was used for bells in medieval and modern Europe and is still used to make modern cymbals in the West and gongs in the Philippines and other places in East Asia (Goodway and Conklin 1987).

High-tin bronze can be made by adding smelted tin to copper and is also easily produced by smelting cassiterite and copper ore. Pigott (1989) noted, "Early in this century W. Gowland demonstrated this point by smelting 15 pounds of malachite with 10 pounds of cassiterite, 7.5 pounds of limestone (flux), and 10 pounds of charcoal.... The resulting bronze had a tin content of 22 percent." Artifacts of high-tin bronze can be created in two ways that have very different results microscopically and mechanically (see TAM 2B, chapter 4 for discussion of microscopic changes). The first treatment is to cast an artifact and allow it to cool in the mold. A high-tin bronze can be cast like any other bronze, and, as the tin content rises beyond 20%, the color becomes more and more silvery with polishing. This alloy is too brittle to work at room temperature. Often high-tin alloys cast and subsequently slow cooled in molds contain several percent of lead for better casting.

The second treatment is to cast the artifact, allow it to cool in the mold, reheat it above 550°C, perform hot work if further shaping is desired, then quench it by cooling it rapidly from a temperature of 550°C or above by plunging it into water or oil. This cycle can be repeated several times. This treatment renders the metal relatively malleable at room temperature. Instead of the silvery color of unquenched high-tin bronze, quenched and polished high-tin bronze appears golden or pale yellow in color (Rajpitak and Seeley 1979) and is extraordinarily lustrous. (See Srinivasan [2010, and especially figure 7] for an ethnographic description of manufacturing high-tin bronze bowls in India and a photograph of the resulting color.) High-tin alloys intended for quenching very seldom contain lead, because the addition of the lead can lower the melting temperature of the alloy to below that needed to produce the desired malleability changes upon quenching (Scott 1991:27).

How hot the object was just before it was quenched can have considerable effect on the post-quenching hardness. Matsuda (1928) conducted a series of experiments on the effects of quenching on hardness in high-tin alloys with different levels of tin (Fig. 7.6). He discovered that quenching the object when it was cooler than 520°C

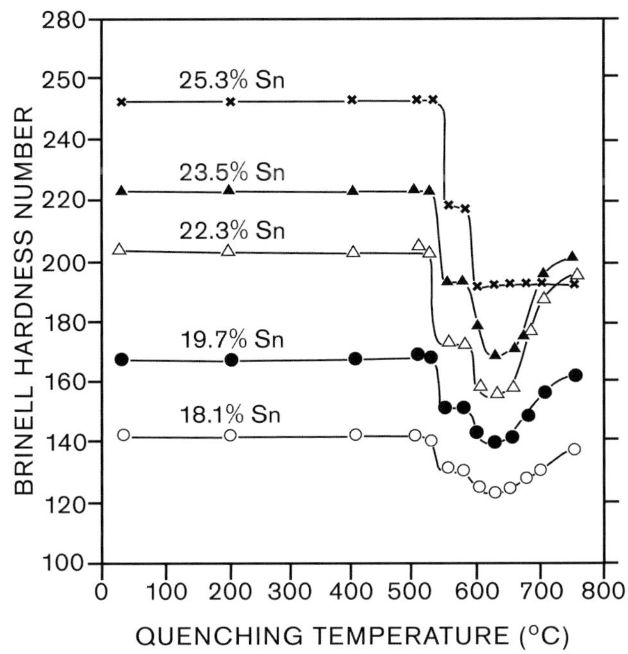

Figure 7.6 Hardness of quenched high-tin bronze as a function of tin (Sn) percentage and temperature at start of quenching (Goodway and Conklin 1987, adapted from Matsuda 1928:86).

did not affect hardness, i.e., the metal remained too hard to work at room temperature. A rapid drop in hardness occurred at a quenching temperature of 520°C (i.e., when the object was at 520°C just before it was quenched), with a further drop in hardness if the quenching took place when the metal was around 590°C. At quenching temperatures higher than these, the metal began to harden again, except when the alloy contained as much as 25.33% tin, where a higher temperature produced no change in hardness. Nonetheless, quenching even as high as 750°C still left the metal softer and more malleable than if it had been cooled in air.

Quenching Other Bronze Alloys: Ottaway and Wang's (2004) experimental casting program has demonstrated that quenching a copper-base artifact has a considerable effect on its hardness and microstructure even when the alloy contains less than 20% tin. Quenching improves the malleability of even 6% tin alloys. Quenching also increased the hardness of the alloys, though not to a large degree; differences in microhardness are still mostly dependent on composition and degree of cold working.

Iron and Steel

Humans had had a limited awareness of the existence of iron, mostly in the form of meteoritic iron, for millennia before smelted iron came into widespread use. Although native iron (iron of terrestrial origin) is extremely rare—mostly occurring in Greenland—meteoritic iron was known and worked into sheets at least as early as 3200 B.C. in Egypt (Rehren et al. 2013). Meteoritic iron is metallic iron that was deposited on earth from meteor showers. It can be identified by its high germanium (Rehren et al. 2013) and nickel contents (4–26% nickel). This high nickel content makes it prone to cracking and difficult to work, according to Tylecote (1992:3), though Rapp (2002:164) wrote that it is easily worked. Artifacts containing high-nickel iron can be found as late as the 1st millennium B.C. in Europe (Tylecote 1992:3), as well as in Shang and Western Zhou contexts (ca. 1500–771 B.C.) in China (Bronson 1999; Gettens et al. 1971; Wagner 1999). Finds of low-nickel iron, some dating to the 5th millennium B.C., have been assumed to be of smelted origin, causing some to postulate precocious iron-smelting technology, but Craddock pointed out that nickel can easily leach out of corroding iron (Craddock 1995:256). Waldbaum's discussion of the earliest iron artifacts showed the difficulty of establishing the origins of ironworking (Waldbaum 1999).

For most of the 2nd millennium, iron was used almost entirely for ornaments and prestige gifts, such as the gold-decorated iron dagger from the tomb of Tutankhamen (14th century B.C.), which was recently determined to be of meteoritic origin (Comelli et al. 2016). It was only at the end of the 2nd millennium and the beginning of the 1st millennium B.C., especially by the 8th century B.C., that iron manufacture became widespread in the Old World (Craddock 1995:258; Waldbaum 1999); though finds of actual iron smelting or smithing sites earlier than 500 B.C. are very rare (Veldhuijzen and Rehren 2007).

The discovery of iron smelting may have been due either to the use of copper ores that contained a great deal of iron, or from the use of iron oxide as a flux in copper smelting. In either case, the iron minerals present may have been accidentally reduced to metallic iron (Killick 2014; Pigott and Marder

1984). Because of the ubiquity of iron in the earth's crust, and the ease with which iron ores can form by sedimentary evaporation, precipitation, and leaching metallogenic processes (see Chapter 6), iron ore is widely distributed over the world. As such, there was seldom need for the elaborate trading networks required in much of the world to obtain copper and tin. In many pre-state contexts, the blacksmiths (iron smiths) smelted the iron themselves. Iron has a number of useful performance characteristics, in addition to the material properties inherent in all metals—ductility, hardness, thermal conductivity, reusability, and ability to be mixed with other materials. Iron can, if treated appropriately in ways to incorporate carbon, be made far harder than bronze, yet retain ductility and toughness, and pieces of iron can be welded together by heat and hammering, as copper cannot be. Its major disadvantages are that it is far more prone to corrosion than is copper or bronze and more difficult to extract from the ore.

The *chaînes opératoires* of iron production and the manufacture of iron objects have similarities with and differences from those of copper-base objects (see Table 7.2). The basic processes are the same—prospecting, mining, ore beneficiation and preparation, and smelting—but the greatest difference is that copper smelting produces molten copper, whereas bloomery iron smelting (the dominant form of pre-industrial iron smelting) produces iron in the solid-state, i.e., the reduced iron never achieves the molten state. The melting point of pure iron is 1550°C. In the later 1st millennium B.C., only the Chinese were able to reliably and consistently produce a furnace heat of this temperature in the manufacture of cast iron (Wagner 1993; although see Rehder [1986] for a view that early charcoal furnaces could attain a very high heat). The iron used in the rest of the world was smelted in a *bloomery* furnace, where the iron ore was reduced at temperatures around 1200°C, in an atmosphere rich in carbon monoxide, to yield a spongy *bloom*, a mixture of iron, carbon, and slag. This bloom was pulled from the furnace and forged—heated red- or white-hot and hammered so that the glassy slag became molten and was squeezed out.

The most important archaeological evidence for manipulation of iron—whether extraction from ore, forging of the bloom, or manufacturing iron artifacts—is slag, a hard, once-molten, siliceous mixture of mineral waste material. Some slags are the result of burnt mud-brick, vitrified fuel ash, or glass manufacture, but most are the result of metallurgical process. They can be distinguished from non-metallurgical slags by their composition (ferrous silicates, even if from copper smelting), density (heavy), streak (dark brown to black), color (dark), porosity (tend to be quite vesicular), or embedded fragments of metal or rust (Bachmann 1982:2–3). Distinguishing slag of copper smelting from iron smelting by the naked eye is harder; often it can be done only if there are telltale prills of copper or green stains on a fresh surface. Distinguishing iron smelting slags from smithing slags can be even more difficult.

The *Chaîne Opératoire* of Bloomery Iron Smelting

PROSPECTING

Iron ores vary widely in appearance but can easily be recognized by the experienced eye. Hematite and limonite ores are often red, orange, and yellow; magnetite and certain varieties of hematite are darkly lustrous, with a gunmetal sheen (Rostoker and Bronson 1990:42), and other varieties of iron ore are heavier than ordinary rocks. Other metal-bearing ores, such as tin, lead, and manganese, have the same appearance and weight as the dark iron ores, but, as Rostoker and Bronson (1990:42) pointed out, iron ore is so much more common than these other ores that in most areas, "any reddish, yellowish, or lustrous gray-black rock that was also heavy was likely to contain appreciable amounts of iron." (This may not have been true in the Burma-Malay tin-rich metallogenic belt [Fig. 6.4], with its copious supplies of placer cassiterite, which can be both heavy and dark.) Laterite is a limonite-hematite ore created in iron-rich tropical soils by sunlight and the action of rainfall washing other soil components to a deeper level and tends to be nodular and distinctively reddish. It can be collected or extracted from the surface without difficulty. Laterite nodules are commonly found at iron period sites on the Khorat Plateau. In some parts of the world, including Southeast Asia, laterites have been used as a source of iron, but the few analyses conducted on laterite samples suggest that most were too low in iron to be usable in bloomery smelting (Cawte and Boyd 2010; Pryce and Natapintu 2010).

Table 7.2 Archaeological Evidence for Iron Primary Metal Production and Finished Artifact Manufacture; adapted from Northover (1989b), Miller (2007:Figure 4.12), and Rostoker and Bronson (1990)

Process/activity	Archaeological Evidence	Notes
Prospecting	None.	
Mining	Bits or piles of ore fragments; pits; lenses or piles of finely crushed rock. Most iron ore mined on the surface, or in open pits, not with shafts or adits.	
Techniques		
Metal picks, chisels, sledgehammers	Long thin scratches, often parallel, on mine walls.	
Shovels and hoes	Shovels and hoes found.	
Hydraulic mining	Uses water to wash through the ore deposit to concentrate the heavier iron ore downstream. No archaeological evidence, but process can be inferred if the probable ore source consisted of magnetite sands. The sands could also be concentrated by manual panning.	
Charcoal Preparation	None.	
Ore Preparation		
Techniques		
Ore crushing and sizing	For shaft furnace smelting, ore bits should be of uniform size. Revealed by uniformly sized debris.	
Beneficiation	Bits of discarded ore; thick deposits of uniformly sized, finely crushed host rock; small hand-held hammerstones with pecking; shallow depressions in flat rocks used as anvils.	Crushed rock should be examined to determine efficiency of separation.
Ore roasting	Often done to dry the ore and make it more porous. Temperature should be ~600°C. Traditionally done in heaps, low enclosures, or holes dug into hillsides (of which traces might remain).	
Ore sizing	For shaft furnace smelting, ore bits should be of uniform size.	
Smelting	Primary evidence is slag, bits of ore, flux.	
Techniques		
Furnaces	Large quantities of dense, dark-colored slag; bowl-shaped or hole-like burned depressions; remains of clay tuyeres; pieces of thick ceramic furnace pieces, sometimes vitrified and slagged. Charcoal in soil. Furnaces with slag-tapping will produce slags with smooth or ropy top surfaces; non-tapping furnaces will produce rough and blocky slag.	Metal produced from slagging furnaces has a much higher iron content than crucible metal.
Smithing/forging	Hammerscale, spheroids, slags lighter and more vesicular than smelting slags. Charcoal. Occasionally anvils or hammers.	
Carburization	Metallographic (visual) analysis showing the presence of carbon.	

Finding a suitable-looking ore source was not enough, however; every source would need to be tested. As noted in Chapter 6, the quality of the ore lies not only in its iron content, but also the percentage of irreducible elements, the gangue, which would need to be removed during the smelt. The slagging process would use up energy and valuable iron, as slag becomes fluid at bloom-forming temperatures only if iron oxides are added as a flux (Rostoker and Bronson 1990:44–45).

MINING

Iron ore miners in pre-industrial contexts seldom used the elaborate system of adits and shafts necessary to follow veins of copper, silver, and gold ores (as at Phu Lon; see TAM 2C, chapter 2), probably because the ore was not rare enough to be worth the effort (Rostoker and Bronson 1990:47). Also, large deposits were often on the surface and easily mined. Most mining was carried out on the surface or in open pits with simple tools: picks and hammers to loosen the ore, pried out with various designs of crowbars, gathered with hoes and shovels, and carried out, probably in baskets on the backs of humans or animals. Fire-setting was used in a few places but was not common (Rostoker and Bronson 1990:48). Ethnographic accounts of iron mining in early-20th-century Africa note that mining was performed by women and children as well as slaves (Goucher and Herbert 1996).

If the ore source was composed of very fine particles, such as black or magnetite sand, it could be mined and concentrated by water washing over the matrix and carrying away the deposits to where the heavy iron sands could be caught by mats or small dams. Equally well, the fine iron deposits could have been panned in baskets by hand or in small troughs (Rostoker and Bronson 1990:49–50).

ORE PREPARATION

As a rule, the ore was extracted from the earth in lumps, which had to be crushed to a uniform size for smelting in a furnace. The experimental work of Tylecote et al. (1971) demonstrated that 12 mm was the maximum size of porous ore that would allow acceptable metal recovery; non-porous ore such as hematite needed to be less than 3 mm. The crushing of the ore would also allow beneficiation, the hand separation of iron-rich fragments from the gangue. This process would have been the same as for copper ores. From ethnographic evidence, it is known that frequently the work of beneficiating ores of various sorts was performed by women and children (Henderson 2000:220); in any case, the process would have required several people.

Frequently, according to historical and ethnographic accounts, the iron ore was then roasted to dry it, to reduce the size of the fragments, and, by driving off gases such as carbon dioxide and water vapor, to produce small cracks in the ore fragments and thereby increase their porosity and ease of smelting. The roasting would have taken place in heaps on the ground, in small stone-sided trenches or enclosures, or in pits scooped into a hillside (Craddock 1995:168). The burn needed to reach at least 600°C to be successful, but ordinary wood could be used as a fuel (Rostoker and Bronson 1990:52).

The roasting would also have helped remove sulfur from the ore, though there is no evidence in the ethnographic literature that metalworkers were aware of this (Rostoker and Bronson 1990:52). Even low percentages of sulfur in iron produce "hot-shortness": the tendency to crack when forged at high temperatures.

CHARCOAL PREPARATION

As with copper smelting, charcoal has been in prehistoric and most pre-industrial times the preferred fuel for iron smelting and smithing. A great deal of fuel is required for smelting iron, and historically, the cost of fuel has been the major determining factor in the economics of iron production. According to Rostoker and Bronson (1990:63), in England during the 16th to 18th centuries A.D., making and transporting charcoal accounted for 55–75% of the cost of iron production. As with beneficiation, ethnographic evidence from West Africa suggested that, in traditional smelting, women were the charcoal makers (Goucher and Herbert 1996).

IRON SMELTING IN PRE-INDUSTRIAL CONTEXTS

The exact details and precise technical choices followed by prehistoric ironmakers in Thailand are not clear. Compared to the Middle East and Europe, relatively little technical work has been done on the prehistoric iron industry in Thailand (though see Bennett 2013a, 2013b, 2015; Biggs et al. 2013; Cawte and Boyd 2010; Glover and Bellina 2011; Pigott and Marder 1984; Pryce 2014; Pryce and Natapintu 2010; Suchitta 1983, 1992), and much has rested on morphological artifact studies. There are numerous technological requirements in iron smelting, including a high temperature, reducing atmosphere, and a steady air supply. To achieve these, most smelting has been done in enclosed furnaces, which contain and magnify the heat. High temperatures are not enough; to produce iron (or any other metal smelting) certain chemical reactions must take place. The most important is reduction, where the iron oxide is separated into elemental iron and an oxygen-containing gas. The separation is performed by a reagent with a strong affinity for oxygen, strong enough to pull oxygen out of the metal oxide when intensely heated. Carbon monoxide is a suitable reagent, and carbon monoxide and carbon dioxide are both produced when a carbon-containing fuel like charcoal is used to heat the furnace (Rostoker and Bronson 1990:25).

Furnace design is difficult, however, as carbon monoxide is not automatically produced from a hot fire. Most fires heated with ample supplies of air produce carbon dioxide, not monoxide. A smelting furnace must have enough oxygen injected to achieve a high temperature, but not enough that all the carbon converts to carbon dioxide. There are numerous ways to solve this technical problem, and a wide variety of pre-industrial approaches can be seen, particularly in ethnoarchaeological work in Africa (Childs and Killick 1993; Schmidt 1996). For a fuller discussion of the chemistry of iron smelting, see Rostoker and Bronson (1990) and Rehder (2000). Crucible iron smelting, where the iron ore and a carbonaceous material is encased in crucibles placed inside a furnace, has been documented in pre-industrial times in China, southern India, Sri Lanka, the Fergana Valley in Central Asia (Rehren and Papakhristu 2000), Southwest Asia, and Western Europe (Rostoker and Bronson 1990:27). Usually, however, the crucible process was not used for actual smelting, but for an optional subsequent step of carburization, the process of infusing carbon into the iron (Craddock 1995:276); the crucible steel process in the Fergana Valley was an exception (Rehren and Papakhristu 2000).

Furnace Types

Most bloomery iron smelting took place in furnaces with a closed superstructure and without crucibles. As in copper-smelting furnaces, usually only furnace bottoms and/or fragments of the furnace walls scattered amid the slag heaps are found in the archaeological record. Often, though not always, the furnace would have to be destroyed to remove the iron bloom, rendering reconstruction by the archaeologist difficult. The high heat requirements of iron smelting necessitated a larger air supply than in copper smelting, so any bloomery furnace needed multiple tuyere holes punched through the furnace walls about two-thirds of the way down from the top of the superstructure; air was normally supplied by bellows, though in exceptional cases natural draft would have been sufficient (see Juleff 1998). As with copper smelting, two generalized types of bloomery furnace existed: bowl furnaces (Fig. 7.3) and shaft furnaces (Fig. 7.4), though, as Craddock noted (1995:146), a better distinction might be between slag-tapping and non-tapping furnaces. The functional division in copper smelting is between slagging furnaces and non-slagging hearths; in contrast, it would be almost impossible to conduct a non-slagging iron smelt. The slag liquefies at a temperature of 1200°C or above, and, as it liquefies and drips down, it shields the newly reduced iron from reoxidizing as air is pushed into the furnace. Impurities such as fuel ash, bits of ceramic lining, and gangue also concentrate in the slag and leave the metal. Slags are mostly ferrous silicate, so some of the iron in the ore must be used in the slag formation process (Pryce and Natapintu 2010).

A slag-tapping furnace that allowed slag to escape via holes in the furnace side during the smelt could run for hours, but a furnace with no way to let slag escape would require stopping and starting periodically to prevent the slag from rising so high that the tuyere holes would be choked. The distinction between the two kinds of furnaces can easily be seen in the slag; tapped slag has a smooth, flowing texture on top, and sometimes shows the impression of the dirt onto which it flowed, whereas slag from a non-tapping furnace will be rougher and blocky. A furnace, once constructed, would have been preheated by letting the fuel burn inside prior to charging, or adding layers of finely crushed and beneficiated ore and charcoal at the top. Often the ore contained enough silica to be self-fluxing, i.e., to produce a slag. The amount of time taken for the bloom to consolidate enough to be workable varied enormously, from a few hours to the 45 hours recorded in Africa (Goucher and Herbert 1996). Little work has been done on correlating smelting variables to time, but it seems to have depended on the temperature achieved, the amount of air delivered, the structure of the furnace, and the quality and quantity of the ore. At a minimum, the process took several hours. After the master metalworker judged that the process was complete, the smelting furnace was allowed to cool and the bloom removed.

Eiji Nitta (1991, 1997) excavated a site called Ban Dong Phlong in Buriram Province, northeast Thailand, which held the remains of 17 iron-smelting furnaces dated by ^{14}C to the 1st through the 3rd centuries B.C. Some were destroyed beyond interpretation, but several were the remains of shaft furnaces, some with slag-tapping pits and some without. In one furnace Nitta discovered a number of half-melted iron nodules. The description was not clear, but he appeared to be referring to laterite nodules. He also found tuyeres in a dump pit. It seems that the iron smelters at Ban Dong Phlong were sophisticated enough to use shaft furnaces with slag-tapping capabilities but did not always choose to do so.

FURNACES AND REFRACTORY MATERIALS

Bloomery furnaces and hearths were normally constructed of clay, as were tuyeres and crucibles, though stone blocks could be used. The greatest demand was on furnace walls; they needed to maintain their structure during a high-temperature smelt, but also resist both the weight of the charge—a column of ore and charcoal—and the corrosive effects of liquid slag. The material also needed to have low thermal conductivity; but, as Rostoker and Bronson (1990:57) noted, often the material most capable of structural integrity was not the most thermally nonconductive. Hence the furnace could be built of multiple materials; "the walls of a furnace were often built of a refractory clay or stone, and sand, gravel, and even earth were mounded around the outside" (Rostoker and Bronson 1990:57). These layers of sand or gravel from a collapsed furnace might be archaeologically visible.

LABOR ORGANIZATION IN IRON SMELTING

Ethnographic and historical accounts of iron smelting show that there was a great deal of variability in how labor was organized, but all show that the process involved numerous participants. Ore must be extracted and carried, charcoal must be made, ore must be beneficiated, roasting pens must be erected and tended, the furnace must be constructed, air must be supplied, and the smelt process must be supervised by a master metalworker. Much of this labor could be done by untrained assistants, as in the African accounts where women and children, who were excluded from the actual smelt, made the charcoal and carried out beneficiation. Unlike the copper-smelting process, the quantity of air needed required the use of bellows (or, in rare cases, natural draft) to provide forced air. Historically, in Southeast Asia piston bellows were the dominant forced air delivery system for iron smelting; an ethnographic account of iron smelting among the Shan of eastern Burma in the 1890s described their system of smelting iron in pits in the ground, using 10–12 piston bellows at once (Rostoker and Bronson 1990:76). In this situation, at a minimum, simply providing air would have required five people working at once. Another account of iron smelting in eastern Burma, written in 1890, described a two-man operation using bellows, where one smelt was run per day for four days and the fifth day "reserved for marketing" (Bronson and Charoenwongsa 1986:14).

RITUAL

Ethnographic accounts of African iron smelting (e.g., Schmidt 1996) emphasize the close intertwining of ritual and technology; from the viewpoint of the smelters and other members of the community, the ritual actions and beliefs were as essential to the success of the smelt as the physical actions involved in mining, roasting, and furnace construction. It is likely that this was true in most prehistoric or pre-industrial smelting, so an account of the *chaîne opératoire* of iron smelting that excluded these rituals and ritual actors is of necessity incomplete. It is noteworthy, however, that Bronson and Charoenwongsa reported, in their collection of 19th- and early-20th-century ethnohistorical accounts of early mining and smelting in mainland Southeast Asia (1986), that none of their sources mentioned any particular rituals performed during smelting or smithing, which "may imply that they were less conspicuous and perhaps less essential than in other regions" (Bronson and Charoenwongsa 1986:23).

GENDER

In Africa, whose pre-industrial iron-making and iron-working practices have been extensively investigated, no women were ever smelters or smiths, though the rituals surrounding these practices are laden with female-based symbolism and ritual, and though women might have performed much of the mining and ore preparation (Herbert 1993:27). Murdock's 1937 analysis of 224+ societies around the world concluded that in none of these societies were women the metalworkers, which was the most extreme sexual division of labor found in a long list of activities from weapon making to grain grinding. In contrast to Murdock's findings and practices in Africa, no taboos against the participation of women are found in Bronson and Charoenwongsa's collection of ethnohistorical accounts, and, in fact, among the Lawa of northern Thailand, women did much (one source said all) of the mining and smelting (Bronson and Charoenwongsa 1986:22). Other cases of Thai women engaging in iron forging are cited by Suchitta (1983:29–31).

Smithing the Bloom

The bloom is not workable iron, but instead a rough mass of iron and slag. Once removed from the furnace, it was then reheated to 1000°C–1100°C in the smith's forge and hammered while orange-yellow-hot (Craddock 1995:247) to squeeze out the pasty slag and consolidate the iron. This process reduced the weight of the mass by at least 50% (Rostoker and Bronson 1990:94) and as much as 75% (Crew 1991) as the surface metal oxidized and flaked off as hammerscale. Under prehistoric conditions, inevitably some slag stringers remained in the metal (see TAM 2C, fig. 4.18 for an example). The presence of numerous slag stringers in the metal is evidence that the metal was produced by the bloomery process. Craddock (1995:247) cited the experimental work of Peter Crew (1991) to demonstrate that a bloom produced in a slag-tapping furnace contained considerably less slag than one produced in a non-tapping furnace, substantially reducing the labor required to consolidate the bloom.

To consolidate the bloom required a great deal of labor and fuel, because the bloom needed to be reheated again and again to keep the slag molten. Hitting the bloom when the temperature was below red hot and the slag was solid would shatter the bloom. In Crew's experiments with a bloom from a non-tapping furnace, it took 45 reheatings to convert a bloom to usable iron (Craddock 1995:248). A bloom from a slag-tapping furnace would take less effort to consolidate, as some slag would have been drained off, but the number of reheatings would depend on the smelting conditions. Blooms from the furnace were usually consolidated at the furnace site, though in places such as India blooms were sold in an unworked state. In Africa, broken pieces of bloom could be passed out to individual members of the smelting teams. More often, the metalworkers worked the bloom into a tradable form, called in Europe "bar iron" (Rostoker and Bronson 1990:96).

LABOR ORGANIZATION IN IRON FORGING

The actual consolidation of the bloom, particularly if it was not large, could have been carried out by one person, though it would have been easier to have multiple hammerers. At least one other helper would have been required to work the bellows. Again, this would have needed charcoal, as well as hammers and an anvil.

MAKING IRON PRODUCTS

The same equipment, though not necessarily the same workers, used to consolidate the bloom was used to turn the iron into products. It should be emphasized that, although in some areas the artifact smiths could be the same as the smelters (Bronson and Charoenwongsa 1986:14; Goucher and Herbert 1996), in others, the ironworking system could be divided into two distinct entities with different organizations and degree of craft specialization. Ehrenreich examined the socio-technical systems of pre-Roman iron making and working in Britain and concluded that the iron production community had achieved a high degree of craft specialization, with most iron in Britain coming from sites in three iron-producing regions. Ironworking—the production and repair of actual utensils and objects—operated at a much less elaborate community level, with evidence for small-scale smithing found at most village sites (Ehrenreich 1991b). He suggested that the reason was the high degree of complex knowledge required to produce large quantities of smelted iron, contrasted with the much less specialized knowledge needed to forge and forge-weld iron tools.

There has been much less research on prehistoric smithing than on smelting, even in areas such as Africa that are rich in ethnographic evidence (Childs and Dewey 1996). One study has made it clear that the actions of smiths can carry the same symbolic and ritual weight as those of smelters; "The evidence suggests that the transformative actions of repeated hammering, annealing, folding, and welding at the forge were metaphorically linked to the growth and maturation of individual members of society and to the reproduction of society in general. This is partly indicated by some of the vocabulary associated with iron smithing. The verb kusumga…for example, means both to fabricate a weapon and to carry out a ritual associated with a rite of passage" (Childs and Dewey 1996:154–155).

Iron and Steel

A bloom as it comes from the furnace will contain not only iron and slag, but also varying amounts of carbon from the fuel. Some of the carbon is lost during the forging of the bloom, as the surface is exposed to oxygen, but as a rule, some will be retained in an irregular pattern. An iron with 0.1–1.9% carbon is called a steel. Pure iron from the furnace is relatively soft, tough, and easy to work. It is well suited to being made into nails, armor, and farm tools, but is less suited for cutting tools. If a harder iron was desired, the metal could, like bronze, be cold worked, though with less impressive results. A cold-worked bronze is considerably harder than a cold-worked iron. Instead of cold working, the best way to harden iron is to combine it with carbon, in a process called carburization, and then quench it from above 695°C. How widespread the knowledge of this process was in the prehistoric world, however, is debated. Most iron made was uncarburized and in most carburized steels that are found, no attempt at heat treatment can be seen (Craddock 1995:238). As Craddock noted, however, an uncarburized iron may be softer than steel, but it can be repaired easily by hammering it back into shape and thus may have been preferred for most tools.

As with copper, iron objects are formed either by working (forging) or casting. In prehistory, with the exception of China, almost all iron artifacts were formed by forging. Forging can be done either hot or cold, though hot forging is by far the most common and effective method. One great advantage of iron over bronze is that iron pieces can be welded together by hot forging. Iron when red-hot (700°C–900°C), or better, white-hot (1000°C and higher) is much less resistant to shaping than when cold. The piece to be worked is held in a forge until it glows at the desired color, then is pulled out by tongs onto an anvil and hammered until it cools to the point when hammering has little effect (600°C–700°C). The piece is then thrust into the fire again to repeat the cycle. Because forges are usually open fires, they are oxidizing, so every time the piece is put back into the forge, the surface turns into a layer of iron oxide. This oxide, or hammerscale, is knocked off when the piece is hammered again, the same process that so severely reduces the size of a bloom. As a result, the piece becomes progressively smaller and rougher with each cycle of heating and hammering. Small objects and iron sheet must be shaped by cold forging because they lose heat too rapidly and because they lose metal by oxidation. Cold forging is also used, along with grinding and polishing, to smooth the surface of larger pieces.

Iron artifacts, both ornamental and utilitarian, appear in the Ban Chiang area during the first half of the 1st millennium B.C. Analysis demonstrates that these artifacts were forged, not cast, and came from bloomery iron. Pigott and Marder (1984) suggested that the technology for smelting and working iron developed indigenously in Southeast Asia. Newer evidence suggests that technological knowledge and perhaps raw iron were transmitted from India and probably China, even though the iron was worked using fairly rudimentary skills (Biggs et al. 2013; Pryce 2014).

Summary

This chapter provides a brief introduction to the technical specifics of extracting metal from ore and manufacturing metal products in pre-industrial contexts, with specific reference to the metals and alloys—copper, tin bronze, high-tin bronze, low-carbon iron—used in prehistoric Southeast Asia. It is only by knowing the technological requirements of manipulating metals that one can properly interpret the archaeological remains of metallurgical processes. The intent of this chapter is to establish the foundation of knowledge necessary to understand the analytical details presented in TAM 2B and 2C, not simply as a record of prehistoric technological behavior devoid of cultural significance, but as a way to obtain insight into the socio-technical system—techniques, time needed, the social coordination of labor, and the social meaning of technological activities (Pfaffenberger 1992:502)—that place the technological acts in their social context.

Appendix: Glossary

This glossary contains brief and relatively non-technical definitions of terms used in this volume (TAM 2A). Terms used in Volumes 2B–2D will appear in glossaries in each of those volumes. The definitions focus on ancient metalworking, not modern industrial processes. For instance, beneficiation today can be carried out by drying, flotation, or magnetic separation. In antiquity, only roasting and hand-picking were used, so the definition mentions only those processes.

Added value (used in economic discussions): increase in the *value* of a resource, product, or service as the result of particular processes. For example, in a segmented production system, communities may add value to copper-base metal by casting ingots that can be easily traded.

Adit: a horizontal or nearly horizontal passage driven into a mine from the side of a hill in order to reach the ore or to drain water.

Adoption: the social process whereby a new technological system becomes widespread in a society.

Adze/axes: implements with cutting edges usually on a splayed flat head with a blade in alignment with the axis of the hafting element. All examples found at the four sites are socketed, with both winged and plain, symmetrically splayed blades. Some of these may be tillers such as hoes or digging stick tips; hence the full name for the artifact class is "adze/axe/tillers," but the shortened "adze/axe" is used in this volume. An adze has its blade mounted with blade perpendicular to the haft, and an axe has its blade mounted parallel to the haft. Without a haft (handle), it can be difficult to distinguish a stone or metal adze from a stone or metal axe.

Alloy: a metallic material that is composed of at least two separate metals, such as copper and tin (or a metal and metalloid element, such as arsenic or antimony). In this publication, each constituent must make up at least 2% by weight of the substance. This 2% figure, while perhaps arbitrary, was chosen following Stech (1999) in order to separate accidental metallic admixtures from deliberately produced alloys. Adding even a small percentage of another element can significantly alter the properties of a metal; for example, adding arsenic or tin to copper lowers the melting point and greatly increases the hardness and hardenability of the metal. Steel and cast iron (iron and carbon) are frequently described as alloys, but strictly speaking, this is not true, because carbon is not a metal.

Alluvial placer: in mining, a placer formed by wind or running water, such as a streambed or alluvial fan. cf. *eluvial*

Amorphous: as used in this volume, amorphous artifacts are irregular lumps of metal. Some may once have been part of an object, but most were probably casting splashes or debris resulting from metal smelting or melting; i.e., most amorphous artifacts were not deliberately shaped.

Annealing: heating a metal or alloy to homogenize and recrystallize the grains, softening it for further shaping.

Anthropology of technology: the sub-branch of anthropology that focuses on the social context of technology, technological transmission, and technological practice.

Archaeometallurgy: the study of metals found in archaeological contexts. The study is carried out in the field, through excavation of mines, metal extraction sites, working sites, and sites where the final products were made, used, or discarded; and also in the laboratory, where a variety of techniques are used to examine metal artifacts, ores, crucibles, slag, and the remains of furnaces and other processing debris.

Arsenical copper (sometimes called arsenical bronze): copper alloy containing over 2% weight of arsenic.

Artifact: any usually portable object created directly or indirectly by human actions. In this study, artifacts include broken pieces of products and manufacturing byproducts like casting spillings and slag.

As-cast: a cast artifact that has not been modified by any post-casting treatment such as annealing or hammering. Most as-cast copper-base artifacts have a characteristic microstructure; see *dendrites*.

Bangles: curved lengths of variable cross section with intact examples comprising a closed or nearly closed ring. The curvature and scale are such that the object could have fit around a human body part. Fragmentary examples are arcs of the original circles. This class includes bracelets, anklets, necklaces, and finger rings.

Behavioral chain (life history sequence): the sequence of physical activities that produced a material or artifact, used the material or artifact, and discarded or reused the material or artifact.

Bells: in this study collection, small round rattle-type bells (clapper is loose within a globe), often with a small loop for attachment and a slit in the body. Also termed jingle bells.

Beneficiation: the process of crushing ore and its associated minerals (gangue) usually with stone hammers, hand-picking the rich ore pieces, and discarding the remaining pieces of gangue. The ore may be roasted to increase its friability, reduce the water content, and drive off sulfur. Also called "ore dressing."

Bimetallic: an artifact composed of two metals, such as a spear point with a bronze hilt cast on to a forged iron blade.

Bivalve mold: a two-piece casting mold in stone, ceramic, or metal, often reusable. Pulling the two pieces apart allows the cast product to be removed easily.

Blades: implements whose primary working areas are cutting edges along a longitudinal side parallel to the axis of the hafting element, if present. Cutting edges can be straight or curved.

Bloom: a spongy mass of solid iron and slag produced in a bloomery furnace. To produce usable iron, the slag must be squeezed out of the bloom by many cycles of heating and forging (hammering).

Bloomery furnace: a smelting furnace whose end product is a solid but vesicular mix of iron and slag, the bloom. The bloom contains varying but often low percentages of carbon.

Brass: an alloy of copper and zinc. In antiquity, the zinc content was almost never higher than 28%.

Bronze: in antiquity, the alloy of copper and tin.

Carburization: the process of heating iron in the presence of a carbon-containing material such as charcoal so that the carbon will be absorbed into the surface of the iron and, by diffusion, form a layer of steel with a concentration gradient of carbon.

Cassiterite: SnO_2, the most important ore of tin.

Cast iron: a mixture of iron and 2%–4% carbon. Without further treatment it is very hard and brittle, so must be cast into shape. It is relatively easy to cast because the large amount of carbon lowers the melting temperature considerably below the melting point of pure iron and steel.

Castability: the ability of a molten metal to be cast easily into molds. Good castability means that the metal is fluid enough to fill well and take fine impressions in a mold. Adding lead to bronze greatly increases the castability of the alloy.

Casting: the process of pouring molten metal into a mold and allowing it to cool until solid; also, the object resulting from a casting operation.

Casting debris or casting splashes: the spillage, splashes, and other accidental metal remains of a casting operation.

Casting flash: the small ridge of metal left on a cast artifact where the two pieces of a bivalve mold met.

Chaîne opératoire: the portion of the life history of an artifact that relates to the process of manufacture. The term refers to the sequence of activities used to make an artifact, from raw material acquisition to finishing the final product. This analysis, together with technical analysis of the performance characteristics of the materials, enables the analyst to discern the technical choices—choices among a variety of available materials and processes—made by the maker or makers.

Charcoal: a lightweight black substance that is the result of the incomplete combustion of wood. It is the preferred fuel in bloomery furnaces both because it burns hotter and cleaner than wood and because burning charcoal produces the carbon monoxide essential to the reduction of the ore.

Cold working: the plastic deformation of a metal at a temperature lower than its recrystallization temperature, usually at room temperature. In most metals and alloys, cold work, often by hammering, causes an increase in hardness and tensile strength and a decrease in ductility because the deformation causes dislocations in the atomic planes to accumulate at the edges of the crystals.

Community of practice: a social group that shares know-how, resources, and other practices and activities in manufacturing products. The communities such as villages, workshops, neighborhoods, or guilds usually share a common identity that may crosscut other identity groupings such as lineages within their society. The concept is linked to "learning framework" and "learning network," as these structure how knowledge and know-how are passed along (transmitted) within the community of practice (Wenger 1998).

Community specialization: a particular genre of craft specialization whereby autonomous individual or household-based production units are aggregated within a single community. This community of skilled practitioners produces for unrestricted regional consumption (Costin 1991:8–9).

Comparanda: things which may be compared usually based on one or more similarities.

Complex technological system: in this volume, a system involving suites of interacting *technologies* comprising more than one material and requiring multiple production steps and social cooperation for implementation. The system consists of a set of tightly interacting artifacts, labor organization, raw material procurement strategies, and social factors required to manufacture a particular product. A complex technological system engages at the societal level, not just at the individual level. Adapted from Schiffer (2005).

Complex technology: technologies like the bow and arrow that require know-how of several interacting materials and technologies.

Compositional analysis: (see *Elemental analysis*)

Constraint: a restrictive condition imposed by social forces, material properties, or nature; e.g., a small village population constrains the number of people available for a mining party and, in turn, the amount of ore that can be mined in a given timeframe. Or the monsoons constrain the ability to smelt or melt metal all year round.

Contact metasomatism: a deposit formed by high-temperature magmatic emanations along an igneous contact. Contact metasomatism is a mass change in the composition of rocks in contact with invading magma, from which fluid constituents are carried out to combine with some of the country-rock [host rock] constituents to form a new suite of minerals (definition from the Dictionary of Mining, Mineral, and Related Terms, 1996 U.S. Bureau of Mines).

Conventional Paradigm: in this volume, this phrase refers to 20th century explanations of prehistoric and pre-industrial metallurgical technology and the role of metallurgical technology in ancient societies. This paradigm relies on the assumptions that technology develops in a linear and progressive sequence in all societies, that technological changes primarily result from the search for greater efficiency in performance, that technological change is a prime mover in social change, that levels of technology tend to be associated with levels of social complexity, and that certain technological stages can be reified, such as the "Bronze Age," and so act as a signifier for a host of other social and organizational traits. (See also Pfaffenberger's [1992] "Standard View of Technology.")

Corrosion: deterioration of metal as a result of chemical attack.

Co-smelting: the accidental or deliberate smelting together of a mixture of ores, particularly oxide/carbonate copper ores and copper sulfide ores, resulting in the direct one-step production of copper. In Thailand evidence suggests co-smelting was frequently performed in a crucible.

Critical performance characteristic (cultural/technological): a performance characteristic that is essential for the technology to function in the desired manner.

Crucible: the heat-resistant (refractory) ceramic vessel, either open or closed, used to contain molten metal. Casting crucibles hold metals that are heated until molten in order to cast them. In Thailand, as well as other parts of the world, smelting was also frequently carried out in crucibles rather than in furnaces.

Cultural transmission: (see *Transmission*)

Disseminated: an ore deposit in which the metallic minerals occur as scattered particles in the rock in sufficient quantity to make the deposit worth exploiting, but only with modern mining methods. Some disseminated deposits are very large.

Drawing: the process of pulling a wire through successively smaller holes in a plate (die) of hard material, with the objective of making the wire smaller and smaller in diameter. This technique of wire production has not been documented thus far in prehistoric Southeast Asia.

Dross: the mineral and impurity-rich layer that forms on the surface of a molten metal. It can result both from oxidation of the molten surface and from the rising of impurities to the surface.

Ductility: the ability of a metal or other substance to deform plastically without cracking or breaking. This property can be quantified by bend tests and other measures of elongation, reduction of area, and torsion.

Elemental analysis (compositional analysis): The process of determining which elements make up a material, usually expressed as weight percentage. Elemental analysis can be performed by a variety of techniques and can be qualitative (detecting only whether an element is present); semi-quantitative (detecting a relative or rough percentage); and quantitative (the percentages are expressed as exact numbers, the accuracy of which must be assessed by knowledge of the level of corrosion, the tendency of certain elements to segregate in the matrix, and the strengths and weaknesses of the particular technique).

Eluvial deposit: a placer deposit formed by the disintegration of a rock in place, rather than deposited by running water or wind. cf. *alluvial deposit*

Essentialize: to attribute and assert that a phenomenon has certain characteristics on the basis that the phenomenon has been assigned to a type assumed to have intrinsic essences.

Fabric: the matter of a fired ceramic, including the clay and inclusions; also called paste.

Fines: very finely crushed ore and host rock.

Finished products: the final aim of the production process; the end, usable objects.

Fire-setting: a mining technique where pieces of rock and ore are broken off from the host rock by building a fire against the host rock. The fire heats the moisture in the rock so that it expands, causing chunks of ore/rock to detach. Often water is thrown on the heated host rock to speed the process of breakup.

Flux: a material such as limestone or hematite added to a smelt or refining operation to help remove impurities while lowering the melting point of the slag and making it more free flowing.

Forging: to shape a piece of metal, especially iron, by heating and hammering.

Furnace: a stationary structure, in pre-industrial times often constructed of clay, in which metal is smelted from ore, usually with the addition of charcoal. An enclosed design facilitates control of temperature and oxygen (redox conditions).

Gangue: the non-metallic rock in which ore minerals are embedded. The gangue is discarded during beneficiation and before smelting.

Gossan: an iron-bearing weathered product that caps a copper sulfide deposit. Through exposure to air and water, the sulfide ores are oxidized into oxide and carbonate ores and the sulfur and most metals leach out and are redeposited in deeper zones, leaving behind hydrated iron oxides. The hydrated iron oxides are most often a distinctive red or brown.

Grog: crushed or ground potsherds added to clay as a tempering material.

Haft: the handle of an implement.

Hafting element: The part of an implement that is used to secure or attach an implement to a handle, such as of wood. Sockets and tangs are examples.

Hammerscale: the thin layer of oxidized iron that forms on the surface of a piece of iron while it is being heated in a smithing hearth. The layer is knocked off every time the piece is hammered, so that the iron mass of the piece is lessened slightly with every cycle of heating and forging. Hammerscale is deposited around the hearth, providing archaeologists with an important marker of blacksmithing.

Hammerstones (or mauls): stones used in antiquity to pound ore from mine walls and to crush ores during beneficiation. They have battered ends and in many parts of the world have distinctive hafting grooves or pecking.

Host rock: a body of rock containing other rocks or mineral deposits such as ores.

Hot working: the working of a metal or alloy when the metal is above the temperature of recrystallization. Unlike cold working, hot working does not cause hardening.

Inclusions: non-metallic particles of slag or mineral in the body of a metal.

Innovation: development of a new process or component in a technological system that can affect the manufacturing or use of products. Innovation occurs when preexisting knowledge is recombined in novel ways that result in a new technological configuration.

Interacting technologies: when products having different materials properties and engaging different kinds of expertise must be designed to function together to carry out an objective. Thus, crucibles made of clay plus inclusions and other treatments must have requisite refractory performance characteristics to withstand molten metallic contents of different temperatures depending on the metal. The metal raw material may need to be prepared in particular ways to liquefy effectively with crucibles of specific designs and with particular fuels.

Know-how: practical, experiential knowledge of material behaviors used in a technology and the skills to use that knowledge to accomplish a task; expertise.

Laterite: a reddish-brown residual rock or soil that develops in well-drained humid, subtropical, and tropical areas via a process of intense chemical weathering. Laterites are leached of silica by rainfall and contain concentrations of iron oxides and hydroxides, along with aluminum hydroxides. The concentrations of iron, aluminum, and nickel are sometimes high enough for the soil to serve as an ore of these metals.

Lathe: a machine for shaping a piece of wood or metal by means of a spindle that rotates the piece against a variety of sharp tools.

Lead isotope analysis (LIA): there are four different isotopic forms of lead, and different lead sources and lead-containing ores have different ratios of the four isotopes. Lead isotope analysis can associate the lead in a metal, slag, glass, or glazed artifact with a particular ore source that shares the same isotopic signature as the analyzed sample.

Leaded antimonial copper: a copper-base alloy that contains over 2% by weight of both lead and antimony.

Leaded bronze: an alloy of copper, tin, and over 2% lead. The addition of lead improves the castability of the metal but decreases the hardness of the product.

Learning frameworks: the social frameworks in which knowledge including technological knowledge is acquired and transmitted among persons. These frameworks are sometimes divided into several types: trial and error self-learning, observation and imitation, verbal (oral or written) explanation, and hands-on instruction/practice (Minar and Crown 2001).

Life cycle of a technology: the stages of invention, adoption, application, adaptation to new uses, innovation, and senescence of a particular technology in a particular society.

Life history framework: an interpretive perspective for analyzing archaeological assemblages that incorporates in its purview the entire history of an artifact, from raw material acquisition to product creation, to use, breakage, and discard, and even includes post-depositional events through time in the archaeological record. A contrasting perspective, especially common in art history, might base an interpretation only on intact final products with little to no reference to the full range of conditions, stages, uses, or depositional contexts of material remains from a past society.

Lost wax casting: the process of casting using a wax model. The wax is shaped in the form of the final product and encased in clay. The clay is then heated, allowing the wax to melt and run out, leaving an empty cavity that is then filled with molten metal. Also called *cire perdue* or investment casting.

Malleability: the degree to which a metal can be hammered or worked without cracking.

Manufacturing byproducts: artifacts and features that are created as a byproduct of human productive activity, but are not the intentional result, such as casting drippings, residues, and slag.

Manufacturing equipment: tools and technological features necessary to produce certain products.

Materials properties (properties of materials): the behavioral and other characteristics of physical objects, which can influence their suitability for a particular cultural purpose, e.g., hardness, flexibility, heat resistance, etc.

Matte smelting: a complex reduction and concentration process for smelting low-grade copper sulfide ores. The ores are first roasted to drive off sulfur. The ore is then smelted along with a quartz flux to produce slag. The result is a matte, or solid mixture of copper sulfides and iron sulfides. This matte is broken up and roasted to convert the sulfides to oxides and smelted again to produce an impure copper.

Mauls: (see *Hammerstones*)

Metal production: the mining, smelting, refining and alloying of metallic minerals in order to produce unshaped metal, ready to be made into finished products. Also termed primary metal production and metal extraction.

Metallurgical province: a large region where a network of metalworkers exploited a common set of metallic resources; shared technological know-how, technological systems, and technological styles; and produced related metal products. A province could extend for millions of square kilometers, encompass many archaeological cultures, and last from several hundred years to two or three millennia. A metallurgical province is characterized by morphological similarity of entire artifact categories, in particular those of tools and weapons, as well as uniformity of metallurgical technology. This morphological and technological uniformity is maintained by continuous interactions among skilled practitioners from different societies located within the bounds of the province (adapted from Chernykh 1992:8–9).

Metallurgical tradition: a technological system of metal production and metal product manufacturing that endures over time.

Meteoritic iron: iron that came to the Earth as part of meteorites, as opposed to terrestrial iron. In many places, meteoritic iron was the first iron to be used. Meteoritic iron is almost always high in nickel.

Non-slagging: a copper smelting process that produces little or no slag.

Ore: a naturally occurring deposit of minerals from which one or more desirable components can be extracted. The term refers mostly to sources of metallic minerals.

Orebody: a continuous, bounded mass of rock/ ore that contains enough ore to make extraction economically feasible.

Orogeny: the process by which structures within fold-belt mountainous areas are formed. These processes include thrusting, folding, and faulting in the outer and higher layers, and plastic folding, metamorphism, and plutonism in the inner and deeper layers (definition from the Dictionary of Mining, Mineral, and Related Terms, 1996. U.S. Bureau of Mines).

Oxidation: the process of bringing a material in contact with oxygen, usually oxygen in the air. Also, the process that produces a coating of tarnish or rust on a metal, as the metal combines with oxygen and returns to a lower-energy, more stable state.

Performance characteristic: a behavioral characteristic of an artifact when it is used in an activity, such as flexibility in a bow or heat resistance in a cooking pot, that affects the use of the artifact. Performance characteristics can be mechanically functional (metal sufficiently hard and rigid so that a pointed object made of it can pierce a hide) and/or socially valued, such as the same point with formal modifications that signal the community of practice that manufactured the item.

Performance problem: a performance characteristic that militates against optimal use and may stimulate technological innovations to rectify the perceived problem.

Performance requirement: a critical performance characteristic that is essential for the use of an artifact in a given activity. For example, a crucible must be capable of remaining intact while holding molten metal, but the performance requirements of crucibles for melting lead (melting temperature 327°C) are considerably less stringent than those for melting copper (1083°C).

PIXE: Proton-induced X-ray emission spectroscopy, a method of elemental (compositional) analysis that uses a beam of protons to ionize or eject electrons from inner shells of atoms. Electrons from the outer shells drop down to replace inner shell vacancies, releasing energy in the form of X-rays of the characteristic energy of each element in the sample.

Placer: a deposit of sand and/or gravel that contains small particles of valuable minerals such as gold or tin.

Planishing: the act of finishing the manufacture of a decorative metal artifact such as a bowl by hammering it all over with a light hammer. This hardens the metal and leaves small overlapping facets that reflect light when polished.

Pluton: a body of medium- to coarse-grained igneous rock that formed beneath the earth's surface by crystallization of a magma (definition from the Dictionary of Mining, Mineral, and Related Terms, 1996. U.S. Bureau of Mines).

Points: long, straight objects with a tapered end coming to a point extending from and in alignment with the hafting element. The shape implies that they were used to pierce objects. The cross section of the point can be flattened or round. The hafting elements on complete examples are either sockets or tangs.

Porphyry: an igneous rock of any composition that contains conspicuous large mineral crystals in a fine-grained groundmass.

Post-manufacturing treatments: modifications performed after an artifact is manufactured in order to alter its properties; for example, annealing and working a cast axe to smooth, sharpen, and harden its edge.

Prestige goods: according to Earle's classic definition (1991b:10), prestige goods are either: (1) special craft products produced for elite patrons who controlled the craftspersons and products; or (2) goods acquired from long-distance exchange only accessible by elites. In other words, prestige goods are products whose consumption was restricted to social elites (see also Flad and Hruby 2007:9). cf. *valuables*

Prills: small droplets of metal found trapped in slag or crucible linings after smelting or melting metal.

Primary production of metal: all processes, activities, materials, manufacturing equipment, etc., used to extract metals from ores, including prospecting, mining, ore beneficiation, and smelting.

Producer specialization: a genre of economic specialization where individuals rely on exchanging their products for some or all of their subsistence needs (Flad and Hruby 2007).

Product specialization: a genre of economic specialization where individuals do not rely on exchanging their products for subsistence needs, but nevertheless do produce products (have the expertise, skills, and role) for non-kin use (Flad and Hruby 2007).

Production equipment: tools and physical features needed to produce both raw material and finished objects.

Production sequence: the sequence of steps carried out to produce a material/artifact; *chaîne opératoire*.

Quenching: plunging a red- or white-hot metal into water or oil to cool it quickly. This frequently has the effect of preserving at room temperature phase structures that would normally exist only at high temperatures. The most common examples of this are martensite in high-tin bronzes and in steel.

Recrystallization: by heating a metal or alloy above a certain temperature, a new, strain-free crystal structure is formed after the grains of a metal have been distorted by cold working. The precise recrystallization temperature will vary with the composition and purity of the metal, the length of the heating process, and the amount of prior cold work.

Reducing atmosphere: an atmosphere low in oxygen and high in carbon monoxide. A reducing atmosphere is usually necessary for smelting.

Refractory material: material that retains its strength and integrity at high temperatures, especially ceramic materials that can tolerate high temperatures without melting, usually because they are high in silica and alumina.

Reify: (see *Essentialize*)

Rhyolites: a group of extrusive porphyritic igneous rocks.

Rifting: the process wherein segments of the earth's crust are pulled apart along linear zones. Most rift systems are accompanied by volcanoes. Both significant metallic orebodies and important oil and gas reserves are associated with rifted crust, the ore reserves being created by hydrothermal fluids injected into overlying sedimentary beds by magmatic activity.

Secondary production: manufacturing of metal finished products; all processes, activities, materials, manufacturing equipment, etc., used to create final products from extracted metals.

Shaft: a vertical or steeply inclined passage in a mine. cf. *adit*

Skarn: rock made up primarily of lime-bearing silicates. It originates when nearby intrusive rocks have injected large quantities of silica, aluminum, iron, and magnesium into nearly pure limestones or dolomites. Usually associated with ore deposits including those of copper, lead, zinc, gold, and iron.

Slag: a dense, glassy, or rock-like waste material usually produced during a smelt. Slag is composed of ore minerals, metal oxides, and remains of fluxes and furnace linings. Melting metal in a crucible can also produce slag from residual impurities that adhere to the crucible lining.

Slag stringers: slag inclusions in an iron or steel that have been flattened during forging. While metal will recrystallize upon heating, slag and other minerals will not; hence they will always retain their worked, linear shape.

Slagging process: a smelting process for any metal ore that is based on the deliberate creation of slag as a means to separate the gangue (waste material) from the smelted metal (Craddock 2000:152).

Smelting: the extraction of a metal or alloy from its ore in a furnace or crucible by means of a high-temperature chemical reaction under reducing conditions.

Socio-technical system: the distinctive, interlocking system of techniques, tools, and labor organization that a society uses to produce a set of related artifacts.

Socket: a hollow tube, often circular or oval in cross section, that forms part of an implement such as axes, adzes, blades, or points, and is intended for the insertion of a shaft or haft (handle), usually of wood. Open sockets are open at both ends, so that the haft protrudes from both sides, parallel with the edge of the blade in the case of an adze or axe. An example would be a modern axe. A blind socket ends inside the artifact, so that the socket is perpendicular to the blade in the case of an adze/axe. They are cast using a mold plug, usually of clay.

Solder: an alloy with a low melting temperature used to join two metallic surfaces together.

Spectroscopy: in its broadest sense, spectroscopy refers to the measurement and analysis of spectra that are produced by sending a beam of electromagnetic radiation onto a material sample. All atoms emit light (whether visible or not) when stimulated by such a beam, and this light can be split into spectra. Each element has a characteristic spectral pattern. Spectroscopic techniques include many different methods, all of which measure the absorption or emission of radiant energy on the surface of a sample.

Sprue: the cone-shaped plug of excess metal that fills the funnel through which metal was poured into a mold and remains after the metal has cooled; also, the small funnel-shaped cavity in a mold that allows molten metal to enter the mold.

Steel: a malleable mixture of iron and 0.1–1.9% carbon.

Stockwork: a mineral deposit consisting of a three-dimensional network of veins spaced closely enough so that the entire body can be mined.

Strain lines: microscopic lines caused by the movement of one part of a metallic crystal lattice relative to another part. They are produced by working or other forms of stress.

Suture: the zone created when two pieces of buoyant continental crust collide. Neither plate may subduct under the other; instead, the collision creates a wide fault zone of major deformation, such as a mountain range.

Tailings: the gangue and other refuse material resulting from the beneficiation of an ore. May also include host rock removed during mining.

Tang: the metal projection of a blade or point that is inserted into the handle or haft.

Technical ceramics: ceramics used in the production of metal and metal artifacts, such as crucibles, tuyeres, blowpipe ends, and furnace walls and bases or bottoms.

Technical choice: in this volume, the term is used to refer to specific alternative techniques possible in a production sequence. For example, a metalworker may make a technical choice to cold work a copper-base point to harden the metal.

Technological activities: any human behavior that purposefully modifies the physical world with the objective of providing for human wants and needs. These activities can create artifacts, as well as meanings and social relationships (Pfaffenberger 2001). Technological activities may be considered to be that subset of activities pertinent to manufacturing and maintaining all artifacts in the interacting set of artifacts within a technological system. Archaeologists may find concrete evidence for many technological activities, such as a kiln for the activity of firing pottery, or a cluster of flakes that can be refitted demonstrating the activity of shaping and sharpening stone tools, or a deposit of crushed host rock indicating the task of beneficiation. Technological activities commonly exist within behavioral chains where activities occur in characteristic sequences in manufacturing *chaînes opératoires*.

Technological choice: in this volume, technological choices are the technical choices that are characteristic of a technological system that derive from cultural values and social practices rather than benefits inherent to the technology itself (Lemonnier 1993).

Technological community: a community of practice comprised of individuals who share knowledge and activities in one or more technological systems.

Technological deposits: incidental remains (i.e., not in situ installations) resulting from technological activities, such as slag heaps, mine tailings, or spreads of crushed host rock.

Technological features: discrete immobile installations created for technological activities, such as casting hearths.

Technological style: patterns and preferences for how materials and processes are handled that are specific to a particular culture. The term started out referring to how similar cultural preferences could be demonstrated across more than one technology (Lechtman 1977); Lechtman's example was how the Inca belief that precious materials should be an intrinsic part of an artifact rather than a plating or appliqué influenced both weaving and metallurgy. The term has more recently been used to describe culturally characteristic methods of processing materials in a single technology or a way of handling materials that is characteristic of a particular group or culture (see also Hegmon 1998).

Technological system: the way technological knowledge is implemented in a specific context; "a specific set of procedures, gestures, tools, materials, finished products, and beliefs and attitudes" that produce and use a desired technological product (after Gosselain 2008:152).

Technological tradition: a system of technological knowledge and practice that endures over time.

Technologically modified landscapes: landscapes modified by activities of a complex technological system, including those altered by mining, deforestation from agriculture or fuel production for pyrotechnologies, or reconfiguration of a natural hydrology to provide irrigation.

Technology: the application of knowledge of the material properties of physical things to achieve human purposes.

Temper: (in a ceramic) material such as grog, straw, sand, or chaff added to the clay in a ceramic to modify its properties and performance characteristics.

Tempering: removing some of the hardness and brittleness from a quenched martensitic steel by heating it to between 450°C–650°C.

Terrane: a large (up to hundreds of kilometers wide) section of the earth's crust with a common origin and relatively uniform characteristics (e.g., a block, massif, orogenic belt, or microplate), that has usually been transported a considerable distance via plate tectonics.

Transmission: how knowledge, behaviors, and values are passed from one person to another and from one society to another vertically through time and horizontally through space. The transmission of technology specifically involves the transfer of know-how, or entire systems of mental knowledge and physical practice in the context of learning frameworks.

Tuyere: a nozzle of ceramic or metal inserted through a furnace wall, through which air is forced from a bellows into the fire.

Use-modifications: unintentional changes in materials resulting from use and exposure to technological processes, for example vitrification of ceramic containers/crucibles after having been exposed to very high temperatures from holding molten metal.

Valuables: products recognized as special at least in some contexts that are circulated in social and political transactions (e.g., rites of passage, establishment of alliances) where they serve to articulate and integrate relationships among groups and/or individuals (see Dalton 1977 and Earle 2002:19–42). cf. *prestige goods*

Vesicular: characterized by the presence of small cavities caused by gas bubbles.

Vitrification: the process of becoming glass-like. Because of exposure to high temperatures, the particles within a ceramic or slag fuse and form a homogeneous glass-like body.

Wire/rods: a length of metal, more or less straight or slightly curved, of consistent width or slightly tapered, commonly with either a round or a square cross section. According to the typology used in this volume, wires range from 1–4 mm in thickness and rods are thicker than 4 mm. Context shows that some of the wires were parts of jewelry, but the function of the other examples, both wires and rods, is unclear. Some may have been parts of tools.

Working: shaping an artifact by hammering. In iron, working is called *forging*.

Wrought iron: low carbon iron heated and hammered from a bloom and containing small slag inclusions; this iron can be wrought or forged readily. Wrought metal of any kind is metal that has been shaped by hammering or forging rather than casting.

X-ray fluorescence (XRF): probably the most widely used form of elemental analysis in museums. X-ray fluorescence analyzes a surface by directing a beam of X-rays at the object. After striking the sample, the X-rays are either scattered or absorbed, generating secondary X-rays with energies characteristic of all the elements that have been struck. These secondary X-rays hit a detector, which turns the rays into electrical pulses. The intensity of the pulses varies with the amount of the X-rays of each element, so the analysis can be quantitative. Handheld portable XRF (PXRF) devices are now in common use.

References

Thai Archaeology Monograph Series

TAM 1

Pietrusewsky, M., and M. T. Douglas

2002 *Ban Chiang, A Prehistoric Village Site in Northeast Thailand I: The Human Skeletal Remains.* Thai Archaeology Monograph Series 1. University of Pennsylvania Museum of Archaeology and Anthropology, Philadelphia.

TAM 2A

White, J. C., and E. G. Hamilton (editors)

2018 *Ban Chiang, Northeast Thailand, 2A: Background to the Study of the Metal Remains.* Thai Archaeology Monograph Series 2A. University of Pennsylvania Museum of Archaeology and Anthropology, Philadelphia.

TAM 2B

White, J. C., and E. G. Hamilton (editors)

forthcoming *Ban Chiang, Northeast Thailand, 2B: Metals and Related Evidence from Ban Chiang, Ban Tong, Ban Phak Top, and Don Klang.* Thai Archaeology Monograph Series 2B. University of Pennsylvania Museum of Archaeology and Anthropology, Philadelphia.

TAM 2C

White, J. C., and E. G. Hamilton (editors)

forthcoming *Ban Chiang, Northeast Thailand, 2C: The Metal Remains in Regional Context.* Thai Archaeology Monograph Series 2C. University of Pennsylvania Museum of Archaeology and Anthropology, Philadelphia.

TAM 2D

White, J. C., and E. G. Hamilton (editors)

forthcoming *Ban Chiang, Northeast Thailand, 2D: Catalogs for Metals and Related Remains from Ban Chiang, Ban Tong, Ban Phak Top, and Don Klang.* Thai Archaeology Monograph Series 2D. University of Pennsylvania Museum of Archaeology and Anthropology, Philadelphia.

Adams, R. L., and S. M. King (editors)

2011 *Residential Burial: A Multiregional Exploration.* Archeological Papers of the American Anthropological Association No. 20. American Anthropological Association, Arlington, VA.

Agrawal, D. P.

2000 *Ancient Metal Technology and Archaeology of South Asia: A Pan-Asian Perspective.* Aryan Books International, New Delhi, India.

Alberti, B., S. Fowles, M. Holbraad, Y. Marshall, and C. Witmore

2011 "Worlds Otherwise": Archaeology, Anthropology, and Ontological Difference. *Current Anthropology* 52(6):896–912.

Alden, J. R.

1982 Marketplace Exchange as Indirect Distribution: An Iranian Example. In *Contexts for Prehistoric Exchange*, edited by J. E. Ericson and T. K. Earle, pp. 83–101. Academic Press, New York.

Allan, J. W.

1979 *Persian Metal Technology 700–1300 AD*. Oxford Oriental Monographs No. 2. Ithaca Press for the Faculty of Oriental Studies and the Ashmolean Museum, University of Oxford, London.

Allan, S.

2007 Erlitou and the Formation of Chinese Civilization: Toward a New Paradigm. *Journal of Asian Studies* 66(2):461–496.

Alsop, J.

1975 Rewriting Human History. *Washington Post* 8 September: A26. Washington, DC.

Ames, K. M.

1995 Chiefly Power and Household Production on the Northwest Coast. In *Foundations of Social Inequality*, edited by T. D. Price and G. M. Feinman, pp. 155–188. Fundamental Issues in Archaeology. Plenum Press, New York.

Amzallag, N.

2009 From Metallurgy to Bronze Age Civilizations: The Synthetic Theory. *American Journal of Archaeology* 113:497–519.

2011 A Return to the Dark Ages? Reply to Thornton et al. 2010. *American Journal of Archaeology* 114 (2010):317–329.

Andrews, K., and R. Doonan

2003 *Test Tubes and Trowels: Using Science in Archaeology*. Tempus, Stroud, UK.

Anthony, D. W.

1996 V. G. Childe's World System and the Daggers of the Early Bronze Age. In *Craft Specialization and Social Evolution: In Memory of V. Gordon Childe*, edited by B. Wailes, pp. 47–66. University Museum Symposium Series Vol. VI, University Museum Monograph 93. University of Pennsylvania Museum of Archaeology and Anthropology, Philadelphia.

2007 *The Horse, the Wheel and Language: How Bronze-Age Riders from the Eurasian Steppes Shaped the Modern World*. Princeton University Press, Princeton, NJ.

2010 The Rise and Fall of Old Europe. In *The Lost World of Old Europe: The Danube Valley, 5000–3500 B.C.*, edited by D. W. Anthony, pp. 28–57. Institute for the Study of the Ancient World at New York University, New York, and Princeton University Press, Princeton, NJ.

Appadurai, A. (editor)

1986 *The Social Life of Things: Commodities in Cultural Perspective*. Cambridge University Press, Cambridge, UK.

Arnold, J. E.

2007 Credit Where Credit Is Due: The History of the Chumash Oceangoing Plank Canoe. *American Antiquity* 72(2):196–209.

2011 Detecting Apprentices and Innovators in the Archaeological Record: The Shell Bead-Making Industry of the Channel Islands. *Journal of Archaeological Method and Theory*. Electronic document, http://www.springerlink.com/index/10.1007/s10816-011-9108-1, accessed November 11, 2011.

Artioli, G., I. Angelini, G. Kaufmann, C. Canovaro, G. Dal Sasso, and I. M. Villa

2017 Long-distance Connections in the Copper Age: New Evidence from the Alpine Iceman's Copper Axe. *PLoS ONE* 12(7): e0179263. https://doi.org/10.1371/journal.pone.0179263.

Babu, T. M.

2003 Advent of the Bronze Age in the Indian Subcontinent. In *Mining and Metal Production through the Ages*, edited by P. Craddock and J. Lang, pp. 174–180. British Museum Press, London.

Bachmann, H.-G.

1982 *The Identification of Slags from Archaeological Sites*. Occasional Publication No. 6. Institute of Archaeology. University of London, London.

Bacus, E. A.

2004 The Archaeology of the Philippine Archipelago. In *Southeast Asia: From Prehistory to History*, edited by I. Glover and P. S. Bellwood, pp. 257–281. RoutledgeCurzon, New York.

Bailey, D. W.

2000 *Balkan Prehistory: Exclusion, Incorporation and Identity*. Routledge, London.

Balfet, H.

1991 Des chaines opératoires, pour quoi faire? In *Observer l'action techniques*, edited by H. Balfet, pp. 11–20. Centre national de la recherche scientifique, Paris.

Banton, M. (editor)

1966 *The Social Anthropology of Complex Societies*. ASA Monographs 4. Tavistock Publications, London.

Barber, A. J., M. F. Ridd, and M. J. Crow

2011 The Origin, Movement and Assembly of the Pre-Tertiary Tectonic Units of Thailand. In *The Geology of Thailand*, edited by M. F. Ridd, A. J. Barber, and M. J. Crow, pp. 507–538. The Geological Society, London.

Barber, E. W.

1994 *Women's Work: The First 20,000 Years: Women, Cloth, and Society in Early Times*. W. W. Norton and Company, Inc., New York.

Barnard, N.

1980 The Origins and Development of Metallurgy in China, with Special Reference to the Crucible. In *The Diffusion of Material Culture*, edited by H. H. E. Loofs-Wissowa, pp. 215–263. Asian Pacific Archaeology Series No. 9. The University Press of Hawaii, Honolulu, HI.

Barnes, I. L., W. T. Chase, L. L. Holmes, E. C. Joel, P. Meyers, and E. V. Sayre

1988 The Technical Examination, Lead Isotope Determination, and Elemental Analysis of Some Shang and Zhou Dynasty Bronze Vessels. In *The Beginning of the Use of Metals and Alloys: Papers from the Second International Conference on the Beginning of the Use of Metals and Alloys, Zhengzhou, China, 21–26 October 1986*, edited by R. Maddin, pp. 296–306. MIT Press, Cambridge, MA.

Barnes, J. W.

1988 *Ores and Minerals: Introducing Economic Geology*. Open University Press, Milton Keynes.

Barr, S. M., and P. Charusiri

2011 Volcanic Rocks. In *The Geology of Thailand*, edited by M. F. Ridd, A. J. Barber, and M. J. Crow, pp. 415–439. Geological Society, London.

Barrett, J. C.

2001 Agency, the Duality of Structure, and the Problem of the Archaeological Record. In *Archaeological Theory Today*, edited by I. Hodder, pp. 141–164. Polity Press, Cambridge, UK.

2012 Agency: A Revisionist Account. In *Archaeological Theory Today*, edited by I. Hodder, pp. 146–166. 2nd ed. Polity Press, Cambridge, UK.

Barthelemy, R. E.

1938 Primitive Tin Metallurgy in Laos. *Mining and Metallurgy* 19(377):244–246.

Bar-Yosef, O., and P. Van Peer

2009 The *Chaîne Opératoire* Approach in Middle Paleolithic Archaeology. *Current Anthropology* 50(1):103–131.

Basalla, G.

1988 *The Evolution of Technology*. Cambridge History of Science Series. Cambridge University Press, Cambridge, UK.

Bauvais, S., and P. Fluzin

2009 Archaeological and Archaeometrical Approaches of the *Chaîne Opératoire* in Iron and Steelmaking: Methodology for a Regional Evolutionary Study. In *Techniques and People: Anthropological Perspectives on Technology in the Archaeology of the Proto-Historic and Early Historic Periods in the Southern Levant*, edited by S. A. Rosen and V. Roux, pp. 159–180. Mémoires et travaux du Centre de recherche français à Jérusalem: archéologie et sciences de l'Antiquité et du Moyen-âge 9. De Boccard, Paris.

Bayard, D. T.

1971 *Non Nok Tha: The 1968 Excavation Procedure, Stratigraphy and Summary of the Evidence*. University of Otago Studies in Prehistoric Anthropology Vol. 4. University of Otago, Dunedin, New Zealand.

1972 Early Thai Bronze: Analysis and New Dates. *Science* 176:1411–1412.

1979 The Chronology of Prehistoric Metallurgy in North-East Thailand: *Silabhumi* or *Samrddhabhumi*? In *Early South East Asia: Essays in Archaeology, History and Historical Geography*, edited by R. B. Smith and W. Watson, pp. 15–32. Oxford University Press, New York.

1984a Rank and Wealth at Non Nok Tha: The Mortuary Evidence. In *Southeast Asian Archaeology at the XV Pacific Science Congress; the Origins of Agriculture, Metallurgy, and the State in Mainland Southeast Asia*, edited by D. T. Bayard, pp. 87–122. University of Otago Studies in Prehistoric Anthropology Vol. 16. University of Otago, Dunedin, New Zealand.

1984b A Tentative Regional Phase Chronology for Northeast Thailand. In *Southeast Asian Archaeology at the XV Pacific Science Congress; the Origins of Agriculture, Metallurgy, and the State in Mainland Southeast Asia*, edited by D. T. Bayard, pp. 161–168. University of Otago Studies in Prehistoric Anthropology Vol. 16. University of Otago, Dunedin, New Zealand.

1987 Chronology, Evolution, and Diffusion in the Later Southeast Asian Cultural Sequence: Some Comments on Higham's Recent Revision. *Bulletin of the Indo-Pacific Prehistory Association* 7:118–140.

1996–97 Bones of Contention: The Non Nok Tha Burials and the Chronology and Context of Early Southeast Asian Bronze. In A*ncient Chinese and Southeast Asian Bronze Age Cultures. The Proceedings of a Conference Held at the Edith and Joy London Foundation Property, Kioloa, NSW, 8–12 February 1988,* Vol. II, edited by F. D. Bulbeck and N. Barnard, pp. 889–940. SMC Publishing, Taipei.

Bayard, D. T., and P. Charoenwongsa

1983 The Development of Metallurgy in Southeast Asia: Reply to Loofs-Wissowa. *Journal of Southeast Asian Studies* 14(1):12–17.

Bayard, D. T., and W. G. Solheim, II

2009 Archaeological Excavations at Non Nok Tha, Northeastern Thailand, 1965–1968, Part 1. Issuu. http://www.guampedia.com/archaeological-excavations-at-non-nok-tha-thailand, accessed April 14, 2014.

Bellwood, P. S.

2004 The Origins and Dispersals of Agricultural Communities in Southeast Asia. In *Southeast Asia: From Prehistory to History*, edited by I. Glover and P. Bellwood, pp. 21–40. Routledge-Curzon, New York.

Bellwood, P., M. Oxenham, B. C. Hoang, N. K. Dzung, A. Willis, C. Sarjeant, P. Piper, H. Matsumura, K. Tanaka, N. Beavan-Athfield, T. Higham, N. Q. Manh, and D. N. Kinh

2013 An Son and the Neolithic of Southern Vietnam. *Asian Perspectives (2011)* 50(1–2):144–175, doi:10.1353/asi.2011.0007.

Benedict, B.

1966 Sociological Characteristics of Small Territories and their Implications for Economic Development. In *The Social Anthropology of Complex Societies*, edited by M. Banton, pp. 23–35. ASA Monographs 4. Tavistock Publications, London.

Bennett, A.

2013a The Importance of Iron–Its Development and Complexity in the Southeast Asian Iron Age. In *Unearthing Southeast Asia's Past: Selected Papers from the 12th International Conference of the European Association of Southeast Asian Archaeologists,* Vol. 1, edited by M. J. Klokke and V. Degroot, pp. 107–121. NUS Press, Singapore.

2013b Protohistoric Iron Weapons and Tools from a Burial Site in West Central Thailand. In *The World of Iron*, edited by J. Humphries and T. Rehren, pp. 371–379. Archetype, London.

2015 Manufacture, Use, and Trade of Late Prehistoric Iron Billhooks from Mainland Southeast Asia. In *Metals and Civilizations: Proceedings of the Seventh International Conference on the Beginnings of the Use of Metals and Alloys (BUMA VII)*, edited by S. Srinivasan, S. Ranganathan, and A. Giumlia-Mair, pp. 158–166. National Institute of Advanced Studies, Bangalore, India.

Bentley, R. A., M. W. Hahn, and S. J. Shennan

2004 Random Drift and Culture Change. *Proceedings of the Royal Society B: Biological Sciences* (271):1443–1450.

Bentley, R. A., M. Pietrusewsky, M. T. Douglas, and T. C. Atkinson

2005 Matrilocality during the Prehistoric Transition to Agriculture in Thailand? *Antiquity* 79(306):865–881.

Berg, I.

2007 Meaning in the Making: The Potter's Wheel at Phylakopi, Melos (Greece). *Journal of Anthropological Archaeology* 26(2):234–252.

Berthoud, T., S. Bonnefuss, M. Dechoux, and J. Français

1980 Data Analysis: Towards a Model of Chemical Modification of Copper from Ores to Metal. In *Scientific Studies in Early Mining and Extractive Metal*, edited by P. T. Craddock, pp. 87–102. British Museum Occasional Paper No. 20, British Museum, London.

Betancourt, P. P., and S. C. Ferrence (editors)

2011 *Metallurgy: Understanding How, Learning Why. Studies in Honor of James D. Muhly*. Prehistory Monographs 29. INSTAP Academic Press, Philadelphia.

Bettinger, R. L.

1982 Aboriginal Exchange and Territoriality in Owens Valley, California. In *Contexts for Prehistoric Exchange*, edited by J. E. Ericson and T. K. Earle, pp. 103–127. Academic Press, New York.

Bettinger, R. L., and J. Eerkens

1999 Point Typologies, Cultural Transmission, and the Spread of Bow-and-Arrow Technology in the Prehistoric Great Basin. *American Antiquity* 64(2):231–242.

Biggs, L., B. Bellina, M. Martinón-Torres, and T. O. Pryce

2013 Prehistoric Iron Production Technologies in the Upper Thai-Malay Peninsula: Metallography and Slag Inclusion Analyses of Iron Artefacts from Khao Sam Kaeo and Phu Khao Thong. *Archaeological and Anthropological Sciences* 5:311–329.

Bijker, W. E.

1995a Sociohistorical Technology Studies. In *Handbook of Science and Technology Studies*, edited by S. Jasanoff, G. E. Markle, J. C. Petersen, and T. Pinch, pp. 229–256. Rev. ed. Sage Publications, Thousand Oaks, CA.

1995b *Of Bicycles, Bakelites, and Bulbs*. MIT Press, Cambridge, MA.

Binford, L. R.

1980 Willow Smoke and Dogs' Tails: Hunter-Gatherer Settlement Systems and Archaeological Site Formation. *American Antiquity* 45:4–20.

Blanton, R. E.

1998 Beyond Centralization: Steps Toward a Theory of Egalitarian Behavior in Archaic States. In *Archaic States*, edited by G. Feinman and J. Marcus, pp. 135–172. School of American Research Press, Santa Fe, NM.

Blanton, R. E., G. M. Feinman, S. A. Kowalewski, and P. N. Peregrine

1996 A Dual-Processual Theory for the Evolution of Mesoamerican Civilization. *Current Anthropology* 37(1):1–14.

Bleed, P.

2001 Artifice Constrained: What Determines Technological Choice? In *Anthropological Perspectives on Technology*, edited by M. B. Schiffer. Amerind Foundation New World Studies Series No. 5. University of New Mexico Press, Albuquerque, NM.

Bode, M., A. Hauptmann, and K. Mezger

2009 Tracing Roman Lead Sources Using Lead Isotope Analyses in Conjunction with Archaeological and Epigraphic Evidence—a Case Study from Augustan/Tiberian Germania. *Archaeological and Anthropological Sciences* 1(3):177–194.

Boileau, M.-C.

2005 *Pottery Production and Distribution in Third Millennium Northeast Syria*. Éditions Épistèmes, Paris.

Booth, J., and N. Sattayarak

2011 Subsurface Carboniferous-Cretaceous Geology of Northeast Thailand. In *The Geology of Thailand*, edited by M. F. Ridd, A. J. Barber, and M. J. Crow, pp. 185–222. The Geological Society, London.

Boudin, M., P. Boeck, P. Vandenabeele, and M. Van Strydonck

2013 Improved Radiocarbon Dating for Contaminated Archaeological Bone Collagen, Silk, Wool and Hair Samples via Cross-Flow Nanofiltrated Amino Acids. *Rapid Communications in Mass Spectrometry* 27(18):2039–2050, doi:10.1002/rcm.6652.

Bourdieu, P.

1977 *Outline of a Theory of Practice*. Cambridge Studies in Social Anthropology No. 16. Cambridge University Press, Cambridge, UK.

Bourgarit, D.

2007 Chalcolithic Copper Smelting. In *Metals and Mines: Studies in Archaeometallurgy*, edited by S. La Niece, D. Hook, and P. Craddock, pp. 3–14. Archetype Publications and the British Museum, London.

Bowser, B. J.

2000 From Pottery to Politics: An Ethnoarchaeological Study of Political Factionalism, Ethnicity, and Domestic Pottery Style in the Ecuadorian Amazon. *Journal of Archaeological Method and Theory* 7(3):219–248.

Boyd, R., and P. J. Richerson

1985 *Culture and the Evolutionary Process*. University of Chicago Press, Chicago.

Boyd, W. E., R. J. McGrath, and C. F. W. Higham

1999 The Geoarchaeology of the Prehistoric Ditched Sites of the Upper Mae Nam Mun Valley N.E. Thailand, II: Stratigraphy and Morphological Sections of the Encircling Earthworks. *Indo-Pacific Prehistory Association Bulletin* 18:169–179.

Brock, F., V. Geoghegan, B. Thomas, K. Jurkschat, and T. F. G. Higham

2013 Analysis of Bone "Collagen" Extraction Products for Radiocarbon Dating. *Radiocarbon* 55(2–3):445–463, doi:10.2458/azu_js_rc.55.16304.

Bronk R., C., M. Scott, and H. van der Plicht

2013 Calibration for Archaeological and Environmental Terrestrial Samples in the Time Range 26–50 ka cal BP. *Radiocarbon* 55(4): 2021–2027.

Bronson, B.

1992 Patterns in the Early Southeast Asian Metals Trade. In *Metallurgy, Trade and Urbanism in Early Thailand and Southeast Asia*, edited by I. Glover, P. Suchitta, and J. Villiers, pp. 63–113. White Lotus, Bangkok.

1999 The Transition to Iron in Ancient China. In *The Archaeometallurgy of the Asian Old World*, edited by V. C. Pigott, pp. 177–98. University Museum Symposium Series Vol. VII, University Museum Monograph 89, MASCA Research Papers in Science and Archaeology Vol. 16. University Museum of Archaeology and Anthropology, Philadelphia.

Bronson, B., and P. Charoenwongsa

1986 *Eyewitness Accounts of the Early Mining and Smelting of Metals in Mainland South East Asia*. Thailand Academic Publishing Co., Bangkok.

Bronson, B., and J. White

1992 Southeast Asia. In *Chronologies in Old World Archaeology*, edited by R. W. Ehrich, pp. 491–503 (Vol. 1), pp. 475–515 (Vol. 2). 3rd ed. University of Chicago Press, Chicago.

Brumfiel, E. M., and T. K. Earle

1987 Specialization, Exchange, and Complex Societies: An Introduction. In *Specialization, Exchange, and Complex Societies*, edited by E. M. Brumfiel and T. K. Earle, pp. 1–9. Cambridge University Press, Cambridge, UK.

Budd, P., D. Gale, A. M. Pollard, R. G. Thomas, and P. A. Williams

1992 The Early Development of Metallurgy in the British Isles. *Antiquity* 66(252):677–686.

Budd, P., and T. Taylor

1995 The Faerie Smith Meets the Bronze Industry: Magic Versus Science in the Interpretation of Prehistoric Metal-making. *World Archaeology* 27(1):133–143.

Buffetaut, E., and R. Ingavat

1985 The Mesozoic Vertebrates of Thailand. *Scientific American* 253(2):80–87.

Buffetaut, E., and V. Suteethorn

1998 The Biogeographical Significance of the Mesozoic Vertebrates from Thailand. In *Biogeography and Geological Evolution of SE Asia*, edited by R. Hall and J. D. Holloway, pp. 83–90. Backhuys Publishers, Leiden, the Netherlands.

Bunopas, S., and P. Vella

1992 Geotectonics and Geologic Evolution of Thailand. In *Proceedings of a National Conference on Geologic Resources of Thailand: Potential for Future Development, 17–24 November 1992, Bangkok, Thailand*, edited by C. Piancharoen pp. 209–228. Department of Mineral Resources, Bangkok.

Cai, J. X., X. C. Tang, Q. G. Wei, and B. Li

2008 Comment on "Was the Indosenian Orogeny a Triassic Mountain Building or a Thermotectonic Reactivation Event?" by A. Carter and P. D. Clift [C. R. Geoscience 340 (2008) 83–93]. *Comptes Rendus Geoscience*, 340(12):893–895, doi:10.1016/j.crte.2008.09.004.

Carbonnel, J.-P., and G. Delibrias

1968 Premières datations absolues de trois gisements néolithiques Cambodgiens. *Comptes Rendus de l'Académie des Sciences, Séries D* 267:1432–1434.

Carneiro, R.

1981 The Chiefdom: Precursor of the State. In *The Transition to Statehood in the New World*, edited by G. D. Jones and R. R. Kautz, pp. 37–79. New Directions in Archaeology. Cambridge University Press, Cambridge, UK.

Carriveau, G., and G. Harbottle

1983 Ban Chiang Pottery: Thermoluminescence Dating Problems. *Antiquity* 57(219):56–58.

Cartailhac, É.

1877 L'age de la pierre dans l'Indochine. *Matériaux pour l'Histoire Primitive et Naturelle de l'Homme* 8:98–101.

1890 Les bronzes préhistoriques du Cambodge et les recherches de M. Ludovic Jammes. *L'Anthropologie* 1:641–650.

Castro, P. V., S. Gili, V. Lull, R. Micó, C. Rihuete, R. Risch, and M. E. Sanahuja Yll

1998 Teoría de la producción de la vida social: mecanismos de explotación en el sudeste Ibérico. *Boletín de Antropoligía Americana* 33:25–77.

Cattin, F., I. M. Villa, and M. Besse

2009 Copper Supply during the Final Neolithic at the Saint-Blaise/Bains des Dames Site (Neuchâtel, Switzerland). *Archaeological and Anthropological Sciences* 1(3):161–176.

Cavalli-Sforza, L. L., and M. W. Feldman

1981 *Cultural Transmission and Evolution: A Quantitative Approach.* Princeton University Press, Princeton, NJ.

Cawte, H. J., and W. E. Boyd

2010 Laterite Nodules: A Credible Source of Iron Ore in Iron Age Northeast Thailand? *Geoarchaeology* 25(5):626–644.

Champion, S.

1985 Production and Exchange in Early Iron Age Central Europe. In *Settlement and Society: Aspects of West European Prehistory in the First Millennium B.C.*, edited by T. C. Champion and J. V. S. Megaw, pp. 133–160. St. Martin's Press, New York.

Chang, K. C.

1980 The Chinese Bronze Age: A Modern Synthesis. In *The Great Bronze Age of China: An Exhibition from the People's Republic of China*, edited by W. Fong, pp. 35–50. The Metropolitan Museum of Art, New York.

Chapman, R.

2003 *Archaeologies of Complexity.* Routledge, New York.

Charles, J. A.

1980 The Coming of Copper and Copper-Base Alloys and Iron: A Metallurgical Sequence. In *The Coming of the Age of Iron*, edited by T. A.

Wertime and J. D. Muhly, pp. 151–181. Yale University Press, New Haven, CT.

Chauvet, M. G.

1928 Objets préhistoriques trouvés à Somrong-Sen (Cambodge). *Bulletin Archéologique du Comité des Travaux Historiques et Scientifiques (Année 1927)* 45:291–299.

Chernykh, E. N.

1992 *Ancient Metallurgy in the USSR: The Early Metal Age*. Translated by Sarah Wright. New Studies in Archaeology. Cambridge University Press, Cambridge, UK.

Chernykh, E. N., S. V. Kuz'minykh, and L. B. Orlovskata

2004 Ancient Metallurgy in Northeast Asia: from the Urals to the Saiano-Altai. In *Metallurgy in Ancient Eastern Eurasia from the Urals to the Yellow River*, edited by K. M. Linduff, pp. 15–36. Chinese Studies Vol. 31. The Edwin Mellen Press, Lewiston, NY.

Childe, V. G.

1930 *The Bronze Age*. Cambridge University Press, Cambridge, UK.

1944 Archaeological Ages as Technological Stages. *Journal of the Royal Anthropological Institute* 74:7–24.

1950 The Urban Revolution. *Town Planning Review* 21:3–17.

1951 *Man Makes Himself*. Rev. ed. New American Library, New York.

Childs, S. T.

1998 Social Identity and Craft Specialization among Toro Ironworkers in Western Uganda. In *Craft and Social Identity*, edited by C. L. Costin and R. P. Wright, pp. 109–121. Archeological Papers of the American Anthropological Association No. 8. American Anthropological Association, Arlington, VA, doi:10.1525/ap3a.1998.8.1.109.

Childs, S. T., and W. J. Dewey

1996 Forging Symbolic Meaning in Zaire and Zimbabwe. In *The Culture and Technology of African Iron Production*, edited by P. Schmidt, pp. 145–171. University Press of Florida, Gainesville.

Childs, S. T., and D. Killick

1993 Indigenous African Metallurgy: Nature and Culture. *Annual Review of Anthropology* 22:317–337.

Chilton, E. S.

1998 The Cultural Origins of Technical Choice: Unraveling Algonquian and Iroquoian Ceramic Traditions in the Northeast. In *The Archaeology of Social Boundaries*, edited by M. T. Stark, pp. 132–160. Smithsonian Series in Archaeological Inquiry. Smithsonian Institution Press, Washington, DC.

Ciarla, R.

2007a Rethinking Yuanlongpo: The Case for Technological Links between the Lingnan (PRC) and Central Thailand in the Bronze Age. *East and West* 57(1–4):305–328.

2007b A Preliminary Report of Lo.R.A.P. Archaeological Excavations at Prehistoric Khao Sai On, Lopburi Province, Central Thailand. *East and West* 57(1–4):395–401.

2008 The Thai-Italian 'Lopburi Regional Archaeological Project' (Lo.R.A.P.): Excavation at Khao Sai On-Noen Din 2008 Preliminary Report. *East and West* 58(1–4):313–336.

Clark, G.

1969 *World Prehistory: A New Outline*. 2nd ed. Cambridge University Press, Cambridge, UK.

1977 *World Prehistory in New Perspective*. 3rd ed. Cambridge University Press, Cambridge, UK.

Clark, J. E.

2007 In Craft Specialization's Penumbra: Things, Persons, Action, Value, and Surplus. In *Rethinking Craft Specialization in Complex Societies: Archaeological Analyses of the Social Meaning of Production*, edited by Z. X. Hruby and R. K. Flad, pp. 20–35. Archeological Papers of the American Anthropological Association No. 17. American Anthropological Association, Arlington, VA, doi:10.1525/ap3a.2007.17.1.20.

Clark, J. E., and W. J. Parry

1990 Craft Specialization and Cultural Complexity. *Research in Economic Anthropology* 12:289–346.

Coles, J. M., and A. F. Harding

1979 *The Bronze Age in Europe: An Introduction to the Prehistory of Europe, c. 2000–700 BC.* Methuen, London.

Collet, D. P.

1993 Metaphors and Representations Associated with Precolonial Iron-Smelting in Eastern and Southern Africa. In *The Archaeology of Africa: Food, Metals and Towns*, edited by T. Shaw, P. Sinclair, B. Andah, and A. Okpoko, pp. 499–511. One World Archaeology 20. Routledge, New York.

Comelli, D., M. D'orazio, L. Folco, M. El-Halwagy, T. Frizzi, R. Alberti, V. Capogrosso, A. Elnaggar, H. Hassan, A. Nevin, F. Porcelli, M. G. Rashed, and G. Valentini

2016 The Meteoritic Origin of Tutankhamun's Iron Dagger Blade. *Meteoritics & Planetary Science* 51(7):1301–1309, doi:10.1111/maps.12664.

Condominas, G.

1977 *We Have Eaten the Forest: The Story of a Montagnard Village in the Central Highlands of Vietnam.* (Translation of the 1957 edition in French). Hill and Wang, New York.

Conkey, M. W., and C. A. Hastorf (editors)

1990 *The Uses of Style in Archaeology.* Cambridge University Press, Cambridge, UK.

Connah, G.

2010 *Writing About Archaeology.* Cambridge University Press, Cambridge, UK.

Coote, V. J.

1990 Ancient Base Metal Mines and Mining in Southwest Thailand. In *Southeast Asian Archaeology 1986. Proceedings of the First Conference of the Association of Southeast Asian Archaeologists in Western Europe*, edited by I. C. Glover and E. Glover, pp. 131–138. BAR International Series 561. British Archaeological Reports, Oxford.

Cordy, R. H.

1974 Complex Rank Cultural Systems in the Hawaiian Islands: Suggested Explanations for Their Origins. *Archaeology and Physical Anthropology in Oceania* 9(2):89–109.

Corre, A.

1879a Rapport sur de nouvelles recherches relatives à l'âge de la pierre polie et du bronze en Indo-chine. *Cochinchine Française: Excursions et Reconnaissances* 1:167–190.

1879b Rapport sur les objets de l'âge de la pierre polie et du bronze recueillis a Som-ron-sen (Cambodge). *Cochinchine Française: Excursions et Reconnaissances* 1:84–116.

Costin, C. L.

1991 Craft Specialization: Issues in Defining, Documenting, and Explaining the Organization of Production. In *Archaeological Method and Theory*, Vol. 3, edited by M. B. Schiffer, pp. 1–56. University of Arizona Press, Tucson.

1998 Introduction: Craft and Social Identity. In *Craft and Social Identity*, edited by C. L. Costin and R. P. Wright, pp. 3–16. Archeological Papers of the American Anthropological Association No. 8. American Anthropological Association, Arlington, VA, doi:10.1525/ap3a.1998.8.1.3.

2001 Craft Production Systems. In *Archaeology at the Millennium: A Sourcebook*, edited by G. M. Feinman and T. D. Price, pp. 273–327. Kluwer Academic/Plenum, New York.

2005 Craft Production. In *Handbook of Archaeological Methods*, edited by H. D. G. Maschner and C. Chippindale, pp. 1034–1107. AltaMira Press, Lanham, MD.

2007 Thinking about Production: Phenomenological Classification and Lexical Semantics. In *Rethinking Craft Specialization in Complex Societies: Archaeological Analyses of the Social Meaning of Production*, edited by Z. X. Hruby and R. K. Flad, pp. 143–162. Archeological Papers of the American Anthropological Association No. 17. American Anthropological Association, Arlington, VA, doi:10.1525/ap3a.2007.17.1.143.

Costin, C. L., and R. P. Wright (editors)

1998 *Craft and Social Identity.* Archeological Papers

of the American Anthropological Association No. 8. American Anthropological Association, Arlington, VA.

Courcier, A.

2007 La métallurgie dans les pays du Caucase au Chalcolithique et au début de l'âge du bronze: bilan des études et perspectives nouvelles. In *Les cultures du Caucase (VIe– IIIe millénaires avant notre ère): Leurs relations avec le Proche-Orient*, edited by B. Lyonnet, pp. 199–232. CNRS/Éditions Recherche sur les Civilisations, Paris.

Craddock, P.

1979 The Copper Alloys of the Medieval Islamic World-Inheritors of the Classical Tradition. *World Archaeology* 11(1):68–79.

1989 The Scientific Investigation of Early Mining and Smelting. In *Scientific Analysis in Archaeology*, edited by J. Henderson, pp. 178–212. Oxford University Committee for Archaeology, Monograph No. 19, UCLA Institute of Archaeology, Archaeological Research Tools 5, Oxford.

1995 *Early Metal Mining and Production*. Smithsonian Institution, Washington, DC.

2000 From Hearth to Furnace: Evidences for the Earliest Metal Smelting Technologies in the Eastern Mediterranean. *Paléorient* 26:151–165.

2007 Introduction: Achievements and Challenges. In *Metals and Mines: Studies in Archaeometallurgy*, edited by S. La Niece, D. Hook, and P. Craddock, pp. ix–xii. Archetype Publications and the British Museum, London.

Craddock, P., and J. Lang (editors)

2003 *Mining and Metal Production through the Ages*. British Museum Press, London.

Craddock, P., N. Meeks, and S. Timberlake

2007 On the Edge of Success: The Scientific Examination of the Products of the Early Mines Research Group Smelting Experiments. In *Metals and Mines: Studies in Archaeometallurgy*, edited by S. La Niece, D. Hook, and P. Craddock, pp. 37–45. Archetype Publications and the British Museum, London.

Crew, P.

1991 The Experimental Production of Prehistoric Bar Iron. *Journal of the Historical Metallurgy Society* 25(2):187–204.

Cross, J. R.

1993 Craft Specialization in Nonstratified Societies. *Research in Economic Anthropology* 14:61–84.

Crow, M. J., and K. Zaw

2011 Metalliferous Minerals. In *The Geology of Thailand*, edited by M. F. Ridd, A. J. Barber, and M. J. Crow, pp. 459–492. The Geological Society, London.

Crown, P. L.

2001 Learning to Make Pottery in the Prehispanic American Southwest. *Journal of Anthropological Research* 57(4):451–469.

Crumley, C. L.

1995 Heterarchy and the Analysis of Complex Societies. In *Heterarchy and the Analysis of Complex Societies*, edited by R. M. Ehrenreich, C. L. Crumley, and J. E. Levy, pp. 1–5. Archeological Papers of the American Anthropological Association No. 6. American Anthropological Association, Arlington, VA, doi:10.1525/ap3a.1995.6.1.1.

Dalton, G.

1977 Aboriginal Economics in Stateless Societies. In *Exchange Systems in Prehistory*, edited by T. K. Earle and J. E. Ericson, pp. 191–212. Academic Press, New York.

Damm, H.

2008 Physical Orientation Map of North-East-Thailand, Khorat Plateau. Digital image, https://en.wikipedia.org/wiki/File:Isaanmountains.svg, accessed May 31, 2012.

Daniel, G.

1965 Editorial. *Antiquity* 39:241–246, doi:10.1017/S0003598X00039600.

Darvill, T.

2002 The Three-Age System. In *The Concise Oxford Dictionary of Archaeology*. 2nd ed., p. 430. Oxford University Press, London.

Davey, C. J.

1979 Some Ancient Near Eastern Pot Bellows. *Levant* XI:101–111.

David, N.

2012 Ricardo in the Mandara Mountains: Iron, Comparative Advantage and Specialization. In *Metals in the Mandara Mountains Society and Culture*, edited by N. David, pp. 115–169. Africa World Press, Trenton.

Davidson, J. H. C. S.

1975 Recent Archaeological Activity in Viêt-Nam. *Journal of the Hong Kong Archaeological Society* 6:80–99.

1979 Archaeology in Northern Viêt-Nam since 1954. In *Early South East Asia: Essays in Archaeology, History, and Historical Geography,* edited by R. B. Smith and W. Watson, pp. 98–124. Oxford University Press, New York.

de Barros, P.

2000 Iron Metallurgy: Sociocultural Context. In *Ancient African Metallurgy: The Socio-Cultural Context*, edited by J. O. Vogel, pp. 147–198. AltaMira Press, Walnut Creek, CA.

Delibrias, G., M. Guillier, and J. Labeyrie

1965 Saclay Natural Radiocarbon Measurements II. *Radiocarbon* 7:236–244.

Demeter, F., E. Peyre, and Y. Coppens

2002 The Prehistoric Human Remains of Som Ron Sen, Cambodia. *Bulletin of the Indo-Pacific Prehistory Association* 22:135–140.

Denaire, A.

2009 Radiocarbon Dating of the Western European Neolithic: Comparison of the Dates on Bones and Dates on Charcoals. *Radiocarbon* 51(2):657–674.

Derry, D. R.

1980 *World Atlas of Geology and Mineral Deposits.* Halstead Press, London. John Wiley & Sons, New York.

Dietler, M., and I. Herbich

1998 *Habitus*, Techniques, Style: An Integrated Approach to the Social Understanding of Material Culture and Boundaries. In *The Archaeology of Social Boundaries*, edited by M. Stark, pp. 232–263. Smithsonian Series in Archaeological Inquiry. Smithsonian Institution Press, Washington, DC.

Dobres, M.-A.

2000 *Technology and Social Agency: Outlining a Practice Theory for Archaeology.* Blackwell, Malden, MA.

2001 Meaning in the Making: Agency and the Social Embodiment of Technology and Art. In *Anthropological Perspectives on Technology*, edited by M. B. Schiffer, pp. 47–76. Amerind Foundation New World Studies Series No. 5. University of New Mexico Press, Albuquerque, NM.

Dobres, M.-A., and C. R. Hoffman

1994 Social Agency and the Dynamics of Prehistoric Technology. *Journal of Archaeological Method and Theory* 1(3):211–258.

1999 *The Social Dynamics of Technology: Practice, Politics and World Views.* Smithsonian Institution Press, Washington, DC.

Dolfini, A.

2014 Early Metallurgy in the Central Mediterranean. In *Archaeometallurgy in Global Perspective: Methods and Syntheses*, edited by B. W. Roberts and C. P. Thornton, pp. 423–446. Springer, New York.

Domett, K. M., and N. Tayles

2006 Adult Fracture Patterns in Prehistoric Thailand: A Biocultural Interpretation. *International Journal of Osteoarchaeology* 16(3):185–199.

Domett, K., J. Newton, A. Colbert, N. Change, and S. Halcrow

2016 Frail, Foreign or Favored? A Contextualized Case Study from Bronze Age Northeast Thailand. In *The Routledge Handbook of Bioarchaeology in Southeast Asia and the Pacific Islands,* edited by M. Oxenham and H. R. Buckley, pp. 68–94. Routledge, New York.

Doonan, R., B. Hanks, D. Zdanovich, E. Kupriyanova, D. Pitman, N. Batanina, and J. Johnson

2014 Metals, Society, and Economy in the Late

Prehistoric Eurasian Steppe. In *Archaeometallurgy in Global Perspective: Methods and Syntheses*, edited by B. W. Roberts and C. P. Thornton, pp. 755–784. Springer, New York.

Douglas, I.

2005 The Mekong River Basin. In *The Physical Geography of Southeast Asia*, edited by A. Gupta, pp. 193–218. Oxford Regional Environments. Oxford University Press, New York.

Dunn, F. L.

1966 Radiocarbon Dating of the Malayan Neolithic. *Proceedings of the Prehistoric Society* 32:352–353.

1970 Cultural Evolution in the Late Pleistocene and Holocene of Southeast Asia. *American Anthropologist* 72(5):1041–1054.

Dunnell, R. C.

1988 The Concept of Progress in Cultural Evolution. In *Evolutionary Progress?*, edited by M. H. Nitecki, pp. 169–194. The University of Chicago Press, Chicago.

Earle, T. K.

1981 Comment on P. Rice, Evolution of Specialized Pottery Production: A Trial Model. *Current Anthropology* 22(3):230–231.

1991b The Evolution of Chiefdoms. In *Chiefdoms: Power, Economy and Ideology*, edited by T. Earle, pp. 1–15. Cambridge University Press, Cambridge, UK.

1991c Property Rights and the Evolution of Chiefdoms. In *Chiefdoms: Power, Economy and Ideology*, edited by T. Earle, pp. 71–99. Cambridge University Press, Cambridge, UK.

1994 Wealth Finance in the Inka Empire. *American Antiquity* 59:433–460.

1997 *How Chiefs Come to Power: The Political Economy in Prehistory*. Stanford University Press, Stanford, CA.

2002 *Bronze Age Economics: The Beginnings of Political Economies*. Westview, Boulder, CO.

2008 Cultural Anthropology and Archaeology: Theoretical Dialogues. In *Handbook of Archaeological Theories*, edited by R. A. Bentley, H. D. G. Maschner, and C. Chippindale, pp. 187–202. AltaMira Press, Lanham, MD.

Earle, T. (editor)

1991a *Chiefdoms: Power, Economy and Ideology*. Cambridge University Press, Cambridge, UK.

Earle, T. K., and J. E. Ericson

1977 *Exchange Systems in Prehistory*. Academic Press, New York.

Earle, T., and K. Kristiansen (editors)

2010 *Organizing Bronze Age Societies: The Mediterranean, Central Europe, and Scandinavia Compared*. Cambridge University Press, New York.

Eerkens, J. W.

2000 Practice Makes Within 5% of Perfect: Visual Perception, Motor Skills, and Memory in Artifact Variation. *Current Anthropology* 41:663–668.

Eerkens, J. W., and C. P. Lipo

2005 Cultural Transmission, Copying Errors, and the Generation of Variation in Material Culture and the Archaeological Record. *Journal of Anthropological Archaeology* 24:316–334.

2007 Cultural Transmission Theory and the Archaeological Record: Providing Context to Understanding Variation and Temporal Changes in Material Culture. *Journal of Archaeological Research* 15:239–274.

Ehrenreich, R. M.

1991a Introduction. In *Metals in Society: Theory Beyond Analysis*, edited by R. M. Ehrenreich, pp. 5–6. MASCA Research Papers in Science and Archaeology Vol. 8, pt. 2. University of Pennsylvania Museum of Archaeology and Anthropology, Philadelphia.

1991b Metalworking in Iron Age Britain: Hierarchy or Heterarchy? In *Metals in Society: Theory Beyond Analysis*, edited by R. M. Ehrenreich, pp. 69–80. MASCA Research Papers in Science and Archaeology Vol. 8, pt. 2. University of Pennsylvania Museum of Archaeology and Anthropology, Philadelphia.

1995 Early Metalworking: A Heterarchical Analysis of Industrial Organization. In *Heterarchy and the Analysis of Complex Societies*, edited by R. M. Ehrenreich, C. L. Crumley, and J. E. Levy, pp. 33–39. Archeological Papers of the

American Anthropological Association No. 6. American Anthropological Association, Arlington, VA, doi:10.1525/ap3a.1995.6.1.33.

Ehrenreich, R. M., C. L. Crumley, and J. E. Levy (editors)

1995 *Heterarchy and the Analysis of Complex Societies*. Archeological Papers of the American Anthropological Association No. 6. American Anthropological Association, Arlington, VA.

Ehrhardt, K. L.

2005 *European Metals in Native Hands: Rethinking Technological Change 1640–1683*. University of Alabama Press, Tuscaloosa.

2009 Copper Working Technologies, Contexts of Use, and Social Complexity in the Eastern Woodlands of Native North America. *Journal of World Prehistory* 22(3):213–235.

2014 Copper Working Technologies, Contexts of Use, and Social Complexity in the Eastern Woodlands of Native North America. In *Archaeometallurgy in Global Perspective: Methods and Syntheses*, edited by B. W. Roberts and C. P. Thornton, pp. 303–328. Springer, New York.

Epstein, S. M.

1993 Cultural Choice and Technological Consequences: Constraint of Innovation in the Late Prehistoric Copper Industry on Cerro Huaringa, Peru. Ph.D. dissertation, Department of Anthropology, University of Pennsylvania.

Ericson, J. E., and T. K. Earle (editors)

1982 *Contexts for Prehistoric Exchange*. Academic Press, New York.

Evans, R. K.

1978 Early Craft Specialization: An Example from the Balkan Chalcolithic. In *Social Archeology: Beyond Subsistence and Dating*, edited by C. Redman, M. J. Berman, E. V. Curtin, W. T. Langhorne, Jr., N. M. Versaggi, and J. C. Wanser, pp. 113–129. Studies in Archeology. Academic Press, New York.

Eyre, C. O.

2006 Prehistoric and Proto-Historic Communities in the Eastern Upper Chao Phraya Valley, Thailand: Analysis of Site Chronology, Settlement Patterns, and Land Use. Ph.D. dissertation, Department of Anthropology, University of Pennsylvania, Philadelphia.

2011 Social Variation and Dynamics in Metal Age and Protohistoric Central Thailand: A Regional Perspective. *Asian Perspectives (2010)* 49(1):43–84.

Fagan, B. M.

2008 *World Prehistory: A Brief Introduction*, 7th ed. Pearson, New York.

Fasnacht, W.

2009 7000 Years of Trial and Error in Copper Metallurgy—in One Experimental Life. In *Metals and Societies: Studies in Honour of Barbara S. Ottaway*, edited by T. L. Kienlin and B. W. Roberts, pp. 395–399. Universitätsforschungen zur Prähistorischen Archäologie Bd. 169. Dr. Rudolf Habelt, Bonn.

Feinman, G. M.

1995 The Emergence of Inequality: a Focus on Strategies and Processes. In *Foundations of Social Inequality*, edited by T. D. Price and G. M. Feinman, pp. 255–280. Fundamental Issues in Archaeology. Plenum Press, New York.

Feinman, G. M., and J. Marcus

1998 *Archaic States*. 1st ed. School of American Research Press, Santa Fe, NM.

Feinman, G., and J. Neitzel

1984 Too Many Types: An Overview of Sedentary Prestate Societies in the Americas. In *Advances in Archaeological Method and Theory*, Vol. 7, edited by M. B. Schiffer, pp. 39–102. Academic Press, New York.

Feldman, M. H.

2006 *Diplomacy by Design. Luxury Arts and an "International Style" in the Ancient Near East, 1400–1200 BCE*. University of Chicago Press, Chicago.

Ferrence, S. C.

2011 Introduction. In *Metallurgy: Understanding How, Learning Why: Studies in Honor of James D. Muhly*, edited by P. P. Betancourt and S. C. Ferrence, pp. xxxiii–xxxv. Prehistory Monographs 29. INSTAP Academic Press, Philadelphia.

Fiedel, S., J. R. Southon, R. E. Taylor, Y. V. Kuzmin, M. Street, T. F. G. Higham, J. van der Plicht, M. J. Nadeau, and S. Nawalade-Chavan

2013 Assessment of Interlaboratory Pretreatment Protocols by Radiocarbon Dating an Elk Bone Found Below Laacher see Tephra At Miesenheim IV (Rhineland, Germany). *Radiocarbon* 55:1443–1453, doi:10.02458/azu_js_rc.55.16239.

Finkbeiner, U., M. Novák, and M. Lebeau

2015 Introduction. In *ARCANE: Associated Regional Chronologies for the Ancient Near East and the Eastern Mediterranean. Vol. IV: Middle Eurphrates*, edited by U. Finkbeiner, M. Novák, F. Sakal, and P. Sconzo, pp. 1–16. Brepols, Turnhout, Belgium.

Flad, R.

2007 Rethinking the Context of Production through an Archaeological Study of Ancient Salt Production in the Sichuan Basin, China. In *Rethinking Craft Specialization in Complex Societies: Archaeological Analyses of the Social Meaning of Production*, edited by Z. X. Hruby and R. K. Flad, pp. 108–128. Archeological Papers of the American Anthropological Association No. 17. American Anthropological Association, Arlington, VA, doi:10.1525/ap3a.2007.17.1.108.

Flad, R. K., and Z. X. Hruby

2007 "Specialized" Production in Archaeological Contexts: Rethinking Specialization, the Social Value of Products, and the Practice of Production. In *Rethinking Craft Specialization in Complex Societies: Archaeological Analyses of the Social Meaning of Production*, edited by R. K. Flad and Z. X. Hruby, pp. 1–19. Archeological Papers of the American Anthropological Association, No. 17. American Anthropological Association, Arlington, VA, doi:10.1525/ap3a.2007.17.1.1.

Flanagan, J. G.

1989 Hierarchy in Simple "Egalitarian" Societies. *Annual Review of Anthropology* 18:245–266.

Fowler, C.

2013 *The Emergent Past: A Relational Realist Archaeology of Early Bronze Age Mortuary Practices.* Oxford University Press, Oxford.

Frachetti, M. D.

2009 Differentiated Landscapes and Non-Uniform Complexity among Bronze Age Societies of the Eurasian Steppe. In *Social Complexity in Prehistoric Eurasia: Monuments, Metals, and Mobility*, edited by B. K. Hanks and K. M. Linduff, pp. 19–46. Cambridge University Press, New York.

2012 Multiregional Emergence of Mobile Pastoralism and Nonuniform Institutional Complexity across Eurasia. *Current Anthropology* 53(1):2–38.

Franklin, U. M.

1983 On Bronze and Other Metals in Early China. In *Origins of Chinese Civilization*, edited by D. Keightley, pp. 279–296. University of California Press, Berkeley.

Fried, M. H.

1967 *The Evolution of Political Society: An Essay in Political Anthropology*. Random House, New York.

Fuchs, E.

1882 Station préhistorique de Son-Ron-Sen, au Cambodge, son âge. *Matériaux pour l'Histoire Primitive et Naturelle de l'Homme* 13:356–365.

Gale, N. H., and Z. A. Stos-Gale

1989 Bronze Age Archaeometallurgy of the Mediterranean: the Impact of Lead-Isotope Studies. In *Archaeological Chemistry IV*, edited by R. O. Allan, pp. 159–198. Advances in Chemistry Series 220. American Chemical Society, Washington, DC.

1992 Lead Isotope Studies in the Aegean (the British Academy Project). In *New Developments in Archaeological Science*, edited by A. M. Pollard, pp. 63–108. Proceedings of the British Academy 77. Oxford University Press, Oxford.

Gale, N. H., Z. A. Stos-Gale, A. Raduncheva, I. Panayotov, I. Ivanov, P. Lilov, and T. Todorov

2003 Early Metallurgy in Bulgaria. In *Mining and Metal Production Through the Ages*, edited by P.

T. Craddock and J. Lang, pp. 122–173. British Museum Press, London.

Gallie, W. B.

1955 Essentially Contested Concepts. *Proceedings of the Aristotelian Society* 56 (1955–1956):167–198.

Gardner, L. S.

1967 Antimony Deposits of Thailand. Report of Investigations, *Department of Mineral Resources No. 13,* Bangkok, as cited in Bronson (1992).

Garshol, L. M.

2004 Metadata? Thesauri? Taxonomies? Topic Maps! *Journal of Information Science* 30(4):378–391, https://doi.org/10.1177/0165551504045856.

Gatinsky, Y. G., A. V. Mischina, I. V. Vinogradov, and A. A. Kovalev

1978 The Main Metallogenic Belts of Southeast Asia as the Result of Different Geodynamic Conditions Interference. In *Proceedings of the Third Regional Conference on Geology and Mineral Resources of Southeast Asia*, edited by P. Nutalaya, pp. 313–321. Asian Institute of Technology, Bangkok.

Geselowitz, M. N.

1993 Archaeology and the Social Study of Technological Innovation. *Science, Technology, & Human Values* 18(2):231–246.

Gettens, R., R. S. Clarke, Jr., and W. T. Chase

1971 *Two Early Chinese Bronze Weapons with Meteoritic Iron Blades.* Freer Gallery of Art Occasional Papers Vol. 4, no. 1. Smithsonian Institution, Washington, DC.

Gibson, B.

1996 Death of a Salesman: Childe's Itinerant Craftsman in the Light of Present Knowledge of Late Prehistoric European Craft Production. In *Craft Specialization and Social Evolution: In Memory of V. Gordon Childe*, edited by B. Wailes, pp. 107–119. University Museum Symposium Series Vol. VI, University Museum Monograph 93. University Pennsylvania Museum of Archaeology and Anthropology, Philadelphia.

Gillespie, R., R. Wood, S. Fallon, T. W. Stafford, and J. Southon.

2014 New ^{14}C Dates for Spring Creek and Mowbray Swamp Megafauna: XAD-2 Processing. *Archaeology in Oceania* 50(1):43–48, doi:10.1002/arco.5045.

Gillieson, D.

2005 Karst in Southeast Asia. In *The Physical Geography of Southeast Asia*, edited by A. Gupta, pp. 157–176. Oxford Regional Environments. Oxford University Press, New York.

Gilmour, B.

2009 Specialization in Iron- and Steel-Making in the Early Middle East and Central Asia: Myths, Assumptions and a Reassessment of Early Manuscript Evidence. In *Metallurgy and Civilisation: Eurasia and Beyond: Proceedings of the Sixth International Conference on the Beginnings of the Use of Metals and Alloys (BUMA VI)*, edited by J. Mei and T. Rehren, pp. 134–144. Archetype, London.

Glover, I. C., and B. Bellina

2011 Ban Don Ta Phet and Khao Sam Kaeo: The Earliest Indian Contacts Reassessed. In *Early Interactions between South and Southeast Asia: Reflections on Cross-Cultural Exchange*, edited by P.-Y. Manguin, A. Mani, and G. Wade, pp. 17–46. Nalanda-Sriwijaya Series, Institute of Southeast Asian Publishing, Singapore.

Glover, I. C., and A. Bennett

2012 The High-Tin Bronzes of Thailand. In *Scientific Research on Ancient Asian Metallurgy: Proceedings of the Fifth Forbes Symposium at the Freer Gallery of Art*, edited by P. Jett, B. E. McCarthy, and J. G. Douglas, pp. 101–114. Archetype Publications, London, and the Freer Gallery of Art, Smithsonian Institution, Washington, DC.

Glover, I. C., and B. Syme

1993 The Bronze Age in Southeast Asia: Its Recognition, Dating, and Recent Research. *Man and Environment* 18(2):41–74.

Glumac, P. D.

1991 The Advent of Metallurgy in Prehistoric

Southeast Europe. Ph.D. dissertation, Department of Anthropology, University of California at Berkeley, Berkeley.

Gnecco, C., and C. Langebaek

2014 Introduction: Against Typological Tyranny. In *Against Typological Tyranny in Archaeology: A South American Perspective*, edited by C. Gnecco and C. Langebaek, pp. v–x. Springer, New York.

Golden, J.

2009 New Light on the Development of Chalcolithic Metal Technology in the Southern Levant. *Journal of World Prehistory* 22(3):283–300.

2010 *Dawn of the Metal Age: Technology and Society During the Levantine Chalcolithic*. Equinox Publishing, London.

2014 Who Dunnit? New Clues Concerning the Development of Chalcolithic Metal Technology in the Southern Levant. In *Archaeometallurgy in Global Perspective: Methods and Syntheses*, edited by B. W. Roberts and C. P. Thornton, pp. 559–578. Springer, New York.

Goloubew, V.

1929 L'âge du bronze au Tonkin et dans le Nord-Annam. *Bulletin de l'École Française d'Extrême-Orient* 29:1–46.

1932 Sur l'origine et la diffusion des tambours métalliques. In *Praehistorica Asiae Orientalis, École Française d'Extreme-Orient*, pp. 137–150. Imprimerie d'Extreme Orient, Hanoi.

1937 *L' âge du bronze au Tonkin et des fouilles de Dong-son*. École Française d'Extrême-Orient, Paris.

Gömöri, J.

1988 The Earliest Use of Metals in Hungary. In *The Beginning of the Use of Metals and Alloys: Papers from the Second International Conference on the Beginning of the Use of Metals and Alloys, Zhengzhou, China, 21–26 October 1986*, edited by R. Maddin, pp. 80–91. MIT Press, Cambridge, MA.

Goodway, M., and H. C. Conklin

1987 Quenched High-tin Bronzes from the Philippines. *Archaeomaterials* 2(1):1–27.

Goody, J.

1971 *Technology, Tradition, and the State in Africa*. Oxford University Press, London.

Gorman, C. F.

1969 Hoabinhian: A Pebble-Tool Complex with Early Plant Associations in Southeast Asia. *Science* 163(3868):671–673.

Gorman, C., and P. Charoenwongsa

1976 Ban Chiang: A Mosaic of Impressions from the First Two Years. *Expedition* 8(4):14–26.

Gosselain, O. P.

1992 Technology and Style: Potters and Pottery among the Bafia of Cameroon. *Man* 27(3):559–586.

2000 Materializing Identities: An African Perspective. *Journal of Archaeological Method and Theory* 7(3):187–217.

2008 Mother Bella Was Not a Bella. In *Cultural Transmission and Material Culture: Breaking Down Boundaries,* edited by M. T. Stark, B. J. Bowser, and L. Horne, pp. 150–177. University of Arizona Press, Tucson.

2011 Fine if I Do, Fine if I Don't: Dynamics of Technical Knowledge in Sub-Saharan Africa. In *Investigating Archaeological Cultures: Material Culture, Variability, and Transmission*, edited by B. W. Roberts and M. Vander Linden, pp. 211–227. Springer, New York.

Goucher, C. L., and E. W. Herbert

1996 The Blooms of Banjeli: Technology and Gender in West African Iron Making. In *The Culture and Technology of African Iron Production*, edited by P. Schmidt, pp. 40–57. University Press of Florida, Gainesville.

Gould, R. A.

2001 From Sail to Steam at Sea in the Late Nineteenth Century. In *Anthropological Perspectives on Technology*, edited by M. B. Schiffer, pp. 193–213. Amerind Foundation New World Studies Series No. 5. University of New Mexico Press, Albuquerque, NM.

Graves, M. W.

1994 Kalinga Social and Material Culture

Boundaries: A Case of Spatial Convergence. In *Kalinga Ethnoarchaeology*, edited by W. A. Longacre and J. M. Skibo, pp. 13–49. Smithsonian Institution Press: Washington, DC.

Grayson, D. K.

1983 *The Establishment of Human Antiquity*. Academic Press, New York.

Gregg, S. A. (editor)

1991 *Between Bands and States*. Occasional Paper/Center for Archaeological Investigation No. 9. Center for Archaeological Investigations, Southern Illinois University at Carbondale, Carbondale, IL.

Grotzinger, J., T. H. Jordan, F. Press, and R. Siever

2007 *Understanding Earth*. 5th ed. W. H. Freeman, New York.

Guedes, J. d'A.

2011 Millets, Rice, Social Complexity, and the Spread of Agriculture to the Chengdu Plain and Southwest China. *Rice* 4(3–4):104–113.

Gullapalli, P.

2009 Early Metal in South India: Copper and Iron in Megalithic Contexts. *Journal of World Prehistory* 22(4):439–459.

Gupta, A.

2005 Landforms of Southeast Asia. In *The Physical Geography of Southeast Asia*, edited by A. Gupta, pp. 38–64. Oxford Regional Environments. Oxford University Press, New York.

Gupta, A., and S. C. Liew

2007 The Mekong from Satellite Imagery: A Quick Look at a Large River. *Geomorphology* 85(3–4):259–274.

Hall, R.

1998 Palaeozoic and Mesozoic Geological Evolution of the Southeast Asian Region: Multidisciplinary Constraints and Implications for Biogeography. In *Biogeography and Geological Evolution of SE Asia*, edited by R. Hall and J. D. Holloway, pp. 99–131. Backhuys Publishers, Leiden, the Netherlands.

Hally, D. J.

1986 The Identification of Vessel Function: A Case Study from Northwest Georgia. *American Antiquity* 51(2):267–295.

Hamilton, E.

1991 Metallurgical Analysis and the Bronze Age of Bohemia, or, Are Cultural Alloys Real? *Archaeomaterials* 5:75–89.

1996 *Technology and Social Change in Belgic Gaul: Copper Working at the Titelberg, Luxembourg, 125 B.C.–A.D. 300*. MASCA Research Papers in Science and Archaeology Vol. 13. Museum Applied Science Center for Archaeology, University of Pennsylvania Museum of Archaeology and Anthropology, Philadelphia.

Han, R., and H. Duan

2009 An Early Iron-using Centre in the Ancient Jin State Region (8th–3rd Century B.C.). In *Metallurgy and Civilisation: Eurasia and Beyond: Proceedings of the Sixth International Conference on the Beginnings of the Use of Metals and Alloys (BUMA VI)*, edited by J. Mei and T. Rehren, pp. 99–106. Archetype, London.

Hanks, B., and R. Doonan

2009 From Scale to Practice: A New Agenda for the Study of Early Metallurgy on the Eurasian Steppe. *Journal of World Prehistory* 22(4):329–356.

Hårdh, B., L. Larsson, D. Olausson, and R. Petre

1988 *Trade and Exchange in Prehistory: Studies in Honour of Berta Stjemquist*. Acta Archaeologica Lundensia 16, no. 8. Lunds Universitets Historika Museu, Lund, Sweden.

Hauptmann, A.

2007 *The Archaeometallurgy of Copper: Evidence from Faynan, Jordan*. Springer, Berlin.

Hayden, B.

1995 Pathways to Power: Principles for Creating Socioeconomic Inequalities. In *Foundations of Social Inequality*, edited by T. D. Price and G. M. Feinman, pp. 15–86. Fundamental Issues in Archaeology. Plenum Press, New York.

1998 Practical and Prestige Technologies: The

Evolution of Material Systems. *Journal of Archaeological Method and Theory* 5(1):1–55.

2001a The Dynamics of Wealth and Poverty in the Transegalitarian Societies of Southeast Asia. *Antiquity* 75(289):571–581.

2001b Richman, Poorman, Beggarman, Chief: The Dynamics of Social Inequality. In *Archaeology at the Millennium: A Sourcebook*, edited by G. M. Feinman and T. D. Price, pp. 231–272. Kluwer Academic/Plenum, New York.

2013 Traditional Corporate Group Economics in Southeast Asia: An Ethnographic Study with Archaeological Implications. *Asian Perspectives (2011)* 50 (1–2):1–23.

Hedges, R. E. M., P. B. Pettitt, C. Bronk Ramsey, and G. J. van Klinken

1997 Radiocarbon Dates from the Oxford AMS System: Archaeometry Datelist 23. *Archaeometry* 39:247–262, doi:10.1111/j.1475-4754.1997.tb00803.x.

Hedges, R. E. M., C. Tiemei, and R. A. Housley

1992 Results and Methods in the Radiocarbon Dating of Pottery. *Radiocarbon* 34(3):906–915.

Heeb, J.

2009 Thinking Through Technology—An Experimental Approach to the Copper Axes from Southeastern Europe. In *Metals and Societies: Studies in Honour of Barbara S. Ottaway*, edited by T. L. Kienlin and B. W. Roberts, pp. 415–420. Universitätsforschungen zur Prähistorischen Archäologie Bd. 169. Dr. Rudolf Habelt, Bonn.

Heeb, J., and B. S. Ottaway

2014 Experimental Archaeometallurgy. In *Archaeometallurgy in Global Perspective: Methods and Syntheses*, edited by B. W. Roberts and C. P. Thornton, pp. 161–192. Springer, New York.

Hegmon, M.

1992 Archaeological Research on Style. *Annual Review of Anthropology* 21:517–536.

1998 Technology, Style, and Social Practices: Archaeological Approaches. In *The Archaeology of Social Boundaries*, edited by M. T. Stark, pp. 265–279. Smithsonian Series in Archaeological Inquiry. Smithsonian Institution Press, Washington, DC.

Heine-Geldern, R.

1935 The Archaeology and Art of Sumatra. *Weiner Beiträge zur Kulturgeschichte und Linguistik* 3:305–331.

1937 L'art prébouddhique de la Chine et de l'Asie du Sud-Est et son influence en Océanie. *Revue des Arts Asiatiques* 11(4):177–206.

1946 Research on Southeast Asia; Problems and Suggestions. *American Anthropologist* 48(2):149–175.

Henderson, J.

2000 *The Science and Archaeology of Materials: An Investigation of Inorganic Materials*. Routledge, London and New York.

Heng, S.

2007 A Study of Polished Stone Tools from Samrong Sen, Cambodia: the French Museum Collections. Master's thesis, Department of Prehistory, Museum National d'Histoire Naturelle, Paris.

Henrich, J.

2001 Cultural Transmission and the Diffusion of Innovations: Adoption Dynamics Indicate that Biased Cultural Transmission is the Predominate Force in Behavioral Change. *American Anthropologist* 103(4):992–1013.

2004 Demography and Cultural Evolution: How Adaptive Cultural Processes Can Produce Maladaptive Losses—the Tasmanian Case. *American Antiquity* 69(2):197–214.

Herbert, E. W.

1984 *Red Gold of Africa: Copper in Precolonial History and Culture*. University of Wisconsin Press, Madison.

1993 *Iron, Gender, and Power*. Indiana University Press, Bloomington, IN.

Herbich, I., and M. Dietler

2008 The Long Arm of the Mother-in-Law. In *Cultural Transmission and Material Culture*, edited by M. T. Stark, B. J. Bowser, and L. Horne, pp. 223–244. University of Arizona Press, Tucson.

Higham, C. F. W.

1984a The Ban Chiang Culture in Wider Perspective. *Proceedings of the British Academy* 69:229–261.

1984b The Social Structure of the Ban Na Di Prehistoric Population. In *Southeast Asian Archaeology at the XV Pacific Science Congress. The Origins of Agriculture, Metallurgy, and the State in Mainland Southeast Asia*, edited by D. T. Bayard, pp. 72–86. University of Otago Studies in Prehistoric Anthropology Vol. 16. University of Otago, Dunedin, New Zealand.

1987 Chronology, Evolution and Diffusion in the Later Southeast Asian Cultural Sequence: Further Comments. *Bulletin of the Indo-Pacific Prehistory Association* 7:141–147.

1988 Ban Chiang and Charcoal in Hypothetical Hindsight: A Comment. *Bulletin of the Indo-Pacific Prehistory Association* 8:75–78.

1989 *The Archaeology of Mainland Southeast Asia: From 10,000 B.C. to the Fall of Angkor*. Cambridge University Press, Cambridge, UK.

1996a *The Bronze Age of Southeast Asia*. Cambridge University Press, Cambridge, UK.

1996b Archaeology and Linguistics in Southeast Asia: Implications of the Austric Hypothesis. *Bulletin of the Indo-Pacific Prehistory Association* 14:110–118.

1996–97 The Social and Chronological Contexts of Early Bronze Working in Southeast Asia. In *Ancient Chinese and Southeast Asian Bronze Age Cultures. The Proceedings of a Conference Held at the Edith and Joy London Foundation Property, Kioloa, NSW, 8–12 February, 1988,* Vol. II, edited by F. D. Bulbeck and N. Barnard, pp. 821–888. SMC Publishing, Taipei.

2002 *Early Cultures of Mainland Southeast Asia*. River Books, Bangkok.

2004 Mainland Southeast Asia from the Neolithic to the Iron Age. In *Southeast Asia: From Prehistory to History*, edited by I. Glover and P. Bellwood, pp. 41–67. RoutledgeCurzon, New York.

2009 The Excavation of Square Y. In *The Origins of the Civilization of Angkor. Volume Three: The Excavation of Ban Non Wat, Part One: Introduction*, edited by C. F. W. Higham and A. Kijngam, pp. 79–138. The Fine Arts Department of Thailand, Bangkok.

2011a The Bronze Age of Southeast Asia: New Insight on Social Change from Ban Non Wat. *Cambridge Archaeological Journal* 21(3):365–389.

2011b The Prehistory of Southeast Asia: A Retrospective View of 40 Years of Research. *Antiquity* 85:639–653.

2012 Conclusions. In *The Origins of the Civilization of Angkor. Volume Five: The Excavation of Ban Non Wat, Part Three: The Bronze Age*, edited by C. F. W. Higham and A. Kijngam, pp. 505–536. The Fine Arts Department of Thailand, Bangkok.

2014 *Early Mainland Southeast Asia: From First Humans to Angkor*. River Books, Bangkok.

2015a Debating a Great Site: Ban Non Wat and the Wider Prehistory of Southeast Asia. *Antiquity* 89(347):1211–1220.

2015b Social Change during the Bronze and Iron Ages of Northeast Thailand. In *Advancing Southeast Asian Archaeology 2013: Selected Papers from the first SEAMEO SPAFA Conference on Southeast Asian Archaeology*, edited by N. H. Tan, pp. 270–278, 329–331. SEAMEO SPAFA Regional Center for Archaeology and Fine Arts, Bangkok.

2015c Reply to Responses. *Antiquity* 89(347): 1233–1237.

Higham, C. F. W., K. Douka, and T. F. G. Higham

2015 A New Chronology for the Bronze Age of Northeastern Thailand and its Implications for Southeast Asian Prehistory. *PLoS ONE* 10(9):e0137542, doi:10.1371/journal.pone.0137542.

Higham, C., and T. Higham

2009a The Chronology of Ban Non Wat. In *The Origins of the Civilization of Angkor. Volume Three: The Excavation of Ban Non Wat, Part One: Introduction*, edited by C. F. W. Higham and A. Kijngam, pp. 17–25. The Fine Arts Department of Thailand, Bangkok.

2009b A New Chronological Framework for Prehistoric Southeast Asia, Based on a Bayesian Model from Ban Non Wat. *Antiquity* 83(319):125–144.

2012 A New Chronological Framework for Ban Chiang, Northeast Thailand. In *Crossing Borders: Selected Papers from the 13th International Conference of the European Association of*

Southeast Asian Archaeologists, Vol. 1, edited by M. L. Tjoa-Bonatz, A. Reinecke, and D. Bonatz, pp. 164–170. NUS Press, Singapore.

Higham, C., T. Higham, R. Ciarla, K. Douka, A. Kijngam, and F. Rispoli

2011a The Origins of the Bronze Age in Southeast Asia. *Journal of World Prehistory* 24:227–274, doi:10.1007/s10963-011-9054-6.

Higham, C. F. W., T. F. G. Higham, and K. Douka

2014 The Chronology and Status of Non Nok Tha, Northeast Thailand. *Journal of Indo-Pacific Archaeology* 34:61–75.

Higham, C., T. Higham, and A. Kijngam

2011b Cutting a Gordian Knot: The Bronze Age of Southeast Asia: Origins, Timing, and Impact. *Antiquity* 85(328):583–598, http://dx.doi.org/10.1017/S0003598X00067971.

Higham, C. F. W., and A. Kijngam (editors)

2012a *The Origins of the Civilization of Angkor. Volume Five: The Excavation of Ban Non Wat, Part Three: The Bronze Age.* The Fine Arts Department of Thailand, Bangkok.

2012b *The Origins of the Civilization of Angkor. Volume Six: The Excavation of Ban Non Wat, Part Four: The Iron Age, Summary and Conclusions.* The Fine Arts Department of Thailand, Bangkok.

Higham, C. F. W., Y. V. Kuzmin, and G. S. Burr

2010 The AMS ^{14}C Dating of Iron Age Rice Chaff Ceramic Temper from Ban Non Wat, Thailand: First Results and Its Interpretation. *Nuclear Instruments and Methods in Physics Research B* 268:1022–1025.

Higham, C. F., and F. Rispoli

2014 The Mun Valley and Central Thailand in Prehistory: Integrating Two Cultural Sequences. *Open Archaeology* 1:2–28, doi:10.2478/opar-2014-0002.

Higham, C. F. W., and R. Thosarat

1998 *Prehistoric Thailand: From Early Settlement to Sukhothai.* River Books, Bangkok.

2012 *Early Thailand: From Prehistory to Sukhothai.* River Books, Bangkok.

Higham, C. F. W., R. Thosarat, and D. O'Reilly

2004 Conclusions. In *The Origins of the Civilization of Angkor. Volume One: The Excavation of Ban Lum Khao*, edited by C. F. W. Higham and R. Thosarat, pp. 325–333. The Fine Arts Department of Thailand, Bangkok.

Higham, C. F. W., and W. Wiriyaromp

2010a Neolithic Phase One. In *The Origins of Angkor. Volume Four: The Excavation of Ban Non Wat, Part Two: The Neolithic Occupation*, edited by C. F. W. Higham and A. Kijngam, pp. 15–64. The Fine Arts Department of Thailand, Bangkok.

2010b Neolithic 1 Ceramic Motifs. In *The Origins of Angkor. Volume Four: The Excavation of Ban Non Wat, Part Two: The Neolithic Occupation*, edited by C. F. W. Higham and A. Kijngam, pp. 93–106. The Fine Arts Department of Thailand, Bangkok.

Higham, T., and K. Douka

2015 Response to Joyce White's comment on our paper. *PLoS ONE*, doi:10.1371/journal.pone.0137542.

Hirth, K. G.

2009b Housework and Domestic Craft Production: An Introduction. In *Housework: Craft Production and Domestic Economy in Ancient Mesoamerica*, edited by K. G. Hirth, pp. 1–12. Archeological Papers of the American Anthropological Association No. 19. American Anthropological Association, Arlington, VA, doi:10.1111/j.1551-8248.2009.01009.x.

2009c Craft Production, Household Diversification, and Domestic Economy in Prehispanic Mesoamerica. In *Housework: Craft Production and Domestic Economy in Ancient Mesoamerica*, edited by K. G. Hirth, pp. 13–32. Archeological Papers of the American Anthropological Association No. 19. American Anthropological Association, Arlington, VA, doi:10.1111/j.1551-8248.2009.01010.x.

Hirth, K. G. (editor)

2009a *Housework: Craft Production and Domestic Economy in Ancient Mesoamerica.* Archeological Papers of the American Anthropological

Association No. 19. American Anthropological Association, Arlington, VA.

Hodder, I.

2012b *Entangled: An Archaeology of the Relationships Between Humans and Things*. Wiley-Blackwell, Malden, MA.

2012c Introduction: Contemporary Theoretical Debate in Archaeology. In *Archaeological Theory Today*, edited by I. Hodder, pp. 1–14. 2nd ed. Polity Press, Cambridge, UK.

Hodder, I. (editor)

2012a *Archaeological Theory Today*. 2nd ed. Polity Press, Cambridge, UK.

Hodder, I., and C. Cessford

2004 Daily Practice and Social Memory at Çatalhöyük. *American Antiquity* 69:17–40.

Hodges, H.

1976 *Artifacts: An Introduction to Early Materials and Technology*. Humanities Press and John Baker, New Jersey and London.

Hoffman, B. C., and H. M. L. Miller

2009 Production and Consumption of Copper-base Metals in the Indus Civilization. *Journal of World Prehistory* 22(3):237–264.

2014 Production and Consumption of Copper-base Metals in the Indus Civilization. In *Archaeometallurgy in Global Perspective: Methods and Syntheses*, edited by B. W. Roberts and C. P. Thornton, pp. 697–727. Springer, New York.

Högbom, B.

1913–14 Contributions to the Geology and Morphology of Siam. *Bulletin of the Geological Institute* 12:65–127.

Holl, A. F. C.

2009 Early West African Metallurgies: New Data and Old Orthodoxy. *Journal of World Prehistory* 22(4):415–438.

Hosler, D.

1988 The Metallurgy of Ancient West Mexico. In *The Beginning of the Use of Metals and Alloys: Papers from the Second International Conference on the Beginning of the Use of Metals and Alloys Zhengzhou, China, 21–26 October 1986*, edited by R. Maddin, pp. 328–343. MIT Press, Cambridge, MA.

1994 *The Sounds and Colors of Power*. MIT Press, Cambridge, MA.

2009 West Mexican Metallurgy: Revisited and Revised. *Journal of World Prehistory* 22(3):185–212.

2014 Mesoamerican Metallurgy: The Perspective from the West. In *Archaeometallurgy in Global Perspective: Methods and Syntheses*, edited by B. W. Roberts and C. P. Thornton, pp. 329–359. Springer, New York.

Hruby, Z. X.

2007 Ritualized Chipped-Stone Production at Piedra Negras, Guatemala. In *Rethinking Craft Specialization in Complex Societies: Archaeological Analyses of the Social Meaning of Production*, edited by Z. X. Hruby and R. K. Flad, pp. 68–87. Archeological Papers of the American Anthropological Association No. 17. American Anthropological Association, Arlington, VA.

Hruby, Z. X., and R. K. Flad (editors)

2007 *Rethinking Craft Specialization in Complex Societies: Archaeological Analyses of the Social Meaning of Production*. Archeological Papers of the American Anthropological Association No. 17. American Anthropological Association, Arlington, VA.

Hua, J.

2000 On the Origin of Metallurgy in China. In *The Beginnings of Metallurgy in China*, edited by K. M. Linduff, H. Rubin, and S. Shuyun, pp. 51–61. Chinese Studies Vol. 11. The Edwin Mellen Press, Lewiston, NY.

Hughes, T. P.

1987 The Evolution of Large Technical Systems. In *The Social Construction of Technological Systems: New Directions in the Sociology and History of Technology*, edited by W. E. Bijker, T. P. Hughes, and T. Pinch, pp. 51–82. MIT Press, Cambridge, MA.

Hummel, C. L., and P. Phawandon

1967 *Geology and Mineral Deposits of the Phuket*

Mining District, South Thailand. Report of Investigation No. 5, Thai Department of Mineral Resources.

Hurst, B. and B. Lawn

1984 University of Pennsylvania Radiocarbon Dates XXII. *Radiocarbon* 26(2):212–240.

Hutchison, C. S.

1983 *Economic Deposits and Their Tectonic Setting.* John Wiley & Sons, Inc., New York.

1989 *Geological Evolution of South-East Asia.* Clarendon Press, Oxford.

2005 The Geological Framework. In *The Physical Geography of Southeast Asia*, edited by A. Gupta, pp. 3–23. Oxford Regional Environments. Oxford University Press, New York.

Hutchison, C. S., and D. Taylor

1978 Metallogenesis in SE Asia. *Journal of the Geological Society* 135(4):407–428.

Ingold, T.

1990 Society, Nature and the Concept of Technology. *Archaeological Review from Cambridge* 9:5–17.

Inomata, T.

2007 Knowledge and Belief in Artistic Production by Classic Maya Elites. In *Rethinking Craft Specialization in Complex Societies: Archaeological Analyses of the Social Meaning of Production*, edited by Z. X. Hruby and R. K. Flad, pp. 129–141. Archeological Papers of the American Anthropological Association No. 17. American Anthropological Association, Arlington, VA.

Jaang, L.

2015 The Landscape of China's Participation in the Bronze Age Eurasian Network. *Journal of World Prehistory* 28(3):179–213.

Jackson, C.

2009 Experimental Archaeology and Education: Theory Without Practice is Empty; Practice Without Theory is Blind. In *Metals and Societies: Studies in Honour of Barbara S. Ottaway*, edited by T. L. Kienlin and B. W. Roberts, pp. 400–406. Universitätsforschungen zur Prähistorischen Archäologie Bd. 169, Dr. Rudolf Habelt, Bonn.

Janse, O. R. T.

1947 *Archaeological Research in Indo-China.* Harvard University Press, Cambridge, MA.

Jantzen, D., U. Brinker, J. Heinemeier, J. Piek, K. Hauenstein, J. Kruger, G. Lidke, H. Lubke, R. Lampe, S. Lorenz, M. Schult, and T. Terberger

2011 A Bronze Age Battlefield? Weapons and Trauma in the Tollense Valley, North-Eastern Germany. *Antiquity* 85 (328):417–433.

Janusek, J. W.

1999 Craft and Local Power: Embedded Specialization in Tiwanaku Cities. *Latin American Antiquity* 10(2):107–131.

Janz, L., J. K. Feathers, and G. S. Burr

2015 Dating Surface Assemblage Using Pottery and Eggshell: Assessing and Luminescence Techniques in Northeast Asia. *Journal of Archaeological Science* 57:119–129.

Janzon, G.

1988 Early Nonferrous Metallurgy in Sweden. In *The Beginning of the Use of Metals and Alloys: Papers from the Second International Conference on the Beginning of the Use of Metals and Alloys, Zhengzhou, China, 21–26 October 1986*, edited by R. Maddin, pp. 104–117. MIT Press, Cambridge, MA.

John, D. A., R. A. Ayuso, M. D. Barton, R. J. Blakely, R. J. Bodnar, J. H. Dilles, F. Gray, F. T. Graybeal, J. C. Mars, D. K. McPhee, R. R. Seal, R. D. Taylor, and P. G. Vikre

2010 Porphyry Copper Deposit Model, Chap. B of Mineral Deposit Models for Resource Assessment: U.S. Geological Survey Scientific Investigations Report 2010–5070–B. Electronic document, http://pubs.usgs.gov/sir/2010/5070/b/, accessed April 5, 2012.

Johnson, A. W., and T. Earle

2000 *The Evolution of Human Societies: From Foraging Group to Agrarian State.* 2nd ed. Stanford University Press, Stanford, CA.

Jones, A.

2002 *Archaeological Theory and Scientific Practice. Topics in Contemporary Archaeology.* Cambridge University Press, Cambridge, UK.

Jones, M., H. Hunt, E. Lightfoot, D. Lister, X. Liu, and G. Motuzaite-Matuzeviciute

2011 Food Globalization in Prehistory. *World Archaeology* 43(4):665–675.

Jordan, P., and S. Shennan

2003 Cultural Transmission, Language, and Basketry Traditions amongst the California Indians. *Journal of Anthropological Archaeology* 22(1):42–74.

Jovanović, B.

1988 Early Metallurgy in Yugoslavia. In *The Beginning of the Use of Metals and Alloys*, edited by R. Maddin, pp. 69–79. MIT Press, Cambridge, MA.

Juleff, G.

1998 *Early Iron and Steel in Sri Lanka: a Study of the Samanalawewa Area.* P. von Zabern, München, Germany.

Jull, A. J. T., G. S. Burr, and G. W. L. Hodgins

2013 Radiocarbon Dating, Reservoir Effects, and Calibration. *Quaternary International* 299:64–71.

Jungyusuk, N., and S. Khositanont

1992 Volcanic Rocks and Associated Mineralization in Thailand. In *Proceedings of a National Conference on Geologic Resources of Thailand: Potential for Future Development, 17–24 November 1992, Bangkok, Thailand*, edited by C. Piancharoen, pp. 522–538. Thai Department of Mineral Resources, Bangkok.

Källén, A.

2004 *And through Flows the River: Archaeology and the Pasts of Lao Pako.* Studies in Global Archaeology 6. Uppsala University, Uppsala, Sweden.

Kamvong, T., and K. Zaw

2009 The Origin and Evolution of Skarn-Forming Fluids from the Phu Lon Deposit, Northern Loei Fold Belt, Thailand: Evidence from Fluid Inclusion and Sulfur Isotope Studies. *Journal of Asian Earth Sciences* 34:624–633.

Kamvong, T., K. Zaw, S. Meffre, R. Maas, H. Stein, and C.-K. Lai

2014 Adakites in the Truong Son and Loci Fold Belts, Thailand and Laos: Genesis and Implications for Geodynamics and Metallogeny. *Gondwana Research* 26:165–184.

Karlgren, B.

1942 The Date of the Early Dong-So'n Culture. *Bulletin of the Museum of Far Eastern Antiquities* 14:1–28.

Keller, C. M.

2001 Thought and Production: Insights of the Practitioner. In *Anthropological Perspectives on Technology*, edited by M. B. Schiffer, pp. 33–45. Amerind Foundation New World Studies Series No. 5. University of New Mexico Press, Albuquerque, NM.

Kelly, R.

1991 Sedentism, Sociopolitical Inequality, and Resource Fluctuations. In *Between Bands and States*, edited by S. A. Gregg, pp. 135–158. Center for Archaeological Investigations, Occasional Paper No. 9. Southern Illinois University at Carbondale, Carbondale, IL.

Kenoyer, J. M., and H. M. L. Miller

1999 Metal Technologies of the Indus Valley Tradition in Pakistan and Western India. In *The Archaeometallurgy of the Asian Old World*, edited by V. C. Pigott, pp. 107–151. University Museum Symposium Series Vol. VII, University Museum Monograph 89, MASCA Research Papers in Science and Archaeology Vol. 16. University of Pennsylvania Museum of Archaeology and Anthropology, Philadelphia.

Keswani, P. S.

2005 Death, Prestige, and Copper in Bronze Age Cyprus. *American Journal of Archaeology* 109:341–401.

Keyes, C. F.

1977 *The Golden Peninsula: Culture and Adaptation*

in Mainland Southeast Asia. Macmillan, New York.

Kienlin, T. L.

2014 Aspects of Metalworking and Society from the Black Sea to the Baltic Sea from the Fifth to the Second Millennium BC. In *Archaeometallurgy in Global Perspective: Methods and Syntheses,* edited by B. W. Roberts and C. P. Thornton, pp. 447–472. Springer, New York.

Kienlin, T. L., and B. S. Ottaway

1998 Flanged Axes of the North-Alpine Regions: An Assessment of the Possibilities of Use Wear Analysis on Metal Artifacts. In *L'atelier du bronzier en Europe du XXᵉ au VIIIᵉ siècle avant notre ère. Actes du colloque international "Bronze '96", Neuchâtel et Dijon, 1998, Tome II: Du minerai au métal, du métal à l'object,* edited by C. Mordant, M. Pernot, and V. Rychner, pp. 271–286. CTHS, Paris.

Killick, D.

2001 Science, Speculation and the Origins of Extractive Metallurgy. In *Handbook of Archaeological Sciences*, edited by D. R. Brothwell and A. M. Pollard, pp. 483–492. John Wiley & Sons, Chichester, UK.

2004 Social Constructionist Approaches to the Study of Technology. *World Archaeology* 36(4):571–578.

2009 Cairo to Cape: The Spread of Metallurgy Through Eastern and Southern Africa. *Journal of World Prehistory* 22(4):399–414.

2014 From Ores to Metals. In *Archaeometallurgy in Global Perspective: Methods and Syntheses*, edited by B. W. Roberts and C. P. Thornton, pp. 11–45. Springer, New York.

Killick, D., and T. Fenn

2012 Archaeometallurgy: The Study of Preindustrial Mining and Metallurgy. *Annual Review of Anthropology* 41(1):559–575.

Kim, J.

2001 Elite Strategy and the Spread of Technological Innovation: The Spread of Iron in the Bronze Age Societies of Denmark and Southern Korea. *Journal of Anthropological Archaeology* 20:442–478, doi:10.1006/jaar.2001.0386.

King, C. L., R. A. Bentley, N. Tayles, U. S. Viðarsdóttir, G. Nowel, and C. G. Macpherson

2013 Moving Peoples, Changing Diets: Isotopic Differences Highlight Migration and Subsistence Changes in the Upper Mun River Valley, Thailand. *Journal of Archaeological Science* 40(4):1681–1688.

King, C. L., N. Tayles, and K. C. Gordon.

2011 Re-Examining the Chemical Evaluation of Diagenesis in Human Bone Apatite. *Journal of Archaeological Science* 38:2222–2230, doi:10.1016/j.jas.2011.03.023.

Kingery, W. D.

1996a Materials Science and Material Culture. In *Learning from Things: Method and Theory of Material Culture Studies*, edited by W. D. Kingery, pp. 181–203. Smithsonian Institution Press, Washington, DC.

1996b A Role for Materials Science. In *Learning from Things: Method and Theory of Material Culture Studies*, edited by W. D. Kingery, pp. 175–180. Smithsonian Institution Press, Washington, DC.

Klemptner, L. J., and P. F. Johnson

1986 Technology and the Primitive Potter: Mississippian Pottery Development Seen through the Eyes of a Ceramic Engineer. In *Ceramics and Civilization, Vol. II: Technology and Style*. edited by W. D. Kingery, pp. 251–271. American Ceramics Society, Columbus, OH.

Knappett, C.

2005 *Thinking through Material Culture*. University of Pennsylvania Press, Philadelphia.

Knauth, P.

1974 *The Metalsmiths*. Time-Life Books, New York.

Kohler, T. A.

2012 Complex Systems and Archaeology. In *Archaeological Theory Today*, edited by I. Hodder, pp. 93–123. 2nd ed. Polity Press, Cambridge, UK.

Kolb, C. C.

2011 *Chaine Operatoire* and Ceramics: Classifications and Typology, Archaeometry, Experimental Archaeology, and Ethnoarchaeology. In

Archaeological Ceramics: A Review of Current Research, edited by S. Scarcella, pp. 1–4. BAR International Series 2193. Archaeopress, Oxford.

Kopytoff, I.

1986 The Cultural Biography of Things: Commoditization as Process. In *The Social Life of Things: Commodities in Cultural Perspective*, edited by A. Appadurai, pp. 64–91. Cambridge University Press, Cambridge, UK.

Kozar, M. G., G. F. Crandall, and S. E. Hall

1992 Integrated Structural and Stratigraphic Study of the Khorat Basin, Rat Buri Limestone (Permian), Thailand. In *Proceedings of a National Conference on Geologic Resources of Thailand: Potential for Future Development, 17–24 November 1992, Bangkok, Thailand,* edited by C. Piancharoen, pp. 692–736. Thai Department of Mineral Resources, Bangkok.

Krause, R.

2009 Bronze Age Copper Production in the Alps: Organization and Social Hierarchies in Mining Communities. In *Metals and Societies: Studies in Honour of Barbara S. Ottaway*, edited by T. L. Kienlin and B. W. Roberts, pp. 47–66. Universitätsforschungen zur Prähistorischen Archäologie Bd. 169. Dr. Rudolf Habelt, Bonn.

Kristiansen, K.

1987 From Stone to Bronze: The Evolution of Social Complexity in Northern Europe, 2300–1200 B.C. In *Specialization, Exchange, and Complex Societies*, edited by E. M. Brumfiel and T. K. Earle, pp. 30–51. Cambridge University Press, Cambridge, UK.

1991 Chiefdoms, States, and Systems of Social Evolution. In *Chiefdoms: Power, Economy, and Ideology*, edited by T. Earle, pp. 16–43. Cambridge University Press, Cambridge, UK.

Kristiansen, K., and T. Earle

2015 Neolithic versus Bronze Age Social Formations: A Political Economy Approach. In *Paradigm Found: Archaeological Theory Present, Past and Future, Essays in Honour of Evžen Neustupný*, edited by K. Kristiansen, L. Šmejda, and J. Turek, pp. 234–247. Oxbow Books, Oxford.

Kristiansen, K., and T. B. Larsson

2005 *The Rise of Bronze Age Society: Travels, Transmissions, and Transformations.* Cambridge University Press, Cambridge, UK.

Krobicki, M., J. Golonka, and K. T. Hung

2008 Major Tectonic Events and Plates of Northwest Vietnam. In *Proceedings of the International Symposia on Geoscience Resources and Environments of Asian Terranes (GREAT 2008), 4th IGCP 516, and 5th APSEG; November 24–26, 2008, Bangkok, Thailand*, pp. 101–104. Bangkok, Thailand.

Kroeber, A. L.

1940 Stimulus Diffusion. *American Anthropologist* 42(1):1–20.

Kuhn, S.

2004 Evolutionary Perspectives on Technology and Technological Change. *World Archaeology* 36(4):561–570.

Kuhn, T. S.

1962 *The Structure of Scientific Revolutions.* University of Chicago Press, Chicago.

Kuno, Y., K. Yamasaki, M. Murozumi, and M. R. R. Abhakorn

1990 Metallographic and Lead Isotope Studies on Metal Objects Unearthed in the Tak Area, Thailand. *Journal of the Siam Society* 78(2):51–59.

Kuzmin, Y. V., J. van der Plicht, and L. D. Sulerzhitsky

2014 Puzzling Radiocarbon Dates for The Upper Paleolithic Site of Sungir (Central Russian Plain). *Radiocarbon* 56(2):451–459.

Lai, C.-K., S. Meffre, A. J. Crawford, K. Zaw, C.-D. Xue, and J. A. Halpin

2014 The Western Ailaoshan Volcanic Belts and their SE Asia Connection: A New Tectonic Model for the Eastern Indochina Block. *Gondwana Research* 26:52–74.

LaMoreaux, P. E., J. Charaljavanaphet, N. Jalichan, P. P. N. Chiengmai, D. Bunnag, A. Thavisri, and C. Rakprathum

1958 *Reconnaissance of the Geology and Ground Water*

of the Khorat Plateau, Thailand. Geological Survey Water-Supply Paper 1429, United States Department of the Interior. United States Government Printing Office, Washington, DC.

LaMotta, V. M., and M. B. Schiffer

2001 Behavioral Archaeology: Toward a New Synthesis. In *Archaeological Theory Today*, edited by I. Hodder, pp. 14–64. Polity Press, Cambridge, UK.

Larick, R.

1987 The Circulation of Spears Among Loikop Cattle Pastoralists of Samburu District, Kenya. In *Research in Economic Anthropology*, Vol. 9, edited by B. L. Isaac, pp. 143–166. Jai Press, Inc., Greenwich, CT.

Lave, J., and E. Wenger

1991 *Situated Learning: Legitimate Peripheral Participation*. Cambridge University Press, Cambridge, UK.

Layton, R.

1989 Pellaport. In *What's New? A Closer Look at the Process of Innovation*, One World Archaeology 14, edited by S. E. van der Leeuw and R. Torrence, pp. 33–53. Unwin Hyman, London.

Lebeau, M. (editor)

2011 *ARCANE: Associated Regional Chronologies for the Ancient Near East and the Eastern Mediterranean. Vol. I: Jezirah*. Brepols, Turnhout, Belgium.

Lechtman, H.

1977 Style in Technology—Some Early Thoughts. In *Material Culture: Style, Organization and Dynamics of Technology*, edited by H. Lechtman and R. S. Merrill, pp. 3–20. West Publishing, St. Paul, MN.

1980 The Central Andes: Metallurgy without Iron. In *The Coming of the Age of Iron*, edited by T. A. Wertime and J. D. Muhly, pp. 267–334. Yale University Press, New Haven, CT.

1988 Traditions and Styles in Central Andean Metalworking. In *The Beginning of the Use of Metals and Alloys: Papers from the Second International Conference on the Beginning of the Use of Metals and Alloys, Zhengzhou, China, 21–26 October 1986*, edited by R. Maddin, pp. 344–378. MIT Press, Cambridge, MA.

1999 The Production of Copper-Arsenic Alloys (Arsenic Bronze) by Cosmelting: Modern Experiment, Ancient Practice. *Journal of Archaeological Science* 26:497–526.

2014 Andean Metallurgy in Prehistory. In *Archaeometallurgy in Global Perspective: Methods and Syntheses*, edited by B. W. Roberts and C. P. Thornton, pp. 361–446. Springer, New York.

Lefèvre-Pontalis, M.

1904 Note. In *Mission Pavie Indo-Chine 1879–1895. Études diverses III: Recherches sur l'Histoire Naturelle de l'Indo-Chine Orientale*, edited by A. Pavie, pp. 16–27. Ernest Leroux, Paris.

Lehner, J. W., and K. A. Yener

2014 Organization and Specialization of Early Mining and Metal Technologies in Anatolia. In *Archaeometallurgy in Global Perspective: Methods and Syntheses*, edited by B. W. Roberts and C. P. Thornton, pp. 529–577. Springer, New York.

Lemonnier, P.

1986 The Study of Material Culture Today: Toward an Anthropology of Technical Systems. *Journal of Anthropological Archaeology* 5:147–186.

1989 Bark Capes, Arrowheads and Concorde: On Social Representations of Technology. In *The Meaning of Things: Material Culture and Symbolic Expression*, edited by I. Hodder, pp. 156–171. Unwin Hyman, London.

1992 *Elements for an Anthropology of Technology*. Anthropological Paper 88. University of Michigan Museum of Anthropology, Ann Arbor, MI.

Lemonnier, P. (editor)

1993 *Technological Choices: Transformation in Material Cultures since the Neolithic*. Routledge, New York.

Lepvrier, C., H. Maluski, V. V. Tich, A. Leyreloup, P. T. Thi, and N. V. Vuong

2004 The Early Triassic Indosinian Orogeny in Vietnam (Truong Son Belt and Kontum Massif); Implications for the Geodynamic Evolution of Indochina. *Tectonophysics* 393(1–4):87–118.

Leroi-Gourhan, A.

1943 *Évolution et techniques. Vol. I: L'homme et la matière*. Sciences d'aujourd'hui. A. Michel, Paris.

Levi-Strauss, C.

1943 Guerre et commerce chez les Indiens de l'Amerique du Sud. *Renaissance* 1:122–139.

Levy, J. E.

1995 Heterarchy in Bronze Age Denmark: Settlement Pattern, Gender, and Ritual. In *Heterarchy and the Analysis of Complex Societies,* edited by R. M. Ehrenreich, C. L. Crumley, and J. E. Levy, pp. 41–53. Archeological Papers of the American Anthropological Association No. 6. American Anthropological Association, Arlington, VA, doi:10.1525/ap3a.1995.6.1.41.

Lewis, P., and E. Lewis

1984 *Peoples of the Golden Triangle: Six Tribes in Thailand*. Thames and Hudson, London.

Li, X.

2000 On the Origin of Bronze in Ancient China. In *The Beginnings of Metallurgy in China*, edited by K. M. Linduff, H. Rubin, and S. Shuyun, pp. 87–98. Chinese Studies Vol. 11. The Edwin Mellen Press, Lewiston, NY.

Linduff, K. M.

2000 Introduction. In *The Beginnings of Metallurgy in China*, edited by K. M. Linduff, H. Rubin, and S. Shuyun, pp. 1–28. Chinese Studies Vol. 11. The Edwin Mellen Press, Lewiston, NY.

Linduff, K. M. (editor)

2004 *Metallurgy in Ancient Eastern Eurasia from the Urals to the Yellow River*. Chinese Studies Vol. 31. The Edwin Mellen Press, Lewiston, NY.

Linduff, K. M., and J. Mei

2009 Metallurgy in Ancient Eastern Asia: Retrospect and Prospects. *Journal of World Prehistory* 22(3):265–281.

2014 Metallurgy in Ancient Eastern Asia: Retrospect and Prospects. In *Archaeometallurgy in Global Perspective: Methods and Syntheses*, edited by B. W. Roberts and C. P. Thornton, pp. 785–804. Springer, New York.

Liu, L.

2003 "The Products of Minds as Well as of Hands": Production of Prestige Goods in the Neolithic and Early State Periods of China. *Asian Perspectives* 42(1):1–40.

Liu, L., and X. Chen

2003 *State Formation in Early China*. Duckworth Debates in Archaeology. Duckworth, London.

Longacre, W. A.

1974 Models of Cultural Process; Testing Hypotheses: Suggestions from Southwestern Archaeology. In *Reconstructing Complex Societies: An Archaeological Colloquium*, edited by C. B. Moore, pp. 29–40. Supplement to the Bulletin of the American Schools of Oriental Research No. 20. American Schools of Oriental Research, Cambridge, MA.

Loofs-Wissowa, H. H. E.

1983a The Development and Spread of Metallurgy in Southeast Asia: A Review of Present Evidence. *Journal of Southeast Asian Studies* 14(1):1–11.

1983b The Development of Metallurgy in Southeast Asia: Reply to Bayard, Pisit, and Solheim. *Journal of Southeast Asian Studies* 14(1):26–31.

1992 The Rise and Fall of Early Bronze in Thailand. In *Proceedings of the XXXII International Congress for Asian and North African Studies, Hamburg, 25th–30th August 1986*, edited by A. Wezler, pp. 117–128. F. Steiner, Stuttgart.

LoSchiavo, F.

1988 Early Metallurgy in Sardinia. In *The Beginning of the Use of Metals and Alloys: Papers from the Second International Conference on the Beginning of the Use of Metals and Alloys, Zhengzhou, China, 21–26 October 1986*, edited by R. Maddin, pp. 92–103. MIT Press, Cambridge, MA.

Ly, V.

2001 Recent Archaeological Research in the Floodplain of Tonle Sap River: The Shell Midden Site of Samrong Sen and Its Pottery Characteristics. *Journal of Southeast Asian Archaeology* 21:47–81.

Lyman, R. L.

2008 Cultural Transmission in North American Anthropology and Archaeology, ca. 1895–1965. In *Cultural Transmission and Archaeology: Issues and Case Studies*, edited by M. J. O'Brien, pp. 10–20. Society for American Archaeology, Washington, DC.

Lyons, E. and F. Rainey

1982 The Road to Ban Chiang: A Dialogue of Events Leading to the University Museum's Participation in the Expedition. *Expedition* 24(4):5–12.

Maddin, R., J. D. Muhly, and T. Stech

1999 Early Metalworking at Çayönü. In *The Beginnings of Metallurgy: Proceedings of the International Conference "The Beginnings of Metallurgy," Bochum 1995*, edited by A. Hauptmann, E. Pernicka, T. Rehren, and Ü. Yalçin, pp. 37–44. Der Anschnitt Beiheft 9. Deutschen Bergbau-Museum, Bochum, Germany.

Mahias, M.-C.

1993 Pottery Techniques in India: Technical Variants and Social Choice. In *Technological Choices: Transformation in Material Cultures since the Neolithic*, edited by P. Lemonnier, pp. 157–180. Routledge, New York.

Mair, V. H.

1998 Priorities. In *The Bronze Age and Early Iron Age Peoples of Eastern Central China, Volume 1: Archeology, Migration and Nomadism, Linguistics*, edited by V. H. Mair, pp. 4–41. Journal of Indo-European Studies Monograph No. 26. Institute for the Study of Man, Washington, DC, and University of Pennsylvania Museum of Archaeology and Anthropology, Philadelphia.

Maldonado, B.

2009 Metal for the Commoners: Tarascan Metallurgical Production in Domestic Contexts. In *Housework: Craft Production and Domestic Economy in Ancient Mesoamerica*, edited by K. G. Hirth, pp. 225–238. Archeological Papers of the American Anthropological Association No. 19. American Anthropological Association, Arlington, VA, doi:10.1111/j.1551-8248.2009.01022.x.

Malinowski, B.

1922 *Argonauts of the Western Pacific: An Account of Native Enterprise and Adventure in the Archipelagos of Melanesian New Guinea*. Routledge and Kegan Paul, London.

Manaka, T., K. Zaw, and S. Meffre

2008 Geological and Tectonic Setting of Cu-Au Deposits in Northern Lao PDR. In *Proceedings of the International Symposia on Geoscience Resources and Environments of Asian Terranes (GREAT 2008), 4th IGCP 516, and 5th APSEG; November 24–26, 2008, Bangkok, Thailand*, pp. 254–257. Bangkok, Thailand.

Mansuy, H.

1920 L'industrie de la pierre et du bronze dans la région de Luang Prabang, Haut-Laos. Contribution à l'étude de la Préhistoire de l'Indochine. *Bulletin du Service Géologique de l'Indochine* 7(1):1–14.

Marom, A., J. S. O. McCullagh, T. F. G. Higham, and R. E. M. Hedges

2013 Hydroxyproline Dating: Experiments on the ^{14}C Analysis of Contaminated and Low-Collagen Bones Radiocarbon. *Radiocarbon* 55(2–3):698–708.

Marom, A., J. S. O. McCullagh, T. F. G. Higham, A. A. Sinitsyn, and R. E. M. Hedges

2012 Single Amino Acid Radiocarbon Dating of Upper Paleolithic Modern Humans. *Proceedings of the National Academy of Sciences of the USA* 109(18):6878–6881.

Martinón-Torres, M., and T. Rehren

2014 Technical Ceramics. In *Archaeometallurgy in Global Perspective: Methods and Syntheses*, edited by B. W. Roberts and C. P. Thornton, pp. 107–131. Springer, New York.

Maschner, H. D. G.

1991 The Emergence of Cultural Complexity on the Northern Northwest Coast. *Antiquity* 65:924–934.

Massie, M.

1904 Catalogue des objets des âges de la pierre et du bronze recueillis dans la région de Luang-Prabang.

In *Mission Pavie Indo-Chine 1879–1895. Études diverses III: Recherches sur l'Histoire Naturelle de l'Indo-Chine Orientale*, edited by A. Pavie, pp. 10–16. Ernest Leroux, Paris.

Masson, M. A.

2005 Trade and Exchange. In *Handbook of Archaeological Methods*, edited by H. D. G. Maschner and C. Chippindale, pp. 1138–1178. AltaMira Press, Lanham, MD.

Matsuda, T.

1928 On the Quenching and Tempering of Brass, Bronze, and "Aluminum" Bronze. *Journal of the Institute of Metals* 39:67–108.

Mauss, M.

1966 *The Gift: Forms and Functions of Exchange in Archaic Societies*. Cohen and West, London.

McAnany, P.

1995 *Living with the Ancestors: Kinship and Kingship in Ancient Maya Society*. University of Texas Press, Austin.

McDougall, A. J.

2009 Independent Technical Report for Angkor Wat Minerals Limited on Cambodia and Indonesian Exploration Projects. Earth Worx Geological Services Ltd. http://www.sec.gov/Archives/edgar/data/1402371/000127351109000092/report.htm.

Meeks, N.

1993a Patination Phenomena on Roman and Chinese High-Tin Bronze Mirrors and Other Artefacts. In *Metal Plating and Patination: Cultural, Technical and Historical Development*, edited by S. La Niece and P. Craddock, pp. 63–84. Butterworth-Heinemann, Oxford.

1993b Surface Characterization of Tinned Bronze, High-Tin Bronze, Tinned Iron and Arsenical Bronze. In *Metal Plating and Patination: Cultural, Technical and Historical Developments*, edited by S. La Niece and P. Craddock, pp. 247–275. Butterworth-Heinemann, Oxford.

Mei, J.

2003 Cultural Interaction between China and Central Asia during the Bronze Age. *Proceedings of the British Academy* 121:1–39.

Mei, J., P. Wang, K. Chen, L. Wang, Y. Wang, and Y. Liu

2015 Archaeometallurgical Studies in China: Some Recent Developments and Challenging Issues. *Journal of Archaeological Science* 56:221–232, doi:10.1016/j.jas.2015.02.026.

Melikian-Chirvani, A. S.

1974 The White Bronzes of Early Islamic Iran. *Metropolitan Museum Journal* 9:123–151.

Merkel, J.

1985 Ore Beneficiation during the Late Bronze/Early Iron Age at Timna, Israel. *MASCA Journal* 3(5):164–169.

1990 Experimental Reconstruction of Bronze Age Copper Smelting Based on Archaeological Evidence from Timna. In *The Ancient Metallurgy of Copper: Researches in the Arabah, 1959–1984*, edited by B. Rothenberg, pp. 78–122. Institute for Archaeo-Metallurgical Studies, Institute of Archaeology, University College London, London.

Mesoudi, A., and M. J. O'Brien

2008a The Cultural Transmission of Great Basin Projectile-Point Technology: An Experimental Simulation. *American Antiquity* 73(1):3–28.

2008b The Cultural Transmission of Great Basin Projectile-Point Technology II: An Agent-Based Computer Simulation. *American Antiquity* 73(4):627–644.

Messili, L., J.-F. Saliège, J. Broutin, E. Messager, C. Hatté, and A. Zazzo

2013 Direct ^{14}C Dating of Early and Mid-Holocene Saharan Pottery. *Radiocarbon* 55(2–3):1391–1402, doi:10.2458/azu_js_rc.55.16334.

Metcalfe, I.

1990 Allochthonous Terrane Processes in Southeast Asia. *Philosophical Transactions of the Royal Society of London. Series A, Mathematical and Physical Sciences* 331(1620):625–640.

1998 Palaeozoic and Mesozoic Geological Evolution of the SE Asian Region: Multidisciplinary

Constraints and Implications for Biogeography. In *Biogeography and Geological Evolution of SE Asia*, edited by R. Hall and J. D. Holloway, pp. 25–41. Backhuys Publishers, Leiden, the Netherlands.

Milisauskas, S. (editor)

2011 *European Prehistory: A Survey*. 2nd ed. Springer, New York.

Miller, H. M.-L.

2007 *Archaeological Approaches to Technology*. Elsevier/Academic Press, Amsterdam and Boston.

Minami, M., K. Sakata, M. Takigami, T. Nagaoka, and T. Nakamura

2013 Ultrafiltration Pretreatment for ^{14}C Dating of Fossil Bones from Archaeological Sites in Japan. *Radiocarbon* 55(2–3):481–490.

Minar, C. J., and P. L. Crown

2001 Learning and Craft Production: An Introduction. *Journal of Anthropological Research* 57(4):369–380.

Minghua, X.

2006 Bronze Cowry-containers of the Dian Culture. *Chinese Archaeology* 6(1):168–173.

Moore, C. B. (editor)

1974 *Reconstructing Complex Societies: An Archaeological Colloquium*. Supplement to the Bulletin of the American Schools of Oriental Research Vol. 20. American Schools of Oriental Research, Cambridge, MA.

Morgan, L. H.

2000 (1877) *Ancient Society.* Transaction Publishers, New Brunswick, NJ.

Muhly, J. D.

1976 University Museum-Thai Fine Arts Department Northeast Thailand Archaeological Project. *Expedition* 18(4):11–12.

1980 The Bronze Age Setting. In *The Coming of the Age of Iron*, edited by T. A. Wertime and J. D. Muhly, pp. 25–67. Yale University Press, New Haven, CT.

1988 The Beginnings of Metallurgy in the Old World. In *The Beginning of the Use of Metals and Alloys: Papers from the Second International Conference on the Beginning of the Use of Metals and Alloys, Zhengzhou, China, 21–26 October 1986*, edited by R. Maddin, pp. 2–20. MIT Press, Cambridge, MA.

2003 Archaeology and Archaeometry: Why We Need (and Should Want) to Work Together. In *Metron, Measuring the Aegean Bronze Age: Proceedings of the 9th International Aegean Conference, New Haven Yale University, 18–21 April 2002*, edited by K. P. Foster and R. Laffineur, pp. 17–23. Aegaeum 24. Annales d'archéologie égéennede l'Université de Liège et UT-PASP.

Müller, R., G. Goldenberg, M. Bartelheim, M. Kunst, and E. Pernicka

2007 Zambujal and the Beginnings of Metallurgy in Southern Portugal. In *Metals and Mines: Studies in Archaeometallurgy*, edited by S. La Niece, D. Hook, and P. Craddock, pp. 15–26. Archetype Publications and the British Museum, London.

Murck, B.

2001 *Geology: A Self-Teaching Guide.* Wiley Self-Teaching Guides. John Wiley and Sons, Inc., New York.

Murdock, G. P.

1937 Comparative Data on the Division of Labor by Sex. *Social Forces* 15(4):551–553.

Murillo-Barroso, M., T. O. Pryce, B. Bellina, and M. Martinón-Torres

2010 Khao Sam Kaeo—an Archaeometallurgical Crossroads for Trans-Asiatic Technological Traditions. *Journal of Archaeological Science* 37:1761–1772.

Murowchick, R. E.

1988a An Examination of Some Early Bronzes from Samreng Sen, Kampuchea. *Journal of the Historical Metallurgy Society* 22(1):1–20.

1988b The Development of Early Bronze Metallurgy in Vietnam and Kampuchea: A Reexamination of Recent Work. In *The Beginning of the Use of Metals and Alloys: Papers from the Second International Conference on the Beginning of the Use of Metals and Alloys, Zhengzhou, China, 21–26 October 1986*, edited by R. Maddin, pp. 182–199. MIT Press, Cambridge, MA.

1989 The Ancient Bronze Metallurgy of Yunnan and Its Environs: Development and Implications. Paper presented at the Department of Anthropology, Cambridge University.

Murphy, S. A.

2010 The Buddhist Boundary Markers of Northeast Thailand and Central Laos, 7th–12th Centuries CE: Towards an Understanding of the Archaeological, Religious and Artistic Landscapes of The Khorat Plateau. Ph.D. dissertation, School of Oriental and African Studies, London.

Nalawade-Chavan, S., J. McCullagh, and R. Hedges

2014 New Hydroxyproline Radiocarbon Dates from Sungir, Russia Confirm Early Mid Upper Palaeolithic Burials in Eurasia. *PLoS ONE*, doi:10.1371/journal.pone.0076896.

Natapintu, S.

1988 Current Research on Ancient Copper-Base Metallurgy in Thailand. In *Prehistoric Studies: The Stone and Metal Ages in Thailand*, edited by P. Charoenwongsa and B. Bronson, pp. 107–124. Amarin Printing Group, Bangkok.

1991 Archaeometallurgical Studies in the Khao Wong Prachan Valley, Central Thailand. In *Indo-Pacific Prehistory 1990. Proceedings of the 14th Congress of the Indo-Pacific Prehistory Association*, edited by P. Bellwood, pp. 153–159. Indo-Pacific Prehistory Association, Bangkok.

Needham, J.

1980 The Evolution of Iron and Steel Technology in East and Southeast Asia. In *The Coming of the Iron Age*, edited by T. A. Wertime and J. D. Muhly, pp. 507–541. Yale University Press, New Haven, CT.

Nelson, S. M.

1993 *The Archaeology of Korea*. Cambridge World Archaeology. Cambridge University Press, Cambridge, UK.

Nezafati, N., E. Pernicka, and M. Momenzadeh

2006 Ancient Tin: Old Question and a New Answer. *Antiquity* 80(308). http://www.cais-soas.com/CAIS/Science/tin.htm, accessed on August 30, 2016.

Nguyen, K. S., P. M. Huyen, and T. T. Tin

2004 Northern Vietnam from the Neolithic to the Han Period. In *Southeast Asia: From Prehistory to History*, edited by I. Glover and P. Bellwood, pp. 177–208. RoutledgeCurzon, New York.

Nguyen, Q. M.

2005 Radiocarbon Dating of the Dong Dau Period. *Bulletin of the Indo-Pacific Prehistory Association* 25:113–116.

Nitta, E.

1991 Archaeological Study on the Ancient Iron-Smelting and Salt-Making Industries in the Northeast of Thailand: Preliminary Report of the Excavations of Non Yang and Ban Don Phlong. *Journal of Southeast Asian Archaeology* 11:1–46.

1997 Iron-Smelting and Salt-Making Industries in Northeast Thailand. *Bulletin of the Indo-Pacific Prehistory Association* 16:153–160.

Northover, J. P.

1980 The Analysis of Welsh Bronze Age Metalwork. In *Guide Catalogue to the Bronze Age Collections*, edited by H. N. Savory, pp. 229–243. National Museum of Wales, Cardiff.

1982 The Exploration of the Long-Distance Movement of Bronze in Bronze and Early Iron Age Europe. *Bulletin of the Institute of Archaeology, University of London* 19:45–72.

1989a Properties and Use of Arsenic-Copper Alloys. In *Proceedings of the International Symposium on Old World Archaeometallurgy 1987*, edited by A. Hauptmann, E. Pernicka, and G. A. Wagner, pp. 111–118. Deutschen Bergbau-Museum, Bochum, Germany.

1989b Non-Ferrous Metallurgy in British Archaeology. In *Scientific Analysis in Archaeology*, edited by J. Henderson, pp. 213–236. Oxford University Committee for Archaeology, Monograph No. 19. UCLA Institute of Archaeology, Archaeological Research Tools 5, Oxford.

Noulet, J. B.

1879 L'âge de la pierre polie et du bronze au Campbodge d'après les découvertes de M. J. Moura. *Archives du Musée d'Histoire Naturelle de Toulouse* 1:5–50.

O'Brien, M. J. (editor)

2008 *Cultural Transmission and Archaeology: Issues and Case Studies.* Society for American Archaeology, Washington, DC.

O'Brien, M. J., and R. A. Bentley

2011 Stimulated Variation and Cascades: Two Processes in the Evolution of Complex Technological Systems. *Journal of Archaeological Method and Theory* 18:309–335.

O'Brien, M. J., and R. L. Lyman

2002 *Applying Evolutionary Archaeology: A Systematic Approach.* Kluwer, New York.

O'Brien, M. J., R. L. Lyman, M. Collard, C. J. Holden, R. D. Gray, and S. J. Shennan

2008 Transmission, Phylogenetics, and the Evolution of Cultural Diversity. In *Cultural Transmission and Archaeology: Issues and Case Studies,* edited by M. J. O'Brien, pp. 39–58. Society for American Archaeology, Washington, DC.

O'Brien, W.

1994 *Mount Gabriel: Bronze Age Mining in Ireland.* Galway University Press, Galway, Ireland.

2015 *Prehistoric Copper Mining in Europe 5500–500 BC.* Oxford University Press, Oxford.

O'Reilly, D.

2001 From the Bronze Age to the Iron Age in Thailand: Applying the Heterarchical Approach. *Asian Perspectives (2000)* 39(1–2):1–19.

2003 Further Evidence of Heterarchy in Bronze Age Thailand. *Current Anthropology* 44(2):300–306.

2007a *Early Civilizations of Southeast Asia.* AltaMira Press, New York.

2007b Non Muang Kao in a Regional Perspective. In *The Origins of the Civilization of Angkor. Volume Two: The Excavation of Noen U-Loke and Non Muang Kao,* edited by C. F. W. Higham, A. Kijngam, and S. Talbot, pp. 575–587. The Fine Arts Department of Thailand, Bangkok.

Ottaway, B. S.

2001 Innovation, Production and Specialization in Early Prehistoric Copper Metallurgy. *European Journal of Archaeology* 4(1):87–112.

Ottaway, B. S., and Q. Wang

2004 *Casting Experiments and Microstructure of Archaeologically Relevant Bronzes.* BAR International Series 1331. Archaeopress, Oxford.

Oxenham, M., and N. Tayles (editors)

2006 *Bioarchaeology of Southeast Asia.* Cambridge Studies in Biological and Evolutionary Anthropology 43. Cambridge University Press, New York.

Patte, É.

1936 L'Indochine préhistorique. *Revue Anthropologie* 46:277–314.

Pavie, A.

1904 *Mission Pavie Indo-Chine 1879–1895. Études diverses III: Recherches sur l'Histoire Naturellede l'Indo-Chine Orientale.* Ernest Leroux, Paris.

Pearson, R.

1962 Dong-So'n and Its Origins. *The Bulletin of the Institute of Ethnology Academia Sinica* 13:27–50.

2007 Debating Jomon Social Complexity. *Asian Perspectives* 46(2):361–388.

Peelo, S.

2011 Pottery-Making in Spanish California: Creating Multi-Scalar Social Identity through Daily Practice. *American Antiquity* 76(4):642–666.

Pernicka, E.

1993 Comments III: Comments on P. Budd, D. Gale, A. M. Pollard, R. G. Thomas, and P. A. Williams, "Evaluating Lead Isotope Data: Further Observations". *Archaeometry* 35(2):259–262.

2014 Provenance Determination of Archaeological Metal Objects. In *Archaeometallurgy in Global Perspective: Methods and Syntheses,* edited by B. W. Roberts and C. P. Thornton, pp. 239–268. Springer, New York.

Pétrequin, P.

1993 North Wind, South Wind: Neolithic Technical

Choices in the Jura Mountains, 3700–2400 BC. In *Technical Choices: Transformations in Material Cultures since the Neolithic*, edited by P. Lemonnier, pp. 36–76. Routledge, New York.

Pfaffenberger, B.

1988 Fetished Objects and Humanised Nature: Towards an Anthropology of Technology. *Man (N.S.)* 23:236–252.

1992 Social Anthropology of Technology. *Annual Review of Anthropology* 21:491–516.

1998 Mining Communities, *Chaînes Opératoire*, and Sociotechnical Systems. In *Social Approaches to an Industrial Past: The Archaeology and Anthropology of Mining*, edited by A. B. Knapp, V. C. Pigott, and E. W. Herbert, pp. 291–300. Routledge, London.

2001 Symbols Do Not Create Meanings—Activities Do: Or, Why Symbolic Anthropology Needs the Anthropology of Technology. In *Anthropological Perspectives on Technology*, edited by M. B. Schiffer, pp. 77–86. Amerind Foundation New World Studies Series No. 5. University of New Mexico Press, Albuquerque, NM.

Phillips, D. A.

1979 Material Culture and Trade of the Postclassic Maya. Ph.D. dissertation, Department of Anthropology, University of Arizona, Sierra Vista.

2006 Comment on Harry's Discussion of Ceramic Specialization and Agricultural Marginality in the Prehistoric U.S. Southwest. *American Antiquity* 71(2):397–398.

Phillips, D. A., and W. L. Rathje

1977 Streets Ahead: Exchange Values and the Rise of the Classic Maya. In *Social Process in Maya Prehistory*, edited by N. Hammond, pp. 103–112. Academic Press: New York.

Pietrusewsky, M., and M. T. Douglas

2002 *Ban Chiang, a Prehistoric Village Site in Northeast Thailand I: The Human Skeletal Remains.* Thai Archaeology Monograph Series 1. University of Pennsylvania Museum of Archaeology and Anthropology, Philadelphia.

Piggott, S.

1965 *Ancient Europe, from the Beginnings of Agriculture to Classical Antiquity: A Survey*. Edinburgh University Press, Edinburgh.

Pigott, V. C.

1989 Bronze i. In Pre-Islamic Iran. In *Encyclopædia Iranica*, Vol. IV, fasc. 5, pp. 457–471. Routledge and Kegan Paul, New York. http://www.iranicaonline.org/articles/bronze-i, accessed 16 August 2016.

1996 The Study of Ancient Metallurgical Technology: A Review. *Asian Perspectives* 35(1):89–97.

1998 Prehistoric Copper Mining in the Context of Emerging Community Craft Specialization in Northeast Thailand. In *Social Approaches to an Industrial Past: The Archaeology and Anthropology of Mining*, edited by A. B. Knapp, V. C. Pigott, and E. W. Herbert, pp. 205–225. Routledge, New York.

1999a Reconstructing the Copper Production Process as Practised among Prehistoric Mining/Metallurgical Communities in the Khao Wong Prachan Valley of Central Thailand. In *Metals in Antiquity*, edited by S. M. M. Young, A. M. Pollard, P. Budd, and R. A. Ixer, pp. 10–21. BAR International Series 792. Archaeopress, Oxford.

1999b The Development of Metal Production on the Iranian Plateau: An Archaeometallurgical Perspective. In *The Archeometallurgy of the Asian Old World*, edited by V. C. Pigott, pp. 73–106. University Museum Symposium Series Vol. VII, University Museum Monograph 89, MASCA Research Papers in Science and Archaeology Vol. 16. University of Pennsylvania Museum of Archaeology and Anthropology, Philadelphia.

2011 Sources of Tin and the Tin Trade in Southwest Asia: Recent Research and Its Relevance to Current Understanding. In *Metallurgy: Understanding How, Learning Why: Studies in Honor of James D. Muhly*, edited by P. P. Betancourt and S. C. Ferrence, pp. 273–291. Prehistory Monograph 29. INSTAP Academic Press, Philadelphia.

2012a On Ancient Tin and Tin-Bronze in the Asian Old World: Further Comments. In *Eastern Mediterranean Metallurgy in the Second Millennium BC*, edited by V. Kassianidou and G. Papasavvas, pp. 222–236. Oxbow Books, Oxford.

2012b Looking East: The Coming of Tin Bronze in the Context of the Bronze Age Trans-Eurasian Exchange: An Overview. In *Scientific Research on Ancient Asian Metallurgy: Proceedings of Fifth Forbes Symposium at the Freer Gallery of Art*, edited by P. Jett, B. McCarthy, and J. G. Douglas, pp. 85–100. Archetype Publications, London, and the Freer Gallery of Art, Smithsonian Institution, Washington, DC.

Pigott, V. C., and R. Ciarla

2007 On the Origins of Metallurgy in Prehistoric Southeast Asia: The View from Thailand. In *Metals and Mines: Studies in Archaeometallurgy*, edited by S. La Niece, D. Hook, and P. Craddock, pp. 76–88. Archetype Publications and the British Museum, London.

Pigott, V. C., and A. R. Marder

1984 Prehistoric Iron in Southeast Asia: New Evidence from Northeast Thailand. In *Southeast Asian Archaeology at the XV Pacific Science Congress. The Origins of Agriculture, Metallurgy, and the State in Mainland Southeast Asia*, edited by D. T. Bayard, pp. 278–301. University of Otago Studies in Prehistoric Anthropology Vol. 16. University of Otago, Dunedin, New Zealand.

Pigott, V. C., and S. Natapintu

1988 Archaeological Investigations into Prehistoric Copper Production: The Thailand Archaeometallurgy Project 1984–1986. In *The Beginning of the Use of Metals and Alloys: Papers from the Second International Conference on the Beginning of the Use of Metals and Alloys, Zhengzhou, China, 21–26 October 1986*, edited by R. Maddin, pp. 156–162. MIT Press, Cambridge, MA.

1996–97 Investigating the Origins of Metal Use in Prehistoric Thailand. In *Ancient Chinese and Southeast Asian Bronze Age Cultures. The Proceedings of a Conference Held at the Edith and Joy London Foundation Property, Kioloa, NSW, 8–12 February, 1988*, Vol. II, edited by F. D. Bulbeck and N. Bernard, pp. 787–808. SMC Publishing, Taipei.

Pigott, V. C., S. Natapintu, and U. Theetiparivatra

1992 Research in the Development of Prehistoric Metal Use in Northeast Thailand. In *Early Metallurgy, Trade and Urban Centres in Thailand and Southeast Asia*, edited by I. C. Glover, P. Suchitta, and J. Villiers, pp. 47–62. White Lotus, Bangkok.

Pigott, V. C., H. C. Rogers, and S. K. Nash

2003 Archaeometallurgical Investigations at Tal-e Malyan: The Evidence for Tin-Bronze in the Kaftari Phase. In *Yeki Bud, Yeki Nabud: Essays on the Archaeology of Iran in Honor of William M. Sumner*, edited by N. F. Miller and K. Abdi, pp. 161–175. University of Pennsylvania Museum of Archaeology and Anthropology, Philadelphia.

Pigott, V. C., and G. Weisgerber

1998 Mining Archaeology in Geological Context: The Prehistoric Copper Mining Complex at Phu Lon, Nong Khai Province, Northeast Thailand. *Der Anschnitt* 8:135–162.

Pigott, V. C., A. D. Weiss, and S. Natapintu

1997 The Archaeology of Copper Production: Excavations in the Khao Wong Prachan Valley, Central Thailand. In *South-East Asian Archaeology 1992. Proceedings of the Fourth International Conference of the European Association of South-East Asian Archaeologists. Rome, 28th September–4th October 1992*, edited by R. Ciarla and F. Rispoli, pp. 119–157. Serie Orientale Roma Vol. 77. Istituto Italiano per l'Africa e l'Oriente, Rome.

Pinch, T., and W. E. Bijker

1987 The Social Construction of Facts and Artifacts, Or How the Sociology of Science and the Sociology of Technology Might Benefit Each Other. In *The Social Construction of Technological Systems: New Directions in the Sociology and History of Technology*, edited by W. E. Bijker, T. P. Hughes, and T. Pinch, pp. 17–50. MIT Press, Cambridge, MA.

Pleiner, R.

1980 Early Iron Metallurgy in Europe. In *The Coming of the Age of Iron*, edited by T. A. Wertime and J. D. Muhly, pp. 375–415. Yale University Press, New Haven, CT.

Plog, S.

1995 Equality and Hierarchy: Holistic Approaches to Understanding Social Dynamics in the Pueblo Southwest. In *Foundations of Social Inequality*, edited by T. D. Price and G. M. Feinman, pp. 189–206. Fundamental issues in Archaeology. Plenum Press, New York.

Plourde, A.

2009 Prestige Goods and the Formation of Political Hierarchy: A Costly Signaling Model. In *Pattern and Process in Cultural Evolution*, edited by S. Shennan, pp. 265–276. University of California Press, Berkeley.

Pollard, A. M., and P. Bray

2014 Chemical and Isotopic Studies of Ancient Metals. In *Archaeometallurgy in Global Perspective: Methods and Syntheses*, edited by B. W. Roberts and C. P. Thornton, pp. 217–268. Springer, New York.

Possehl, G. L.

1998 Sociocultural Complexity without the State: the Indus Civilization. In *Archaic States*, edited by G. M. Feinman and J. Marcus, pp. 261–291. School of American Research Press, Santa Fe, NM.

Potter, D. R., and E. M. King

1995 A Heterarchical Approach to Lowland Maya Socioeconomies. In *Heterarchy and the Analysis of Complex Societies*, edited by R. M. Ehrenreich, C. L. Crumley, and J. E. Levy, pp. 17–32. Archeological Papers of the American Anthropological Association No. 6. American Anthropological Association, Arlington, VA, doi:10.1525/ap3a.1995.6.1.41.

Price, T. D., and G. M. Feinman

1995b Foundations of Prehistoric Social Inequality. In *Foundations of Social Inequality*, edited by T. D. Price and G. M. Feinman, pp. 3–13. Fundamental issues in Archaeology. Plenum Press, New York.

Price, T. D., and G. M. Feinman (editors)

1995a *Foundations of Social Inequality*. Fundamental issues in Archaeology. Plenum Press, New York.

Pryce, T. O.

2009 Prehistoric Copper Production and Technological Reproduction in the Khao Wong Prachan Valley of Central Thailand. Ph.D. dissertation, University College London, London.

2012 A Flux That Binds? The Southeast Asian Lead Isotope Project. In *Scientific Research on Ancient Asian Metallurgy: Proceedings of Fifth Forbes Symposium at the Freer Gallery of Art*, edited by P. Jett, B. McCarthy, and J. G. Douglas, pp. 113–121. Archetype Publications, London, and the Freer Gallery of Art, Smithsonian Institution, Washington, DC.

2014 Metallurgy in Southeast Asia. In *Encyclopaedia of the History of Science, Technology, and Medicine in Non-Western Cultures*, edited by H. Selin, pp. 1–17. Springer Netherlands, Dordrecht, the Netherlands.

Pryce, T. O., S. Baron, B. H. M. Bellina, P. S. Bellwood, N. Chang, P. Chattopadhyay, E. Dizon, I. C. Glover, E. Hamilton, C. F. W. Higham, A. A. Kyaw, V. Laychour, S. Natapintu, V. Nguyen, J.-P. Pautreau, E. Pernicka, V. C. Pigott, M. Pollard, C. Pottier, A. Reinecke, T. Sayavongkhamdy, V. Souksavatdy, and J. White

2014 More Questions than Answers: The Southeast Asian Lead Isotope Project 2009–2012. *Journal of Archaeological Science* 42:273–294.

Pryce, T. O., Y. Bassiakos, M. Catapotis, and R. C. Doonan

2007 "De Caerimoniae" Technological Choices in Copper-Smelting Furnace Design at Early Bronze Age Chrysokamino, Crete. *Archaeometry* 49(3):543–557.

Pryce, T. O., M. Brauns, N. Chang, E. Pernicka, A. M. Pollard, C. Ramsey, T. Rehren, V. Souksavatdy, and T. Sayavongkhamdy

2011 Isotopic and Technological Variation in Prehistoric Southeast Asian Primary Copper Production. *Journal of Archaeological Science* 38(12):3309–3322.

Pryce, T. O., and S. Natapintu

2010 Smelting Iron from Laterite: Technical Possibility or Ethnographic Aberration? *Asian Perspectives (2009)* 48(2):249–264.

Pryce, T. O., V. C. Pigott, M. Martinón-Torres, and T. Rehren

2010 Prehistoric Copper Production and Technological Reproduction in the Khao Wong Prachan Valley of Central Thailand. *Archaeological and Anthropological Sciences* 2:237–264.

Rachdawong, S.

1959 Beneficiation of Lead-Zinc Ore, Ban Bo Ngam Deposits, Sri Sawat Sub-District, Kanchanaburi. In *Report of Investigation No. 2*. Royal Department of Mines, Bangkok.

Radivojević, M., and T. Rehren

2016 Paint it Black: The Rise of Metallurgy in the Balkans. *Journal of Archaeological Method and Theory* 23(1):200–237.

Radivojević, M., T. Rehren, and J. Kuzmanovic-Cvetkovic

2013 Tainted Ores and the Rise of Tin Bronzes in Eurasia, c. 6500 Years Ago. *Antiquity* 87:1030–1045.

Radivojević, M., T. Rehren, E. Pernicka, D. Sljivar, M. Brauns, and D. Boric

2010 On the Origins of Extractive Metallurgy: New Evidence from Europe. *Journal of Archaeological Science* 37(11):2775–2787.

Rainey, F.

1992 *Reflections of a Digger: Fifty Years of World Archaeology*. University of Pennsylvania Museum of Archaeology and Anthropology, Philadelphia.

Rajpitak, W., and N. J. Seeley

1979 The Bronze Bowls from Ban Don Ta Phet, Thailand: An Enigma of Prehistoric Metallurgy. *World Archaeology* 11(1):26–31.

Rapp, G., Jr.

1988 On the Origins of Copper and Bronze Alloying. In *The Beginning of the Use of Metals and Alloys*, edited by R. Maddin, pp. 21–27. MIT Press, Cambridge, MA.

2002 *Archaeomineralogy*. Springer, Berlin and New York.

Rapp, G., J. Allert, V. Vitali, Z. Jing, and E. Hendrickson

2000 *Determining Geologic Sources of Artifact Copper: Source Characterization Using Trace Element Patterns*. University Press of America, Lanham, MD.

Raymond, R.

1984 *Out of the Fiery Furnace: The Impact of Metals on the History of Mankind*. The Pennsylvania State University Press, University Park, PA.

Rebay-Salisbury, K. C.

2011 Thoughts in Circles: Kulturkreislehre as a Hidden Paradigm in Past and Present Archaeological Interpretations. In *Investigating Archaeological Cultures: Material Culture, Variability, and Transmission*, edited by B. W. Roberts and M. Vander Linden, pp. 41–59. Springer, New York.

Redmond, E. M., and C. S. Spencer

2012 Chiefdoms at the Threshold: The Competitive Origins of the Primary State. *Journal of Anthropological Archaeology* 3(1):22–37.

Reents-Budet, D.

1998 Elite Maya Pottery and Artisans as Social Indicators. In *Craft and Social Identity*, edited by C. L. Costin and R. P. Wright, pp. 57–89. Archeological Papers of the American Anthropological Association No. 8. American Anthropological Association, Arlington, VA, doi:10.1525/ap3a.1998.8.1.71.

Rehder, J. E.

1986 Primitive Furnaces and the Development of Metallurgy. *Journal of the Historical Metallurgy Society* 20(2):87–92.

2000 *The Mastery and Uses of Fire in Antiquity*. McGill-Queen's University Press, Montreal.

Rehren, T.

2003 Crucibles as Reaction Vessels in Ancient Metallurgy. In *Mining and Metal Production through the Ages*, edited by P. Craddock and J. Lang, pp. 207–215. British Museum Press, London.

Rehren, T., and P. Bedford

2013 The Future of Archaeometallurgy and

Historical Metallurgy. *The Crucible: Historical Metallurgy Society News* 83:6–7.

Rehren, T., T. Belgya, A. Jambon, G. Káli, Z. Kasztovszky, Z. Kis, I. Kovács, B. Maróti, M. Martinón-Torres, G. Miniaci, V. C. Pigott, M. Radivojevi, L. Rosta, L. Szentmiklósi, and Z. Szökefalvi-Nagy

2013 5,000 Years Old Egyptian Iron Beads Made from Hammered Meteoritic Iron. *Journal of Archaeological Science* 40:4785–4792.

Rehren, T., M. F. Charlton, S. Chirikure, J. Humphris, A. Ige, and H. A. Veldhuijzen

2007 Decisions Set in Slag: The Human Factor in African Iron Smelting. In *Metals and Mines: Studies in Archaeometallurgy*, edited by S. La Niece, D. Hook, and P. Craddock, pp. 211–218. Archetype Publications and the British Museum, London.

Rehren, T., and O. Papakhristu

2000 Cutting Edge Technology—the Ferghana Process of Medieval Crucible Steel Smelting. *Metalla* 7(2):55–69.

Reid, A., and R. MacLean

1995 Symbolism and the Social Contexts of Iron Production in Karagwe. *World Archaeology* 27(1):144–161.

Reimer, P. J., E. Bard, A. Bayliss, J. W. Beck, P. G. Blackwell, C. B. Ramsey, C. E. Buck, H. Cheng, R. L. Edwards, M. Friedrich, P. M. Grootes, T. P. Guilderson, H. Haflidason, I. Hajdas, C. Hatté, T. J. Heaton, D. L. Hoffmann, A. G. Hogg, K. A. Hughen, K. F. Kaiser, B. Kromer, S. W. Manning, M. Niu, R. W. Reimer, D. A. Richards, E. M. Scott, J. R. Southon, R. A. Staff, C. S. M. Turney, and J. van der Plicht

2013 IntCal13 and Marine13 Radiocarbon Age Calibration Curves 0–50,000 Years cal BP. *Radiocarbon* 55(4):1869–1887.

Renfrew, C.

1969 The Autonomy of the South-East European Copper Age. *Proceedings of the Prehistoric Society* 35:12–47.

1972 *The Emergence of Civilization: The Cyclades and the Aegean in the Third Millennium B.C.* Methuan & Co., London.

1973 *Before Civilization: The Radiocarbon Revolution and Prehistoric Europe.* 1st American ed. Knopf, New York.

1986 Varna and the Emergence of Wealth in Prehistoric Europe. In *The Social Life of Things: Commodities in Cultural Perspective,* edited by A. Appadurai, pp. 141–167. Cambridge University Press, Cambridge, UK.

Renfrew, C., and P. Bahn (editors)

2005 *Archaeology: The Key Concepts.* Routledge, New York.

Rice, P.

1987 *Pottery Analysis: A Sourcebook.* University of Chicago Press, Chicago.

Ridd, M. F., A. J. Barber, and M. J. Crow

2011a Introduction to the Geology of Thailand. In *The Geology of Thailand*, edited by M. F. Ridd, A. J. Barber, and M. J. Crow, pp. 1–17. Geological Society, London.

Ridd, M. F., A. J. Barber, and M. J. Crow (editors)

2011b *The Geology of Thailand.* Geological Society, London.

Rispoli, F.

1997 Late 3rd–Mid 2nd Millennium B.C. Pottery Traditions in Central Thailand: Some Preliminary Observations in a Wider Perspective. In *South-East Asian Archaeology 1992. Proceedings of the Fourth International Conference of the European Association of Southeast Asian Archaeologists, Rome, 28th September–4th October 1992*, edited by R. Ciarla and F. Rispoli, pp. 59–97. Serie Orientale Roma 77. Istituto Italiano per l'Africa e l'Oriente, Rome.

2007 The Incised & Impressed Pottery Style of Mainland Southeast Asia: Following the Paths of Neolithization. *East and West* 57(1–4):235–304.

Rispoli, F., R. Ciarla, and V. C. Pigott

2013 Establishing the Prehistoric Cultural Sequence for the Lopburi Region, Central Thailand. *Journal of World Prehistory* 26(2):101–171.

Rizvi, U. Z.

2007 Configuring the Space in Between: Redefining

the Ganeshwar Jodhpura Cultural Complex in Northeastern Rajasthan, India. Ph.D. dissertation, Department of Anthropology, University of Pennsylvania, Philadelphia.

Roberts, B. W.

2008 Creating Traditions and Shaping Technologies: Understanding the Earliest Metal Objects and Metal Production in Western Europe. *World Archaeology* 40(3):354–372.

2009a Production Networks and Consumer Choice in the Earliest Metal of Western Europe. *Journal of World Prehistory* 22(4):461–481.

2009b Origins, Transmission and Traditions: Analysing Early Metal in Western Europe. In *Metals and Societies: Studies in Honour of Barbara S. Ottaway*, edited by T. L. Kienlin and B. W. Roberts, pp. 129–142. Universitätsforschungen zur Prähistorischen Archäologie Bd. 169. Dr. Rudolf Habelt, Bonn.

2011 Ancient Technology and Archaeological Cultures: Understanding the Earliest Metallurgy in Eurasia. In *Investigating Archaeological Cultures: Material Culture, Variability, and Transmission*, edited by B. W. Roberts and M. Vander Linden, pp. 137–150. Springer, New York.

2014 Production Networks and Consumer Choice in the Earliest Metal of Western Europe. In *Archaeometallurgy in Global Perspective: Methods and Syntheses*, edited by B. W. Roberts and C. P. Thornton, pp. 423–446. Springer, New York.

Roberts, B. W., and C. P. Thornton (editors)

2014 *Archaeometallurgy in Global Perspective: Methods and Syntheses*. Springer, New York.

Roberts, B. W., C. P. Thornton, and V. C. Pigott

2009 Development of Metallurgy in Eurasia. *Antiquity* 83(322):1012–1022.

Roberts, B. W., and M. Vander Linden (editors)

2011 *Investigating Archaeological Cultures: Material Culture, Variability, and Transmission*. Springer, New York.

Rochette, E. T.

2009 Jade in Full: Prehispanic Domestic Production of Wealth Goods in the Middle Motagua Valley, Guatemala. In *Housework: Craft Production and Domestic Economy in Ancient Mesoamerica*, edited by K. G. Hirth, pp. 205–224. Archeological Papers of the American Anthropological Association No. 19. American Anthropological Association, Arlington, VA, doi:10.1111/j.1551-8248.2009.01021.x.

Rogers, E.

1962 *Diffusion of Innovations*. The Free Press of Glencoe, New York.

Rohl, B., and S. Needham

1998 *The Circulation of Metal in the British Bronze Age: The Application of Lead Isotope Analysis*, British Museum Occasional Paper No. 102. The British Museum, London.

Rostoker, W., and B. Bronson

1990 *Pre-Industrial Iron: Its Technology and Ethnology*. Archeomaterials Monograph No. 1. Archeomaterials, Philadelphia.

Rostoker, W., V. C. Pigott, and J. Dvorak

1989 Direct Reduction to Copper Metal by Oxide-Sulfide Mineral Interaction. *Archeomaterials* 3:69–87.

Rostoker, W., and M. Sadowski

1980 The Carbon Reduction of Fully Oxidized Chalcopyrite (Copper) Ores. *Journal of the Historical Metallurgy Society* 14(1):38–42.

Rousseau, J.

2001 Hereditary Stratification in Middle-range Societies. *Journal of the Royal Anthropological Institute (N.S.)* 7:117–131.

2006 *Rethinking Social Evolution: The Perspective from Middle-Range Societies*. McGill-Queen's University Press, Montreal.

Roux, V.

2003 A Dynamic Systems Framework for Studying Technological Change: Application to the Emergence of the Potter's Wheel in the Southern Levant. *Journal of Archaeological Method and Theory* 10(1):1–30.

2008 Evolutionary Trajectories of Technological Traits and Cultural Transmission: A Qualitative Approach to the Emergence and

Disappearance of the Ceramic Wheel-fashioning Technique in the Southern Levant during the Fifth to Third Millennia BC. In *Breaking Down Boundaries: Anthropological Approaches to Cultural Transmission and Material Culture*, edited by M. T. Stark, B. Bowser, and L. Horne, pp. 82–104. University of Arizona Press, Tucson.

Rovira, S.

2002 Metallurgy and Society in Prehistoric Spain. In *Metals and Society: Papers from a Session Held at the European Association of Archaeologists Sixth Annual Meeting in Lisbon 2000*, edited by B. S. Ottaway and E. C. Wager, pp. 5–20. BAR International Series 1061. British Archaeological Reports, Oxford.

Rovira, S., I. Montero-Ruiz, and M. Renzi

2009 Experimental Co-smelting to Copper-tin Alloys. In *Metals and Societies: Studies in Honour of Barbara S. Ottaway*, edited by T. L. Kienlin and B. W. Roberts, pp. 407–414. Universitätsforschungen zur Prähistorischen Archäologie Bd. 169. Dr. Rudolf Habelt, Bonn.

Rowlands, M. J.

1971 The Archaeological Interpretation of Prehistoric Metalworking. *World Archaeology* 3(2):210–224.

1976 *The Organisation of Middle Bronze Age Metalworking. Part i: Discussion.* British Archaeological Reports 31(i). British Archaeological Reports, Oxford.

1988 A Question of Complexity. In *Domination and Resistance*, edited by D. Miller, M. Rowlands, and C. Tilley, pp. 29–40. Allen and Unwin, London.

Rowlands, M. J., and J.-P. Warnier

1993 The Magical Production of Iron in the Cameroon Grassfields. In *The Archaeology of Africa: Food, Metals and Towns*, edited by T. Shaw, P. Sinclair, B. Andah, and A. Okpoko, pp. 512–550. Routledge, New York.

Ruiz-Taboada, A., and I. Montero-Ruiz

1999 The Oldest Metallurgy in Western Europe. *Antiquity* 73(282):897–903.

Rundquist, D. V.

1984 Metallogenic Belts, Provinces and Epochs of the U.S.S.R. *Episodes* 7(1):28–32.

Rutnin, S.

1988 The Prehistory of Western Udon Thani and Loei Provinces, Northeast Thailand. Ph.D. dissertation, Department of Prehistory and Anthropology, Australian National University, Canberra.

Sahlins, M.

1972 *Stone Age Economics*. Aldine de Gruyter, New York.

Sahlins, M., and E. Service

1960 *Evolution and Culture*. University of Michigan Press, Ann Arbor.

Saitta, D. J.

1999 Prestige, Agency, and Change in Middle-Range Societies. In *Material Symbols: Culture and Economy in Prehistory*, edited by J. E. Robb, pp. 136–149. Occasional Papers 26. Center for Archaeological Investigations, Southern Illinois University, Carbondale, IL.

Sampson, K., and D. Esarey

1993 A Survey of Elaborate Mississippian Copper Artifacts from Illinois. *Illinois Archaeology* 5(1–2):452–480.

Sarjeant, C.

2014 *Contextualising the Neolithic Occupation of Southern Vietnam: The Role of Ceramics and Potters at An Son*. Terra Australis 42. ANU Press, Canberra, Australia.

Sassaman, K. E.

1998 Crafting Cultural Identity in Hunter-Gatherer Economies. In *Craft and Social Identity*, edited by C. L. Costin and R. P. Wright, pp. 93–107. Archeological Papers of the American Anthropological Association No. 8. American Anthropological Association, Arlington, VA, doi:10.1525/ap3a.1998.8.1.93.

Sassaman, K. E., and W. Rudolphi

2001 Communities of Practice in the Early Pottery

Traditions of the American Southeast. *Journal of Anthropological Research* 57(4):407–425.

Saurin, E.

1968 Nouvelles observations prehistoriques a l'est de Saigon. *Bulletin de la Société des Études Indochinoises* 43(1):1–25.

1971 Les recherches préhistoriques au Cambodge, Laos, et Viêt Nam (1877–1966). *Asian Perspectives (1969)* 12:27–41.

Sayavongkhamdy, T., N. Chang, S. Viengko, and H. Cawte

2009 The Archaeology of Sepon, Lao PDR: Archaeometallurgy, Unexploded Bombs and Collaborations. Paper delivered at the 19th Congress of the Indo-Pacific Prehistory Association in Hanoi, Vietnam, 29th November to 5th December 2009.

Sayre, E. V., K. A. Yener, E. C. Joel, and I. L. Barnes

1992 Statistical Evaluation of the Presently Accumulated Lead Isotope Data from Anatolia and Surrounding Regions. *Archaeometry* 34(1):73–105.

Scarry, J. F.

1999 Elite Identities in Apalachee Province: The Construction of Identity and Cultural Change in a Mississippian Polity. In *Material Symbols: Culture and Economy in Prehistory*, edited by J. E. Robb, pp. 342–361. Occasional Papers 26. Center for Archaeological Investigations, Southern Illinois University, Carbondale, IL.

Schauffler, W.

1976 Archaeological Survey and Excavation of Ban Chiang Culture Sites in Northeast Thailand. *Expedition* 18(4):27–37.

Schiffer, M. B.

2001b Toward an Anthropology of Technology. In *Anthropological Perspectives on Technology*, edited by M. B. Schiffer, pp. 1–15. Amerind Foundation New World Studies Series No. 5. University of New Mexico Press, Albuquerque, NM.

2001c The Explanation of Long-Term Technological Change. In *Anthropological Perspectives on Technology*, edited by M. B. Schiffer, pp. 215–235. Amerind Foundation New World Studies Series No. 5. University of New Mexico Press, Albuquerque, NM.

2004 Studying Technological Change: A Behavioral Perspective. *World Archaeology* 36(4):589–585.

2005 The Devil Is in the Details: The Cascade Model of Invention Processes. *American Antiquity* 70(3):485–502.

2008 Transmission Processes: A Behavioral Perspective. In *Cultural Transmission and Archaeology: Issues and Case Studies*, edited by M. J. O'Brien, pp. 102–111. Society for American Archaeology, Washington, DC.

2011 *Studying Technological Change: A Behavioral Approach*. University of Utah Press, Salt Lake City.

Schiffer, M. B. (editor)

2001a *Anthropological Perspectives on Technology*. Amerind Foundation New World Studies Series No. 5. University of New Mexico Press, Albuquerque, NM.

Schiffer, M. B., and J. M. Skibo

1987 Theory and Experiment in the Study of Technological Change. *Current Anthropology* 28:595–622.

1990 The Influence of Surface Treatment on Heating Effectiveness of Ceramic Vessels. *Journal of Archaeological Science* 17(4):373–381.

1997 The Explanation of Artifact Variability. *American Antiquity* 62:27–50.

Schiffer, M. B., J. M. Skibo, T. C. Boelke, M. A. Neupert, and M. Aronson

1994 New Perspectives on Experimental Archaeology: Surface Treatments and Thermal Response of the Clay Cooking Pot. *American Antiquity* 59(2):197–217.

Schlanger, N.

1990 Techniques as Human Action: Two Perspectives. *Archaeological Review from Cambridge* 9(1):18–26.

Schmidt, P. (editor)

1996 *The Culture and Technology of African Iron Production*. University Press of Florida, Gainesville.

Schoop, U.-D.

1995 *Die Geburt des Hephaistos: Technologie und Kulturgeschichte neolithischer Metallverwendung im Vorderen Orient.* Internationale Archäologie No. 24. M. L. Leidorf, Espelkamp, Germany.

Scott, D.

1991 *Metallography and Microstructure of Ancient and Historic Metals.* J. Paul Getty Museum and Archetype Books, Singapore.

Selimkhanov, I. R.

1984 On the Question of Southeast Asian Tin and its Use in the Bronze Metallurgy of Ancient Azerbaijan. In *Southeast Asian Archaeology at the XV Pacific Science Congress. The Origins of Agriculture, Metallurgy, and the State in Mainland Southeast Asia*, edited by D. T. Bayard, pp. 270–277. University of Otago Studies in Prehistory Vol. 16. University of Otago, Dunedin, New Zealand.

1996–97 Remarks on Research in the Ancient Metallurgy of Transcaucasia and the Issue of Southeast Asian Tin. In *Ancient Chinese and Southeast Asian Bronze Age Cultures. The Proceedings of a Conference Held at the Edith and Joy London Foundation Property, Kioloa, NSW, 8–12 February 1988*, Vol. II, edited by F. D. Bulbeck, pp. 949–959. SMC Publishing, Taipei.

Service, E. R.

1962 *Primitive Social Organization: An Evolutionary Perspective.* Random House, New York.

1975 *Origins of the State and Civilization: The Process of Cultural Evolution.* W. W. Norton and Company, Inc., New York.

Shao, W.

2000 The Longshan Period and Incipient Chinese Civilization. *Journal of East Asian Archaeology* 2(1–2):195–226.

Shennan, S.

1986 Central Europe in the Third Millennium B.C.: An Evolutionary Trajectory for the Beginning of the European Bronze Age. *Journal of Anthropological Archaeology* 5:115–146.

1999 Cost, Benefit and Value in the Organization of Early European Copper Production. *Antiquity* 73(280):352–363.

2008 Evolution in Archaeology. *Annual Review of Anthropology* 37:75–91.

Shennan, S. J., and J. R. Wilkinson

2001 Ceramic Style Change and Neutral Evolution: A Case Study from Neolithic Europe. *American Antiquity* 66(4):577–593.

Sherratt, A.

1994a The Transformation of Early Agrarian Europe: The Later Neolithic and Copper Ages 4500–2500 BC. In *The Oxford Illustrated Prehistory of Europe*, edited by B. Cunliffe, pp. 167–201. Oxford University Press, Oxford.

1994b The Emergence of Elites: Earlier Bronze Age Europe, 2500–1300 BC. In *The Oxford Illustrated Prehistory of Europe*, edited by B. Cunliffe, pp. 244–276. Oxford University Press, Oxford.

2006 The Trans-Eurasian Exchange: The Prehistory of Chinese Relations with the West. In *Contact and Exchange in the Ancient World: Perspectives on the Global Past*, edited by V. Mair, pp. 30–61. University of Hawaii Press, Honolulu.

Shimada, I., S. Epstein, and A. K. Craig

1982 Batán Grande: A Prehistoric Metallurgical Center. *Science* 216:952–959.

Sillar, B., and M. S. Tite

2000 The Challenge of "Technological Choices" for Materials Science Approaches in Archaeology. *Archaeometry* 42:2–20.

Sitthithaworn, E., and P. Wasuwanich

1992 Metallogenic Map of Thailand. In *Proceedings of a National Conference on Geologic Resources of Thailand: Potential for Future Development, 17–24 November 1992, Bankok, Thailand*, edited by C. Piancharoen pp. 1–15. Department of Mineral Resources, Bangkok, Thailand.

Skibo, J. M.

1999 Pottery and People. In *Pottery and People: A Dynamic Interaction* edited by J. M. Skibo and G. M. Feinman, pp. 1–8. Foundations of Archaeological Inquiry. University of Utah Press, Salt Lake City.

Skibo, J. M., and M. B. Schiffer

2001 Understanding Artifact Variability and Change: A Behavioral Framework. In *Anthropological Perspectives on Technology*, edited by M. B. Schiffer, pp. 139–149. Amerind Foundation New World Studies Series No. 5. University of New Mexico Press, Albuquerque, NM.

Smith, C. S.

1981 Art, Technology, and Science: Notes on their Historical Interaction. In *A Search for Structure*, edited by C. S. Smith, pp. 191–241. MIT Press, Cambridge, MA.

Smith, R. B.

1979 A Comment on the Non Nok Tha Dates. In *Early South East Asia: Essays in Archaeology, History, and Historical Geography*, edited by R. B. Smith and W. Watson, pp. 39–41. Oxford University Press, New York.

Smith, S., V. C. Pigott, and N. Chang

2012 The Distribution of Copper and Tin Deposits in Mainland Southeast Asia and Their Relevance to the Development of Metallurgy in Prehistory. Paper presented at the 14th International Conference of the European Association for Southeast Asian Archaeologists, 18–21 September 2012, Dublin, Ireland.

Sofaer Derevenski, J.

2000 Rings of Life: The Role of Early Metalwork in Mediating the Gendered Life Course. *World Archaeology* 31:389–406.

Solheim, W. G., II

1968 Early Bronze in Northeastern Thailand. *Current Anthropology* 9:59–62.

1972 Northern Thailand, Southeast Asia, and World Prehistory. *Asian Perspectives (1970)* 13:145–162.

1982 A Look at "L'Art Prébouddhique de la Chine et de l'Asie du Sud-Est et Son Influence en Océanie" Forty Years After. *Asian Perspectives (1979)* 22(2):165–205.

1983 The Development of Metallurgy in Southeast Asia: Another Reply to Loofs-Wissowa. *Journal of Southeast Asian Studies* 14(1):18–25.

Solometo, J.

2006 The Dimensions of War: Conflict and Culture Change in Central Arizona. In *The Archaeology of Warfare: Prehistories of Raiding and Conquest*, edited by E. N. Arkush and M. W. Allen, pp. 23–65. University Press of Florida, Gainesville.

Sone, M., and I. Metcalfe

2008 Parallel Tethyan Sutures in Mainland Southeast Asia: New Insights for Palaeo-Tethys Closure and Implications for the Indosinian Orogeny. *Comptes Rendus Geoscience* 340(2–3):166–179, doi:10.1016/j.crte.2007.09.008.

Sørensen, L.

2014 Axes from Central European Agrarian Societies and Their Imitations in South Scandinavia. *Acta Archaeologica* 85(1):126–177, doi:10.1111/j.1600-0390.2014.00931.x.

Sørensen, P.

1967 *Archaeological Excavations in Thailand. Volume II: Ban-Kao, Neolithic Settlements with Cemeteries in the Kanchanaburi Province, Part One: The Archaeological Material from the Burials.* Munksgaard, Copenhagen.

Soressi, M., and J.-M. Geneste

2011 The History and Efficacy of the *Chaîne Opératoire* Approach to Lithic Analysis: Studying Techniques to Reveal Past Societies in an Evolutionary Perspective. *PaleoAnthropology* 2011:334–335, doi:10.4207/PA.2011.ART63.

Spielmann, K.

1998 Ritual Craft Specialists in Middle Range Societies. In *Craft and Social Identity*, edited by C. L. Costin and R. P. Wright, pp. 153–159. Archeological Papers of the American Anthropological Association No. 8. American Anthropological Association, Arlington, VA, doi:10.1525/ap3a.1998.8.1.153.

Spriggs, M. J. T.

1996–97 The Dating of Non Nok Tha and the "Gakushuin Factor." In *Ancient Chinese and*

Southeast Asian Bronze Age Cultures. The Proceedings of a Conference Held at the Edith and Joy London Foundation Property, Kioloa, NSW, 8–12 February, 1988, Vol. II, edited by F. D. Bulbeck and N. Barnard, pp. 941–948. SMC Publishing, Taipei.

Srinivasan, S.

2010 Megalithic High-Tin Bronzes and Peninsular India's "Living Prehistory." In *50 Years of Archaeology in Southeast Asia: Essays in Honour of Ian Glover*, edited by B. Bellina, E. A. Bacus, T. O. Pryce, and J. W. Christie, pp. 261–271. River Books, Bangkok.

Srinivasan, S., and I. Glover

1995 Wrought and Quenched, and Cast High-Tin Bronzes in Kerala State, India. *Journal of the Historical Metallurgy Society* 29(2):69–88.

Stark, M. T.

1991 Ceramic Production and Community Specialization: A Kalinga Ethnoarchaeological Study. *World Archaeology* 23(1):64–78.

2015 Looking Forward by Studying the Past in East and Southeast Asian Archaeology: The Next 50 Years. *Journal of Indo-Pacific Archaeology* 35:67–75.

Stark, M. T. (editor)

1998 *The Archaeology of Social Boundaries*. Smithsonian Series in Archaeological Inquiry. Smithsonian Institution Press, Washington, DC.

Stark, M. T., B. J. Bowser, and L. Horne (editors)

2008 *Cultural Transmission and Material Culture: Breaking Down Boundaries*. University of Arizona Press, Tucson.

Stech, T.

1999 Aspects of Early Metallurgy in Mesopotamia and Anatolia. In *The Archaeometallurgy of the Asian Old World*, edited by V. C. Pigott, pp. 59–71. University Museum Symposium Series Vol. VII, University Museum Monograph 89, MASCA Research Papers in Science and Archaeology, Vol. 16. University of Pennsylvania Museum of Archaeology and Anthropology, Philadelphia.

Stech, T., and R. Maddin

1988 Reflections on Early Metallurgy in Southeast Asia. In *The Beginning of the Use of Metals and Alloys: Papers from the Second International Conference on the Beginning of the Use of Metals and Alloys, Zhengzhou, China, 21–26 October 1986*, edited by R. Maddin, pp. 163–174. MIT Press, Cambridge, MA.

Stech Wheeler, T., and R. Maddin

1976 The Techniques of the Early Thai Metalsmith. *Expedition* 18(4):38–47.

Stein, G. J.

1996 Producers, Patrons and Prestige: Craft Specialists in Early Mesopotamia. In *Craft Specialization and Social Evolution: In Memory of V. Gordon Childe*, edited by B. Wailes, pp. 25–38. University Museum Symposium Series Vol. VI, University Museum Monograph 93. University Museum of Archaeology and Anthropology, Philadelphia.

Steponaitis, V. P.

1978 Location Theory and Complex Chiefdoms: A Mississipian Example. In *Mississippian Settlement Patterns*, edited by B. D. Smith, pp. 417–453. Academic Press, New York.

Steward, J. H.

1955 *Theory of Culture Change: The Methodology of Multilinear Evolution*. University of Illinois Press, Urbana, IL.

Stöllner, T. R.

2014 Methods of Mining Archaeology (Montanarchäologie). In *Archaeometallurgy in Global Perspective: Methods and Syntheses*, edited by B. W. Roberts and C. P. Thornton, pp. 133–159. Springer, New York.

Stos-Gale, Z. A., and N. H. Gale

2009 Metal Provenancing Using Isotopes and the Oxford Archaeological Lead Isotope Database (OXALID). *Archaeological and Anthropological Sciences* 1:195–213.

Strahm, C., and A. Hauptmann

2009 The Metallurgical Development Phases in the

Old World. In *Metals and Societies: Studies in Honour of Barbara S. Ottaway*, edited by T. L. Kienlin and B. W. Roberts, pp. 116–128. Universitätsforschungen zur Prähistorischen Archäologie Bd. 169. Dr. Rudolf Habelt, Bonn.

Suchitta, P.

1983 The History and Development of Iron Smelting Technology in Thailand. Ph.D. dissertation, Department of Anthropology, Brown University, Providence, RI.

1992 Early Iron-Smelting in Thailand and Its Implications. In *Early Metallurgy, Trade and Urban Centres in Thailand and Southeast Asia*, edited by I. C. Glover, P. Suchitta, and J. Villiers, pp. 115–122. White Lotus, Bangkok.

Suthakorn, P.

1992 The Distribution of Tin and Associated Minerals in Thailand. In *Proceedings of a National Conference on Geologic Resources of Thailand: Potential for Future Development, 17–24 November 1992, Bangkok, Thailand*, edited by C. Piancharoen, pp. 450–485. Department of Mineral Resources, Bangkok.

Suthiragsa, N.

1979 The Ban Chieng Culture. In *Early South East Asia: Essays in Archaeology, History and Historical Geography*, edited by R. B. Smith and W. Watson, pp. 42–52. Oxford University Press, New York.

Sylvain, R.

2014 Essentialism and the Indigenous Politics of Recognition in Southern Africa. *American Anthropologist* 116(2):251–264.

Tantisukrit, C.

1978 Review of the Metallic Mineral Deposits of Thailand. In *Proceedings of the Third Regional Conference on Geology and Mineral Resources of Southeast Asia*, edited by P. Nutalaya, pp. 783–793. Asian Institute of Technology, Bangkok.

Tayles, N., S. Halcrow, and A. Clark

2015 Ban Non Wat: Current Research on late Prehistoric People in the Upper Mun River Valley, Northeast Thailand. In *Advancing Southeast Asian Archaeology 2013: Selected Papers from the first SEAMEO SPAFA Conference on Southeast Asian Archaeology*, edited by N. H. Tan, pp. 279–288. SEAMEO SPAFA Regional Center for Archaeology and Fine Arts, Bangkok.

Taylor, R. E., and O. Bar-Yosef

2014 *Radiocarbon Dating: An Archaeological Perspective*, 2nd ed. Left Coast Press, Walnut Creek, CA.

Taylor, S. J., and C. A. Shell

1988 Social and Historical Implications of Early Chinese Iron Technology. In *The Beginning of the Use of Metals and Alloys: Papers from the Second International Conference on the Beginning of the Use of Metals and Alloys, Zhengzhou, China, 21–26 October 1986*, edited by R. Maddin, pp. 205–221. MIT Press, Cambridge, MA.

Taylor, T.

1999 Envaluing Metal: Theorizing the Eneolithic "Hiatus". In *Metals in Antiquity*, edited by S. M. M. Young, A. M. Pollard, P. Budd, and R. A. Ixer, pp. 22–32. BAR International Series 792. Archaeopress, Oxford.

Theunissen, R. G.

2003 Agate and Carnelian Beads and the Dynamics of Social Complexity in Iron Age Mainland Southeast Asia. Ph.D. dissertation, University of New England, Armidale, Australia.

Thomas, R., and R. McLauchlan

2006 Dating Vietnamese Prehistory: Towards the Establishment of a Secure Database for Archaeological ^{14}C Measurements. In *Uncovering Southeast Asia's Past: Selected Papers from the 10th International Conference of the European Association of Southeast Asian Archaeologists*, edited by E. A. Bacus, I. C. Glover, and V. C. Pigott, pp. 184–195. National University of Singapore Press, Singapore.

Thornton, C. P.

2009a The Emergence of Complex Metallurgy on the Iranian Plateau: Escaping the Levantine Paradigm. *Journal of World Prehistory* 22(3):301–327.

2009b Archaeometallurgy: Evidence of a Paradigm Shift? In *Metals and Societies: Studies in Honour of Barbara S. Ottaway*, edited by T. L. Kienlin and B. W. Roberts, pp. 25–33. Universitätsforschungen zur Prähistorischen Archäologie Bd. 169. Dr. Rudolf Habelt, Bonn.

2012 Archaeometallurgy in the 21st century. *Reviews in Anthropology* 41(2):173–187.

2014 The Emergence of Complex Metallurgy on the Iranian Plateau. In *Archaeometallurgy in Global Perspective: Methods and Syntheses*, edited by B. W. Roberts and C. P. Thornton, pp. 665–696. Springer, New York.

Thornton, C. P., J. M. Golden, D. J. Killick, V. C. Pigott, T. H. Rehren, and B. W. Roberts

2010 A Chalcolithic Error: Rebuttal to Amzallag 2009. *American Journal of Archaeology* 114:305–315.

Thornton, C. P., and B. W. Roberts

2009 Introduction: The Beginnings of Metallurgy in Global Perspective. *Journal of World Prehistory* 22(3):181–184.

2014 Introduction. In *Archaeometallurgy in Global Perspective: Methods and Syntheses*, edited by B. W. Roberts and C. P. Thornton, pp. 1–10. Springer, New York.

Thorp, R. L.

2006 *China in the Early Bronze Age: Shang Civilization*. Encounters with Asia. University of Pennsylvania Press, Philadelphia.

Tilley, C. (editor)

1990 *Reading Material Culture: Structuralism, Hermaneutics and Post-Structuralism*. Basil Blackwell, Oxford.

Timberlake, S.

2007 The Use of Experimental Archaeology/Archaeometallurgy for the Understanding and Reconstruction of Early Bronze Age Mining and Smelting Technologies. In *Metals and Mines: Studies in Archaeometallurgy*, edited by S. La Niece, D. Hook, and P. Craddock, pp. 27–36. Archetype Publications and the British Museum, London.

Torrence, R.

1986 *Production and Exchange of Stone Tools*. Cambridge University Press, Cambridge.

Tosi, M.

1984 The Notion of Craft Specialization and Its Representation in the Archaeological Record of Early States in the Turanian Basin. In *Marxist Perspectives in Archaeology*, edited by M. J. T. Spriggs, pp. 22–52. New Directions in Archaeology. Cambridge University Press, Cambridge, UK.

Tran, T. H., A. E. Izokh, G. V. Polyakov, A. S. Borisenk, T. T. Anh, P. A. Balykin, N. T. Phuong, S. N. Rudnev, V. V. Van, and B. A. Nien

2008 Permo-Triassic Magmatism and Metallogeny of Northern Vietnam in Relation to the Emeishan Plume. *Russian Geology and Geophysics* 49:480–491.

Trigger, B. G.

1986 The Role of Technology in V. Gordon Childe's Archaeology. *Norwegian Archaeological Review* 19(1):1–14.

2006 *A History of Archaeological Thought*. 2nd ed. Cambridge University Press, Cambridge, UK.

Tringham, R.

1974 The Concept of "Civilization" in European Archaeology. In *The Rise and Fall of Civilizations: Modern Archaeological Approaches to Ancient Cultures*, edited by J. A. Sabloff and C. C. Lamberg-Karlovsky, pp. 470–485. Cummings Publishing Co., Menlo Park, CA.

Tripathi, V.

2015 Metallurgical Innovations and Pattern of Adaptation of Iron in Early Cultures of India. In *Metals and Civilizations: Proceedings of the Seventh International Conference on the Beginnings of the Use of Metals and Alloys (BUMA VII)*, edited by S. Srinivasan, S. Ranganathan, and A. Giumlia-Mair, pp. 95–102. National Institute of Advanced Studies, Bangalore, India.

Tucci, A., T. Sayavongkhamdy, N. Chang, and V. Souksavatdy

2014 Ancient Copper Mining in Laos: Heterarchies, Incipient States or Post-State Anarchists?

Journal of Anthropology and Archaeology 2(2):1–15.

Tylecote, R. F.

1974 Can Copper Be Smelted in a Crucible? *Journal of Historical Metallurgy* 8(1):54.

1976 *A History of Metallurgy*. 1st. ed. Institute of Materials, London.

1992 *A History of Metallurgy*. 2nd. ed. Institute of Materials, London.

Tylecote, R. F., J. M. Austin, and A. E. Wraith

1971 The Mechanism of the Bloomery Process in Shaft Furnaces. *Journal of the Iron and Steel Institute* 209:342–363.

Underhill, A.

1996 Craft Production and Social Evolution During the Longshan Period of Northern China. In *Craft Specialization and Social Evolution: In Memory of V. Gordon Childe*, edited by B. Wailes, pp. 133–150. University Museum Symposium Series Vol. VI, University Museum Monograph 93. University of Pennsylvania Museum of Archaeology and Anthropology, Philadelphia.

United Nations

1968 *Atlas of Physical, Economic and Social Resources of the Lower Mekong Basin*. United Nations.

1990 *Atlas of Mineral Resources of the ESCAP Region, Vol. 7, Lao People's Democratic Republic*. Economic and Social Commission for Asia and the Pacific. United Nations.

Upham, Steadman

1987 A Theoretical Consideration of Middle Range Societies. In *Chiefdoms in the Americas*, edited by R. D. Drennan and C. A. Uribe, pp. 345–367. University Press of America, New York.

1990a Decoupling the Processes of Political Evolution. In *The Evolution of Political Systems: Sociopolitics in Small-Scale Sedentary Societies*, edited by S. Upham, pp. 1–17. School of American Research Advanced Seminar Series. Cambridge University Press, New York.

Upham, Steadman (editor)

1990b *The Evolution of Political Systems: Sociopolitics in Small-Scale Sedentary Societies*. School of American Research Advanced Seminar Series. Cambridge University Press, New York.

U.S. Bureau of Mines

1996 Dictionary of Mining, Mineral, and Related Terms. 2nd ed. Department of the Interior. http://www.infomine.com/dictionary/, accessed 3 November 3, 2015.

U.S. Geological Survey

2005 Mineral Resources Data System: U.S. Geological Survey, Reston, Virginia. http://mrdata.usgs.gov/mrds/, accessed January 2012.

U.S. Geological Survey Geologic Names Committee

2010 Divisions of Geologic Time—Major Chronostratigraphic and Geochronologic Units: U.S. Geological Survey Fact Sheet 2010–3059.

van der Leeuw, S.

1993 Giving the Potter a Choice: Conceptual Aspects of Pottery Techniques. In *Technological Choices: Transformation in Material Culture since the Neolithic*, edited by P. Lemonnier, pp. 238–288. Routledge, New York.

Van Heekeren, H. R.

1958 The Bronze-Iron Age of Indonesia. *Verhandelingen van het Koninklijk Instituut voor Taal-, Land- en Volkenjunde,* Vol. 22. Martinus Nijhoff, The Hague, the Netherlands.

Vandkilde, H.

1996 *From Stone to Bronze: The Metalwork of the Late Neolithic and Earliest Bronze Age in Denmark*. Aarhus University Press, Aarhus, Denmark.

Veldhuijzen, H. A., and T. Rehren

2007 Slags and the City: Early Iron Production at Tell Hammeh, Jordan, and Tel Beth-Shemesh, Israel. In *Metals and Mines: Studies in Archaeometallurgy*, edited by S. La Niece, D. Hook, and P. Craddock, pp. 189–201. Archetype Publications and the British Museum, London.

Verneau, M.

1904 Les âges de la pierre et du bronze dans les pays des Bahnars, des Sédangs, des Reungaos et dans l'arrondissement de Bienhoa. In *Mission Pavie*

Indo-Chine 1879–1895. Études diverses III: Recherches sur l'Histoire Naturelle de l'Indo-Chine Orientale, edited by A. Pavie, pp. 27–40. Ernest Leroux, Paris.

Vernon, W. W.

1988 Unpublished Field Notes from the TAP Geological Survey of the KWPV, 18–23 February 1988. Notes held in TAP Archives, Penn Museum, Philadelphia.

1996–97 The Crucible in Copper-Bronze Production at Prehistoric Phu Lon, Northeast Thailand: Analyses and Interpretation. In *Ancient Chinese and Southeast Asian Bronze Age Cultures. The Proceedings of a Conference Held at the Edith and Joy London Foundation Property, Kioloa, NSW, 8–12 February, 1988,* Vol. II, edited by F. D. Bulbeck and N. Barnard, pp. 809–820. SMC Publishing, Taipei.

Villa, I. M.

2009 Lead Isotopic Measurements in Archeological Objects. *Archaeological and Anthropological Sciences* 1:149–153.

Voce, E.

1975 Examination of Specimens from the Pitt Rivers Museum. In *Notes on the Prehistoric Metallurgy of Copper and Bronze in the Old World*, edited by H. H. Coghlan, pp. 129–135. 2nd ed. Occasional Papers on Technology Vol. 4. Pitt Rivers Museum, Oxford.

Wade, J. H.

1989 The Context of Adoption of Brass Technology in Northeastern Nigeria and Its Effects on the Elaboration of Culture. In *What's New? A Closer Look at the Process of Innovation*, edited by S. E. van der Leeuw and R. Torrence, pp. 225–244. One World Archaeology 14. Unwin Hyman, London.

Wagner, D. B.

1993 *Iron and Steel in Ancient China*. Brill, Leiden.

1999 The Earliest Use of Iron in China. In *Metals in Antiquity*, edited by S. M. M. Young, A. M. Pollard, P. Budd, and R. A. Ixer, pp. 1–9. BAR International Series 792. Archaeopress, Oxford.

Wailes, B.

1996a V. Gordon Childe and the Relations of Production. In *Craft Specialization and Social Evolution: In Memory of V. Gordon Childe*, edited by B. Wailes, pp. 3–14. University Museum Symposium Series Vol. VI, University Museum Monograph 93. University Museum of Archaeology and Anthropology of the University of Pennsylvania, Philadelphia.

Wailes, B. (editor)

1996b *Craft Specialization and Social Evolution: In Memory of V. Gordon Childe*. University Museum Monograph 93. University Museum Symposium Series Vol. VI, University Museum of Archaeology and Anthropology of the University of Pennsylvania, Philadelphia.

Waldbaum, J. C.

1980 The First Archaeological Appearance of Iron and the Transition to the Iron Age. In *The Coming of the Age of Iron*, edited by T. A. Wertime and J. D. Muhly, pp. 69–98. Yale University Press, New Haven, CT.

1999 The Coming of Iron in the Eastern Mediterranean: Thirty Years of Archaeological and Technological Research. In *The Archaeometallurgy of the Asian Old World*, edited by V. C. Pigott, pp. 27–57. University Museum Symposium Series Vol. VII, University Museum Monograph 89, MASCA Research Papers in Science and Archaeology Vol. 16. University Museum of Archaeology and Anthropology of the University of Pennsylvania, Philadelphia.

Wallaert-Pêtre, H.

2001 Learning How to Make the Right Pots: Apprenticeship Strategies and Material Culture, a Case Study in Handmade Pottery from Cameroon. *Journal of Anthropological Research* 57(4):471–493.

Wang, Q., and J. Mei

2009 Some Observations on Recent Studies of Bronze Casting Technology in Ancient China. In *Metals and Societies: Studies in Honour of Barbara S. Ottaway*, edited by T. L. Kienlin and B. W. Roberts, pp. 383–391. Universitätsforschungen zur Prähistorischen Archäologie Bd. 169. Dr. Rudolf Habelt, Bonn.

Waterbolk, H. T., and J. J. Butler

1965 Comments on the Use of Metallurgical Analysis in Prehistoric Studies. *Helinium* 5:227–251.

Watson, P. J., S. A. LeBlanc, and C. L. Redman

1971 *Explanation in Archaeology: An Explicitly Scientific Approach*. Columbia University Press, New York.

Wattenmaker, P.

1998 Craft Production and Social Identity in Northwest Mesopotamia. In *Craft and Social Identity*, edited by C. L. Costin and R. P. Wright, pp. 47–55. Archeological Papers of the American Anthropological Association No. 8. American Anthropological Association, Arlington, VA, doi:10.1525/ap3a.1998.8.1.47.

Webb, G. E., G. J. Price, L. D. Nothdurft, L. Deer, and L. Rintoul

2007 Cryptic Meteoric Diagenesis in Freshwater Bivalves: Implications for Radiocarbon Dating. *Geology* 35:803–806, doi:10.1130/G23823A.1.

Weeks, L.

2007 Coals to Newcastle, Copper to Magan? Isotopic Analysis and the Persian Gulf Metals Trade. In *Metals and Mines: Studies in Archaeometallurgy*, edited by S. La Niece, D. Hook, and P. Craddock, pp. 89–96. Archetype Publications and the British Museum, London.

Wellbourn, D. A.

1985 Craft Specialization and Complex Societies: A Critique. In *Settlement and Society: Aspects of West European Prehistory in the First Millennium B.C.*, edited by T. C. Champion and J. V. S. Megaw, pp. 123–131. St. Martin's Press, New York.

Wells, P. S.

1996 Location, Organization, and Specialization of Craft Production in Late Prehistoric Central Europe. In *Craft Specialization and Social Evolution: In Memory of V. Gordon Childe*, edited by B. Wailes, pp. 85–98. University Museum Symposium Series Vol. VI, University Museum Monograph 93. University Museum of Archaeology and Anthropology, Philadelphia.

Welsch, R. L., and J. E. Terrell

1998 Material Culture, Social Fields, and Social Boundaries on the Sepik Coast of New Guinea. In *The Archaeology of Social Boundaries*, edited by M. T. Stark, pp. 50–77. Smithsonian Series in Archaeological Inquiry. Smithsonian Institute Press, Washington, DC.

Wenger, E.

1998 *Communities of Practice: Learning, Meaning, and Identity*. Learning in Doing: Social, Cognitive, and Computational Perspectives Series. Cambridge University Press, New York.

Wertime, T. A.

1964 Man's First Encounter with Metallurgy. *Science* 146(3649):1257–1267.

1973 The Beginnings of Metallurgy: A New Look. *Science* 182(4115):875–887.

1980 The Pyrotechnologic Background. In *The Coming of the Age of Iron*, edited by T. A. Wertime and J. D. Muhly, pp. 1–24. Yale University Press, New Haven, CT.

Wertime, T. A., and J. D. Muhly (editors)

1980 *The Coming of the Age of Iron*. Yale University Press, New Haven, CT.

White, A. A.

2008 A Developmental Perspective on Technological Change. *World Archaeology* 40(4):597–608.

White, J. C.

1978 Osteological Reconfigurations in the Indo-Pacific: Indications from Don Klang, Northeast Thailand. Unpublished Manuscript Available at the University of Pennsylvania Museum.

1982 *Ban Chiang: Discovery of a Lost Bronze Age*. The University Museum, University of Pennsylvania and the Smithsonian Institution Traveling Exhibition Service, Philadelphia.

1986 A Revision of the Chronology of Ban Chiang and Its Implications for the Prehistory of Northeast Thailand, Ph.D. dissertation. Department of Anthropology, University of Pennsylvania, Philadelphia.

1988 Early East Asian Metallurgy: The Southern Tradition. In *The Beginning of the Use of Metals and Alloys: Papers from the Second International*

Conference on the Beginning of the Use of Metals and Alloys, Zhengzhou, China, 21–26 October 1986, edited by R. Maddin, pp. 175–181. MIT Press, Cambridge, MA.

1995 Incorporating Heterarchy into Theory on Socio-Political Development: The Case of Southeast Asia. In *Heterarchy and the Analysis of Complex Societies*, edited by R. M. Ehrenreich, C. L. Crumley, and J. E. Levy, pp. 101–123. Archeological Papers of the American Anthropological Association No. 6. American Anthropological Association, Arlington, VA, doi:10.1525/ap3a.1995.6.1.101.

1997 A Brief Note on new Dates for the Ban Chiang Cultural Tradition. *Bulletin of the Indo-Pacific Prehistory Association* 16:103–106.

2002 Series Editor's Preface. In *Ban Chiang, a Prehistoric Village Site in Northeast Thailand I: The Human Skeletal Remains*, edited by M. Pietrusewsky and M. T. Douglas, pp. xvi–xvii. Thai Archaeology Monograph Series 1. University of Pennsylvania Museum of Archaeology and Anthropology, Philadelphia.

2008 Dating Early Bronze at Ban Chiang, Thailand. In *From Homo Erectus to the Living Traditions: Choice of Papers from the 11th International Conference of the European Association of Southeast Asian Archaeologists, Bougon, 25th–29th September 2006*, edited by J. P. Pautreau, A. S. Coupey, V. Zeitoun, and E. Rambault, pp. 91–104. The 11th International Conference of the European Association of Southeast Asian Archaeologists, Chiang Mai.

2013 Ban Non Wat: New Light on the Metal Ages of Southeast Asia. *Antiquity* 87:909–911.

2015a Comment on Higham et al. (2015) re New Chronology for Northeast Thailand. PLoS ONE 10(9): http://journals.plos.org/plosone/article/comment?id= info%3Adoi%2F10.1371%2Fannotation%2F-dac672b9-11aa-4052-afda-08f788831ed1.

2015b Comment on "Debating a Great Site: Ban Non Wat and the Wider Prehistory of Southeast Asia". *Antiquity* 89 (347):1230–1232, doi:10.15184/aqy.2015.109.

2017 Changing Paradigms in Southeast Asian Archaeology. *Journal of Indo-Pacific Archaeology* 41:66–77.

White, J. C., and B. Bouasisengpaseuth

2008 Archaeology of the Middle Mekong: Introduction to the Luang Prabang Province Exploratory Survey. In *Recherches nouvelles sur le Laos*, edited by Y. Goudineau and M. Lorrillard, pp. 37–52. Études Thématiques Vol. 18. École française d'Extrême-Orient, Paris.

White, J. C., and C. O. Eyre

2011 Residential Burial and the Metal Age of Thailand. In *Residential Burial: A Multiregional Exploration*, edited by R. L. Adams and S. M. King, pp. 57–78. Archeological Papers of the American Anthropological Association No. 20. American Anthropological Association, Arlington, VA, doi:10.1111/j.1551-8248.2011.01028.

White, J. C., and E. G. Hamilton

2009 The Transmission of Early Bronze Technology to Thailand: New Perspectives. *Journal of World Prehistory* 22:357–397, doi:10.1016/j.jas.2015.02.019.

2014 The Transmission of Early Bronze Technology to Thailand: New Perspectives. In *Archaeometallurgy in Global Perspective*, edited by B. W. Roberts and C. P. Thornton, pp. 805–852. Springer, New York.

White, J. C., and V. C. Pigott

1996 From Community Craft to Regional Specialization: Intensification of Copper Production in Pre-state Thailand. In *Craft Specialization and Social Evolution: In Memory of V. Gordon Childe*, edited by B. Wailes, pp. 151–175. University Museum Symposium Series Vol. VI, University Museum Monograph 93. University Museum of Archaeology and Anthropology, Philadelphia.

White, L.

1949 *The Science of Culture*. Farrar Straus and Cudahy, New York.

White, R.

1992 Beyond Art: Towards an Understanding of the Origins of Material Representation in Europe. *Annual Review of Anthropology* 21:537–564.

Whitehouse, R.

1974 Later Prehistory. In *The Origins of Europe: Four New Studies in Archaeology and History*, edited by D. Collins, pp. 127–215. George Allen and Unwin, Ltd., London.

Wiessner, P.

1983 Style and Social Information in Kalahari San Projectile Points. *American Antiquity* 48(2):253–276.

Willey, G. R.

1974 Summary of the Complex Societies Colloquium. In *Reconstructing Complex Societies: An Archaeological Colloquium*, edited by C. B. Moore, pp. 145–153. Supplement to the Bulletin of the American Schools of Oriental Research No. 20. American Schools of Oriental Research, Cambridge, MA.

Wilson, L., and A. M. Pollard

2001 The Provenance Hypothesis. In *Handbook of Archaeological Sciences*, edited by D. R. Brothwell and A. M. Pollard, pp. 507–517. John Wiley & Sons, Chichester, UK.

Winegrad, D. P.

1993 *Through Time, Across Continents: A Hundred Years of Archaeology and Anthropology at the University Museum*. University Museum, University of Pennsylvania, Philadelphia.

Wiriyaromp, W.

2010 The Wider Relationships of the Neolithic 1 Ceramics. In *The Origins of the Civilization of Angkor. Volume Four: The Excavation of Ban Non Wat, Part Two: The Neolithic Occupation*, edited by C. F. W. Higham and A. Kijngam, pp. 107–123. The Fine Arts Department of Thailand, Bangkok.

Wood, R.

2015 From Revolution to Convention: The Past, Present and Future of Radiocarbon Dating. *Journal of Archaeological Science* 56:61–72, doi:10.1016/j.jas.2015.02.019.

Woolley, L.

1963 *The Beginnings of Civilization*. History of Mankind, UNESCO, Vol. 1, pt. 2, pp. 357–854. Harper and Row, New York.

Workman, D. R.

1972 *Mineral Resources of the Lower Mekong Basin and Adjacent Areas of Khmer Republic, Laos, Thailand and Republic of Viet-Nam*. Mineral Resources Development Series No. 39. United Nations, Economic Commission for Asia and the Far East, Bangkok.

1977 *Geology of Laos, Cambodia, South Vietnam and the Eastern Part of Thailand*. Overseas Geology and Mineral Resources 50. Institute of Geological Sciences and Stationery Office Books, London.

Worman, E. C., Jr.

1949a The "Neolithic" Problem in the Prehistory of India. *Journal of the Washington Academy of Sciences* 39:181–201.

1949b Somrong Sen and the Reconstruction of Prehistory in Indo-China. *Southwestern Journal of Anthropology* 5(4):318–329.

Wright, E. G.

1974 The Tell: Basic Unit for Reconstructing Complex Societies in the Near East. In *Reconstructing Complex Societies: An Archaeological Colloquium*, edited by C. B. Moore, pp. 123–130. Supplement to the Bulletin of the American Schools of Oriental Research No. 20. American Schools of Oriental Research, Cambridge, MA.

Wright, H.

1977 Recent Research on the Origins of the State. *Annual Review of Anthropology* 6:379–397.

1984 Prestate Political Formations. In *On the Evolution of Complex Societies: Essays in Honor of Harry Hoijer, 1982*, edited by T. K. Earle, pp. 47–77. Undena, Malibu, CA.

1986 The Evolution of Civilizations. In *American Archaeology Past and Future: A Celebration of the Society for American Archaeology 1935–85*, edited by D. J. Meltzer, D. D. Fowler, and J. A. Sabloff, pp. 323–365. Smithsonian Institution, Washington, DC.

Wright, R. P.

1996 Contexts of Specialization: V. Gordon Childe

and Social Evolution. In *Craft Specialization and Social Evolution: In Memory of V. Gordon Childe*, edited by B. Wailes, pp. 123–132. University Museum Symposium Series Vol. VI, University Museum Monograph 93. University Museum of Archaeology and Anthropology, University of Pennsylvania, Philadelphia.

1998 Crafting Social Identity in Ur III Southern Mesopotamia. In *Craft and Social Identity*, edited by C. L. Costin and R. P. Wright, pp. 58–69. Archeological Papers of the American Anthropological Association No. 8. American Anthropological Association, Arlington, VA, doi:10.1525/ap3a.1998.8.1.57.

Wu, G.

2009 From Western Asia to the Tianshan Mountains: On the Early Iron Artefacts Found in Xinjiang In *Metallurgy and Civilisation: Eurasia and Beyond*, edited by J. Mei and T. Rehren, pp. 107–115. Archetype, London.

Yan, W.

2000 A Discussion of the Chalcolithic Age in China. In *The Beginnings of Metallurgy in China*, edited by K. M. Linduff, H. Rubin, and S. Shuyun, pp. 99–115. Chinese Studies Vol. 11. The Edwin Mellen Press, Lewiston, NY.

Yener, K. A.

2000 *The Domestication of Metals: The Rise of Complex Metal Industries in Anatolia*. Brill, Leiden.

Young, S. M. M., A. M. Pollard, P. Budd, and R. A. Ixer (editors)

1999 *Metals in Antiquity*. BAR International Series 792. Archaeopress, Oxford.

Zaw, K., S. Meffre, C.-K. Lai, C. Burrett, M. Santosh, I. Graham, T. Manaka, A. Salam, T. Kamvong, and P. Cromie

2014b Tectonics and Metallogeny of Mainland Southeast Asia—a Review and Contribution. *Gondwana Research* 26:5–30.

Zaw, K., M. Santosh, and I. T. Graham

2014a Tectonics and Metallogeny of Mainland Southeast Asia: Preface. *Gondwana Research* 26:1–4.

Zazzo, A., and J. F. Saliège

2011 Radiocarbon Dating of Biological Apatites: A Review. *Palaeogeography, Palaeoclimatology, Palaeoecology* 310:52–61, doi:10.1016/j.palaeo.2010.12.004.

Zettler, R. L.

1996 Gordon Childe and the Socioeconomic Position of Craft Specialists in Early Mesopotamia. In *Craft Specialization and Social Evolution: in Memory of V. Gordon Childe*, edited by B. Wailes, pp. 17–23. University Museum Symposium Series Vol. VI, University Museum Monograph 93. University Museum of Archaeology and Anthropology, Philadelphia.

Zhang, L.

2012 *Ancient Society and Metallurgy: A Comparative Study of Bronze Age Societies in Central Eurasia and North China*. BAR International Series 2328. Archaeopress, Oxford.

Zhou, B., H. Youyan, and L. Benshan

1988 Ancient Copper Mining and Smelting at Tonglushan, Daye. In *The Beginning of the Use of Metals and Alloys: Papers from the Second International Conference on the Beginning of the Use of Metals and Alloys, Zhengzhou, China, 21–26 October 1986*, edited by R. Maddin, pp. 125–129. MIT Press, Cambridge, MA.

Zhu, S., and T. He

1993 Studies of Ancient Chinese Mirrors and Other Bronze Artifacts. In *Metal Plating and Patination: Cultural, Technical and Historical Development*, edited by S. La Niece and P. Craddock, pp. 50–62. Butterworth-Heinemann, Oxford.

Zwicker, U., H. Greiner, K. H. Hofmann, and M. Reithinger

1985 Smelting, Refining and Alloying of Copper and Copper Alloys in Crucible Furnaces during Prehistoric up to Roman Times. In *Furnaces and Smelting Technology in Antiquity*, edited by P. T. Craddock and M. J. Hughes, pp. 103–115. British Museum Occasional Paper No. 48, British Museum, London.

Index